MODERN PROTEIN CHEMISTRY

Practical Aspects

MODERN PROTEIN CHEMISTRY

Practical Aspects

Edited by
Gary C. Howard
William E. Brown

CRC PRESS

Boca Raton London New York Washington, D.C.

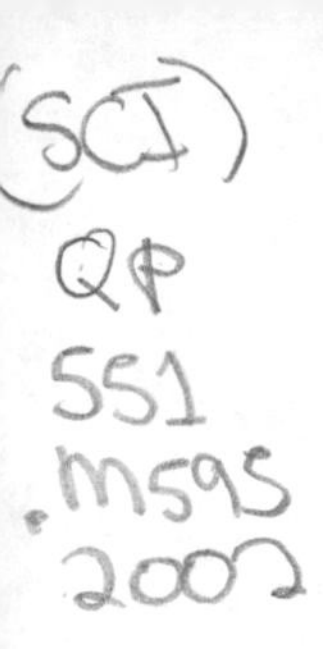

Library of Congress Cataloging-in-Publication Data

Modern protein chemistry : practical aspects / edited by Gary C. Howard and William E. Brown.
p. cm.
Includes bibliographical references and index.
ISBN 0-8493-9453-8 (alk. paper)
1. Proteins--Research--Methodology. I. Howard, Gary C. II. Brown, William E., 1945-

QP551 .M595 2001
572'.6--dc21 2001035684

Direct all inquiries to CRC Press LLC, 2000 N.W. Corporate Blvd., Boca Raton, Florida 33431.

Visit the CRC Press Web site at www.crcpress.com

International Standard Book Number 0-8493-9453-8
Library of Congress Card Number 2001035684
Printed in the United States of America 1 2 3 4 5 6 7 8 9 0
Printed on acid-free paper

Preface

In recent years, interest in proteins has surged. Protein chemistry methods are used by nearly every discipline of the very large field of biomedical research. Many techniques have been used in less traditional ways with good results.

Many fine books describe the techniques of protein chemistry. Some provide step-by-step approaches in the form of protocols. Because the methods presented in this volume are in many cases built on advanced technology, we have sought to describe some of the practical aspects associated with the use of each of the techniques. Thus, while this is not meant to be a user's guide, the authors were asked to address each topic as they would if a colleague or collaborator had asked about using this technology for the first time. The contributors have been selected for their prominence in their specific fields and because they run laboratories that actively collaborate with other scientists (or in some cases, provide services to researchers). Thus, they routinely advise other scientists in these practical matters. Introductory material accompanies each chapter to give a basic understanding of the method discussed.

We hope this book will be useful to researchers and students in biochemistry, molecular biology, immunology, biology, zoology, and other fields of biomedical research. These commentaries will assist researchers, both beginners and experienced practitioners, who are looking for new ideas and who are interested in applying some of these more advanced methods.

We want to thank the contributors to this volume. Without their professional knowledge and enthusiastic efforts, this book would not have been possible. We thank Fequiere Vilsaint from CRC Press for his patience and assistance. We also thank Yale Altman for his encouragement, guidance, and unflagging support throughout the project.

Contributors

Mark E. Bier, Ph.D.
Director
Center for Molecular Analysis
Department of Chemistry
Carnegie Mellon University
Pittsburgh, Pennsylvania

Jan Borén, M.D., Ph.D.
Associate Professor
Department of Medical Biochemistry
Wallenberg Laboratory
Göteborg University
Göteborg, Sweden

William E. Brown, Ph.D.
Professor
Department of Biological Sciences
Carnegie Mellon University
Pittsburgh, Pennsylvania

Chun-Jung Chen, Ph.D.
Postdoctoral Fellow
Department of Biochemistry and Molecular Biology
University of Georgia
Athens, Georgia

Claes M. Gustafsson, M.D., Ph.D.
Associate Professor
Department of Medical Biochemistry
Karolinska Institute
Stockholm, Sweden

Douglas G. Hayes, Ph.D.
Associate Professor
Department of Chemical Engineering
University of Alabama
Huntsville, Alabama

John Hempel, Ph.D.
Professor
Division of Primary Structure
Department of Biological Sciences
University of Pittsburgh
Pittsburgh, Pennsylvania

Jonathan D. Hirst, Ph.D.
Lecturer
The School of Chemistry
University of Nottingham
Nottingham, U.K.

T. Kevin Hitchens
Graduate Student
Department of Biological Sciences
Carnegie Mellon University
Pittsburgh, Pennsylvania

Gary C. Howard, Ph.D.
Editor
Castro Valley, CA

Zhi-Jie Liu, Ph.D.
Assistant Research Scientist
Department of Biochemistry and Molecular Biology
University of Georgia
Athens, Georgia

Jonathan S. Minden, Ph.D.
Associate Professor
Department of Biological Sciences
Carnegie Mellon University
Pittsburgh, Pennsylvania

M. Gary Newton, Ph.D.
Professor
Department of Chemistry
University of Georgia
Athens, Georgia

Frederic M. Richards, Ph.D.
Sterling Professor Emeritus of Molecular Biology and Biophysics
Senior Research Scientist
Department of Molecular Biophysics and Biochemistry
Yale University
New Haven, Connecticut

John P. Rose, Ph.D.
Associate Research Scientist
Department of Biochemistry and Molecular Biology
University of Georgia
Athens, Georgia

Gordon S. Rule, Ph.D.
Associate Professor
Department of Biological Sciences
Carnegie Mellon University
Pittsburgh, Pennsylvania

Mustafa Ünlü
Graduate Student
Department of Biological Sciences
Carnegie Mellon University
Pittsburgh, Pennsylvania

Bi-Cheng Wang, Ph.D.
Professor and Eminent Scholar in Structural Biology
Department of Biochemistry and Molecular Biology
University of Georgia
Athens, Georgia

Table of Contents

Chapter 1
Modern Protein Chemistry 1
Frederic M. Richards, Mark E. Bier, Douglas G. Hayes, John Hempel, Jonathan S. Minden, Gordon S. Rule, William E. Brown, and Gary C. Howard

Chapter 2
Protein Crystallography 7
Chun-Jung Chen, John P. Rose, M. Gary Newton, Zhi-Jie Liu, and Bi-Cheng Wang

Chapter 3
Protein NMR Spectroscopy 37
Gordon S. Rule and T. Kevin Hitchens

Chapter 4
Analysis of Proteins by Mass Spectrometry 71
Mark E. Bier

Chapter 5
An Orientation to Edman Chemistry 103
John Hempel

Chapter 6
Computer Modeling of Protein Structure 123
Jonathan D. Hirst

Chapter 7
Genetic Analysis of Proteins 145
Jan Borén and Claes M. Gustafsson

Chapter 8
How to Employ Proteins in Nonaqueous Environments 179
Douglas G. Hayes

Chapter 9
Proteomics: Difference Gel Electrophoresis 227
Mustafa Ünlü and Jonathan Minden

Index 245

1 Modern Protein Chemistry

Frederic M. Richards, Mark E. Bier, Douglas G. Hayes, John Hempel, Jonathan S. Minden, Gordon S. Rule, William E. Brown, and Gary C. Howard

CONTENTS

Introduction 1
Traditional Bases of Protein Chemistry 2
New Methodologies 2
New Opportunities 4

INTRODUCTION

Approximately 50 years ago, the determination of the structure of DNA ignited the modern era of molecular biology and thrust the determination of molecular structure and function to prominence in the quest to understand the workings of the cell. Beginning in the 1980s, studies in molecular biology accelerated with the development of rapid DNA sequencing methods, restriction enzymes, and cloning. In this era of expanding knowledge, efforts to define the principles involved in regulating cellular growth, development, and differentiation led to a plethora of theoretical models that ultimately required us to push forward with the elucidation of both the sequences of genomes and the structures of the gene products.

Technological advances often lead theory and, for some time, studies of DNA and genomes have reigned as evidenced by Nobel Prizes and millions of dollars in venture capital. In more recent years, interest in proteins has grown as researchers realized that DNA and RNA are only one half of life's equation. This renewed interest in proteins has taken a new name, structural genomics, and is driven by new techniques in protein chemistry and major advances with older techniques. Extensive genetic and biochemical efforts in the last decades have led to a detailed description of many of the molecular pathways that constitute normal and aberrant cellular functions. The recent completion of the DNA sequences of a number of genomes will lead to an unprecedented description of cellular processes at the molecular level.

0-8493-9453-8/02/$0.00+$1.50

TRADITIONAL BASES OF PROTEIN CHEMISTRY

The traditional bases of protein chemistry have been sequencing for primary structure analysis and x-ray diffraction and nuclear magnetic resonance (NMR) spectroscopy for three-dimensional structure determinations. It is clear that a complete understanding of biochemical processes and their regulation will depend on information at the atomic level. Historically, structure determination by x-ray diffraction and NMR, and the characterization of dynamic processes by NMR, have lagged behind the genetic and biochemical characterization of cellular processes. However, new advances in both of these methods promise to accelerate the rate of structure determination and to increase the size and complexity of systems that can be studied by magnetic resonance.

In the case of x-ray diffraction, important advances include the use of robotics to ensure successful growth of crystals, the development of synchrotron sources for macromolecular crystallography, the use of direct methods and anomalous dispersion in phasing, and the development of automated methods of map fitting.

New advances in NMR include the availability of higher magnetic field strengths, development of labeling schemes and relaxation-compensated experimental methods to alleviate the inherent size limitation associated with NMR, and the use of residual dipolar coupling to obtain important long-range structure constraints. These advances will make structural genomics a reality, by permitting the atomic level description of the structure and dynamics of a large number of complex biological assemblies.

NEW METHODOLOGIES

These two basic methods have been joined by new methodologies that have dramatically expanded the study of proteins in recent years. New techniques have enhanced data acquisition, and other computation methods have simplified the data analysis. With ever-faster computers, computer modeling is rapidly advancing. Novel protein engineering approaches have been developed to determine the high-resolution structures of proteins during folding reactions. These have improved the modeling of interactions between multiple proteins and the predictions of structural changes in response to genetically engineered amino acids substitutions. Thus, large protein complexes (e.g., enzymes and ion channels) and the dynamics of their structural and functional rearrangements can now be studied.

While x-ray diffraction and NMR continue to develop, especially NMR of the solid state, one method that is making a significant impact on structure determination is cryo electron microscopy (EM). There have been tremendous advances in recent years and the resolution is continuously improving. The gap between cryo EM and x-ray/NMR is improving. In the foreseeable future, work on both sides of the gap, with effort on the high resolution side largely theoretical, will succeed in being able to go from a three dimensional EM image to an atomic level structure with x-ray/NMR data inserted when available. The ribosome structures provide an excellent example of the present status of the field.

In recent years, x-ray diffraction and NMR spectroscopy have been joined by mass spectroscopy (MS). For many years, use of MS was limited by the lack of ways to put large molecules, such as proteins, into a gas phase. Ionization by fast atom bombardment, secondary ionization MS, and Californium radiation have made the analysis of large proteins by MS possible.

Recent advances in protein analysis by MS are due to the introduction of electrospray ionization (ESI), matrix-assisted laser desorption ionization (MALDI), MSN scan modes, as well as improvements in instrument sensitivity, resolution, and mass accuracy. With these improved techniques, researchers will continue to use MS to help elucidate primary, secondary, and to a lesser extent, tertiary structure of proteins.

The number of human genes was recently estimated as 30,000 rather than the previously estimated number of ~100,000, still believed by other experts. Francis S. Collins, Director of the National Human Genome Research Institute, commented that the complexity of how proteins function in humans probably accounts for the difference. If this functional complexity is the result of protein modifications, MS will play a vital role in determining the position and type of modification (e.g., phosphorylation, liposylation, and glycosylation).

The pursuit of attomole protein identification and peptide sequencing is driven by the need to work at these levels. Although greater confidence will be obtained in MS protein identifications with corrected and enlarged protein databases, identification at reduced amounts is essential. Improvements in mass spectrometers as well as low-level separation techniques are crucial to achieving this goal. Two-dimensional gel analysis of low abundance proteins has limited capabilities, especially for hydrophobic proteins. MS detection limits for peptides have improved by approximately one order of magnitude every 10 years. Detection of zeptomole levels for real samples should soon become a reality.

The desire for rapid MS analysis is currently pushing manufacturers to automate. The simultaneous developments of robotic sample preparation stations, faster mass spectrometers, and small online introduction devices (i.e., chips), will augment this automation. We may someday utilize mass spectrometer arrays analyzing hundreds of different protein samples at once. Rapid MALDI-MS tissue imaging could one day be used to identify and map proteins, and diagnose disease.

MS all too often is viewed by those on the periphery as a replacement for Edman chemistry. Actually, the two complement each other with their nonoverlapping strengths and weaknesses. Sequencing by mass spectrometry is a deductive effort, based on mass differences. With Edman degradation, residue identifications rest solidly on the chromatographic differences of derivatives released at each cycle of the degradation. MS generates such a plethora of ions that it becomes increasingly difficult to determine an unambiguous sequence as the peptide length passes ten residues. Prolines introduce the added complications of preferential fragmentation. Peptides do not fragment as a specific result of Edman coupling, except under extraordinary circumstances (e.g., synthetic peptides containing an unnatural dipeptide with a methylene in place of the carbonyl of the first residue, which creates an internal secondary amine). For this and other reasons, intact proteins cannot be directly sequenced by MS. As a result, Edman degradation is still the only reasonable approach to gaining information from an intact protein with a free N-terminus. Further, with proteins obtained from bacterial and other expression systems, blockage of N-termini seems less a problem than when proteins are obtained from the natural source. As these natural sources for proteins come into wider and wider use, particularly for crystallographic and therapeutic goals, Edman degradation is the most straightforward way of determining that the intended initiation site was used and that, at least at the N-terminus, the preparation is homogeneous. Furthermore, Edman degradation provides the only unambiguous way for most investigators to differentiate isoleucine from leucine and glutamine from lysine, a determination that is difficult to achieve by MS.

Proteins, long thought to be limited to aqueous solutions, are increasingly used in nonaqueous environments (e.g., organic solvents, supercritical fluids, and vapors). The ability of enzymes to function in nonaqueous environments expands the role of enzymes in organic syntheses. Reactions not possible in aqueous media, including those involving nonpolar substrates or those susceptible to hydrolysis, readily occur in low-water media. Reactions of great interest that have been successfully performed in low-aqueous media include peptide synthesis, tailor-made synthesis of cholesterol and lipids, synthesis of beta blockers and other chiral drugs, racemic separation of amino acids, and enzymatic degradation of nerve gas agents. The preparation of the biocatalyst and the composition of the media strongly dictate its performance in nonaqueous media. This work will lead to more stable and active biocatalysts and methods to manipulate the reaction medium and conditions to maximize conversion, selectivity, and productivity.

NEW OPPORTUNITIES

The large quantities of information coming from the genome projects have brought new challenges and opportunities. New questions involving multiple pathways and proteins arise. With reverse genetics, one can take a cloned gene and determine the function of its product. Transgenic animals and other genetic techniques will greatly impact the biological and medical sciences.

The scale of the genome sequencing projects pales in comparison to the goals of the various proteome projects. There are plans to search for protein changes in dozens of different tissues at dozens of different stages of development and dozens of different diseases from hundreds of samples and in the presence of thousands of different drugs and conditions. The number of possible experiments is truly staggering. There is a great need to increase the rate of proteome analysis 10- to 100-fold or more.

The challenges for proteome studies are many. The sensitivity of protein identification must be increased by a factor of 100 to 1000. In addition, systems must automatically detect protein concentrations over a 10,000-fold range, determine and rank protein changes between two or more samples, and identify change signatures.

Difference gel electrophoresis (DIGE) is a good start. This method can rapidly identify protein changes between two or three samples on the same two-dimensional electrophoresis gel. This method relies on fluorescently tagging all proteins in each sample with one of a set of matched fluorescent dyes that do not affect the relative mobility of proteins during electrophoresis. This method greatly reduces the complexity of the search for protein differences from several thousand proteins observed in the protein extract to several dozen candidate proteins that are significantly different between the samples.

It is important to remember that the main mediator of cellular behavior is protein. Understanding how proteins direct cellular changes is at the heart of learning how to control these changes, which will have an important impact on human health. The techniques used in protein chemistry often come from the intersection of diverse disciplines. In the 1970s, who would have predicted that it would be more efficient to sequence proteins by sequencing the DNA that encodes the protein or that enzymes would work in nonaqueous environments? For the past couple of decades, biophysics and chemistry have intersected to great advantage. In the next generation, biophysics might intersect with cell biology itself.

2 Protein Crystallography

Chun-Jung Chen, John P. Rose, M. Gary Newton, Zhi-Jie Liu, and Bi-Cheng Wang

CONTENTS

Introduction 7
Crystals and Diffraction 9
Sample Preparation and Crystallization 11
Sample Preparation 11
Crystallization 11
X-Ray Sources and Detectors 13
X-Ray Sources 13
Detectors 15
Data Collection and Processing 15
Crystal Selection and Mounting 15
Data Collection 17
The Phase Problem 19
Isomorphous Replacement Methods 21
Anomalous Dispersion Methods 24
Molecular Replacement 28
Model Building 29
Refinement 30
Validation 32
Further Reading 33
References 33

INTRODUCTION

x-ray crystallography, as the name implies, is the study of crystals, three-dimensional molecular assemblies, using x-rays. It is a powerful technique that, when successful, can offer a unique unambiguous picture of the molecule in three dimensions. The technique, refined over the past century, works equally well on molecules ranging in size from simple ionic salts (Bragg, 1913) to large assemblies of molecules such as the ribosome (Ban et al., 2000). X-ray crystallography, together with NMR spectroscopy, constitutes the basis of structural biology. Combined, they form a powerful tool with crystallography, providing detailed structural information, and with NMR, providing information on chemical kinetics and dynamics of the system.

0-8493-9453-8/02/$0.00+$1.50

x-rays constitute the region of the electromagnetic spectrum that lies between the ultraviolet and gamma ray regions of the spectrum. X-rays are useful in imaging a molecule because their wavelengths, ranging from 0.01 to 100 Å, are comparable to or smaller than the atoms making up the molecule. This is what is meant by the term atomic resolution. X-rays are scattered by the electrons of the atom or molecule producing a diffraction pattern. Because the amount of x-ray scattering is proportional to the number of electrons in a given atom, the diffraction pattern contains information about the elemental composition of the molecule. However, the x-ray diffraction pattern produced by an isolated atom or molecule is too weak to be measured. Crystals, on the other hand, containing millions of copies of the atom or molecule in an ordered three-dimensional array, can produce a measurable x-ray diffraction pattern. Thus, the x-ray diffraction experiment in its simplest terms requires

1. A crystal
2. An x-ray source
3. A means of recording the x-ray diffraction pattern
4. A means of reconstructing the image of the molecule from the recorded diffraction pattern

This chapter introduces the reader to the x-ray diffraction experiment as applied to macromolecules and describes some of the basics of crystallography. The following topics are covered.

- Crystals and diffraction
- Sample preparation and crystallization
- x-ray sources and detectors
- Data collection and processing
- The phase problem
- Model building
- Refinement
- Validation

The scope of the chapter is limited. The reader can obtain additional information on these topics in the references and World Wide Web links provided in the text. In addition, we offer several texts at the end of the chapter where further information can be found.

There are also several good WWW-based courses on x-ray diffraction and macromolecular crystallography, these include Kevin Cowtan's *Book of Fourier* [*www.yorvic.york.ac.uk/~cowtan/fourier/fourier.html*], Proffen and Neder's Interactive Tutorial about Diffraction [*www.pa.msu.edu/~proffen/teaching/teaching.html*], and the University of California at San Diego (UCSD) x-ray diffraction tutorial [*www.wilson.ucsd.edu/education/pchem/xraydiff*].

CRYSTALS AND DIFFRACTION

A crystal is made up of atoms or molecules that form a pattern that is repeated periodically in three dimensions. The smallest repeat unit found within the crystal is called the unit cell. The unit cell can be described by three lengths, termed *a*, *b*, and *c*, representing the edges or axes of the unit cell and three angles, termed α, β, and γ, representing the angles between two of the corresponding axes. The angle between the *b* and *c* axes is called α, the angle between the *c* and *a* axes is called β, and the angle between the *a* and *b* axes is called γ. These six parameters are referred to as cell constants or lattice parameters. Based on the unit cell parameters, crystals can be grouped into the seven crystal systems (see Table 2.1).

Mathematics (Hassel, 1830) has shown that there are only 32 combinations of symmetry operations (rotation, inversion, and reflection) that are consistent with a three-dimensional crystal lattice. These 32 point groups, or crystal classes, can be grouped into one of the seven crystal systems given in Table 2.1. There are four types of crystal lattices: primitive (*P*), end-centered (*C*, *B*, and *A*), face-centered (*F*), and body-centered (*I*). The primitive lattice contains a lattice point at each corner of the unit cell, the end-centered lattice has an additional lattice point on one of the lattice faces, the face-centered lattice has an extra lattice on each of the lattice faces, and the body-centered lattice has an extra lattice point at the center of the crystal lattice. By combining the seven crystal systems with the four lattice types (*P*, *C*, *I*, *F*), 14 unique crystal lattices, also known as Bravais lattices (Bravais, 1849), are produced.

In addition to the four lattice symmetries described previously, translational symmetry can exist within the crystal lattice. These are unit cell translations, screw axes, and glide planes. Unit cell translations are simply the translation of atoms or molecules at a point in one unit cell to the identical point in an adjacent unit cell. Screw axes are rotational axes combined with translations, while glide planes are reflections across mirror planes combined with translations. This extra symmetry combined with the 14 Bravais lattices and the 32 crystallographic point groups gives the 230 possible crystallographic space groups (see *International Tables of Crystallography*, Vol. A). Every crystal must belong to one of the 230 space groups. However, because proteins are chiral, their crystals belong to one of the 65 enantiomorphic space groups. Space group symmetry when applied to

TABLE 2.1 THE SEVEN CRYSTAL SYSTEMS

System	Lattice Symmetry	Unit Cell Constraints	
Triclinic	1	*a* *b* *c*,	α β γ 90°
Monoclinic	2/m	*a* *b* *c*,	α = γ = 90°, β 90°
Orthorhombic	mmm	*a* *b* *c*,	α = β = γ = 90°
Tetragonal	4/mmm	*a* = *b* *c*,	α = β = γ = 90°
Trigonal/rhombohedral	–3/m	*a* = *b* = *c*,	α = β = γ 90°, γ <120°
Hexagonal	6/mmm	*a* = *b* *c*,	α = β = 90°, γ = 120°
Cubic	m3m	*a* = *b* = *c*,	α = β = γ = 90°

the contents of the asymmetric unit (the part of the unit cell having no symmetry) generates the contents of the entire unit cell. This simplifies the crystal structure determination, because once we know the structure of the asymmetric unit, space group symmetry generates the complete crystal structure.

The x-ray diffraction pattern produced when a crystal is placed in the x-ray beam reflects the arrangement of atoms in the crystal (including space group symmetry) and is governed by Bragg's law. In Bragg's treatment, parallel planes of reflection passing through the crystal lattice points reflect the incoming x-ray beam much like a mirror reflects light, that is, the angle of incidence is equal to the angle of reflection.

$$n\lambda = 2d \sin\theta \qquad \textbf{Bragg's Law}$$

where

λ = the x-ray wavelength
θ_{hkl} = the diffraction angle
d_{hkl} = the lattice spacing
n = the diffraction order, usually 1

The diffraction pattern (Figure 2.1) then simply results from the interference of the reflections from sets of parallel planes within the crystal. The spacing of the lattice planes is determined by the lattice geometry, that is, it is a function of the unit cell parameters. The orientation of the plane with respect to the axes of the unit cell is defined by three integers, h, k, and l (Miller indices) that denoted the points where the plane intersects the three unit cell axes. Miller indices are defined as $h = a/X$, $k = b/Y$, and $l = c/Z$, where X, Y, and Z are the points where the plane intersects the a, b, and c unit cell axes. Thus, the plane intersecting the unit cell at $a/2$, $b/2$, and $c/2$ would have indices 222.

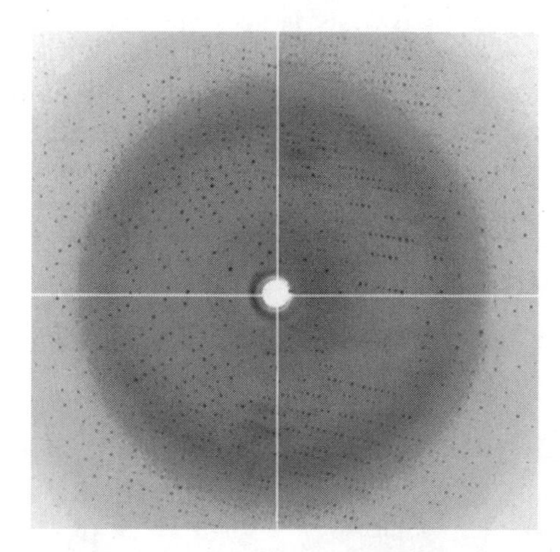

FIGURE 2.1 An x-ray diffraction pattern of a protein crystal recorded on a CCD detector. Each of the black dots (reflections) can be indexed using Miller indices that correspond to a set of reflecting planes in the crystal.

Knowing the unit cell constants and the orientation of the crystal in the x-ray beam, each spot or reflection observed in the diffraction pattern can be defined using the Miller indices corresponding to the set of lattice planes that produced the reflection. There is a reciprocal relationship between the diffraction angle θ_{hkl} and the spacing between the lattice planes d_{hkl} in the crystal. Thus, crystals with larger unit cells, such as proteins, will produce a more dense diffraction pattern than that of salt crystals or other small molecules with much smaller unit cells.

SAMPLE PREPARATION AND CRYSTALLIZATION

SAMPLE PREPARATION

Successful crystallization often depends on the preparation of a pure and homogeneous protein sample. Thus, all factors that create chemical heterogeneity, such as proteolytic cleavage, denaturation, contamination, oxidation, and so on, must be minimized or eliminated. In addition, the proper transport and storage of samples are crucial in preventing sample degradation.

Ideally, several milligrams of pure soluble protein should be available. One- or two-dimensional SDS-PAGE (polyacrylamide gel electrophoresis) and native gel PAGE stained with silver or colloidal gold stains (Merril, 1990) should be used to examine sample purity and protein aggregation before crystallization is attempted. Dynamic light scattering (DLS) provides another powerful tool to characterize protein homology, purity, size, and molecular weight (Hitscherich et al., 2000; Georgalis and Saenger, 1999; and Malkin and McPherson, 1994). Electrospray mass spectrometry (ESI-MS), which can give the mass of the protein to within the mass of a single amino acid residue (Edmonds and Smith, 1990; Scoble and Martin, 1990; Jardine, 1991), can be used to detect protein cleavage and to monitor selenomethionine incorporation when preparing samples for multi-wavelength anomalous dispersion (MAD) experiments. The use of affinity tags, such as the histidine tag, for detection and purification (Sambrook et al., 1989) can affect crystallization. Therefore these tags, if used, should be of minimal length or removed prior to crystallization. Insoluble proteins, such as membrane proteins, need to be solubilized, usually by adding detergent, before crystallization can be attempted.

CRYSTALLIZATION

Protein crystallization can be viewed as a controlled precipitation of the protein from an aqueous media. The solubility of proteins is dependent on the competition between solvent–solute interactions and intermolecular interactions between protein molecules. The balance of these interactions can be modified by several factors.

1. Temperature (affects solubility)
2. pH (affects both solute and solvent)
3. Salts (salt-in or salt-out effects)
4. Hydrogen-bond competitors (urea, guanidiuum salts, etc.)
5. Hydrophobic additives (nonionic detergents, etc.)
6. Organic solvents (modification of dielectric constants)

The goal of the crystallization process is to very slowly drive the protein-solvent system to a state of reduced solubility of the macromolecule by adjusting one or more of the six factors listed previously. This is generally achieved by increasing the concentration of precipitating agents such as salts, polyethylene glycol (PEG) polymers, or organic solvents, using a technique called vapor diffusion, the most common crystallization technique used today.

A simple phase diagram for the vapor diffusion process is shown in Figure 2.2a. Initially the protein is in solution [Figure 2.2(a), point A]. As either the protein or precipitant concentration increases, a point is reached [Figure 2.2(a), point B] where

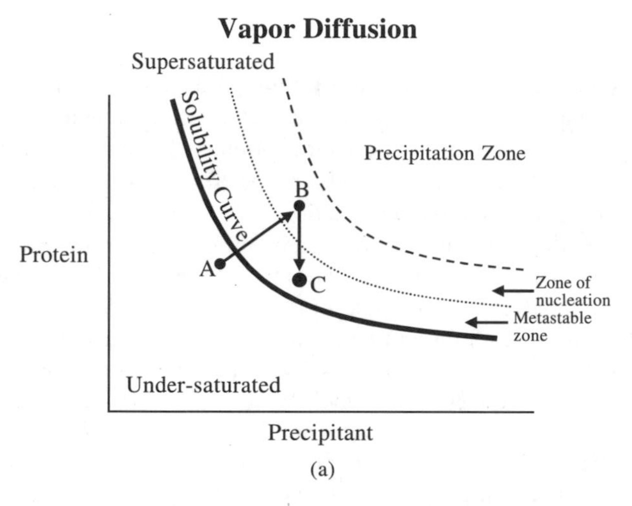

FIGURE 2.2 **(a)** A phase diagram of the vapor diffusion process showing the initial (A), nucleation (B), and final (C) states of the system.

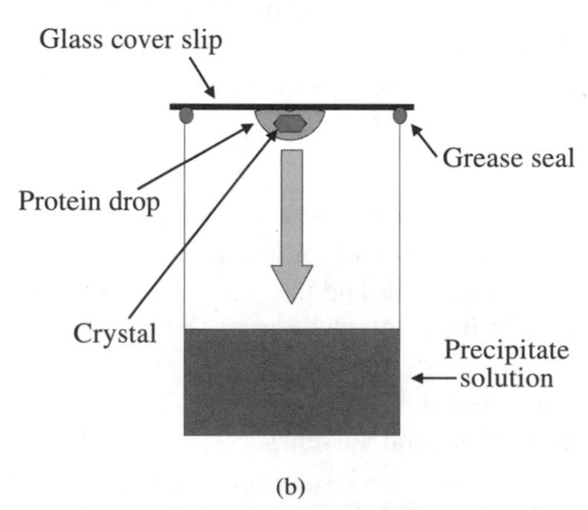

FIGURE 2.2 **(b)** A cartoon showing a typical hanging drop vapor diffusion apparatus. Crystals grow as the concentration of protein and precipitants increase in the drop.

crystal nucleation occurs. Nucleation and crystal growth depletes the local protein concentration and state C is reached, where crystal growth occurs.

In practice, vapor diffusion crystallization trials are generally carried out using 24-well Linbro plates with each well sealed by a microscope cover slip of appropriate size (See Figure 2.2b). Each well of the plate contains ~500 μl of a precipitant solution. A 2- to 4-μl drop, containing equal parts protein solution (~10 mg/ml) and precipitant solution (taken from the well), is then placed on a siliconized microscope cover slip that is carefully inverted and placed over the well. A bead of high vacuum grease previously applied to the lip of each well provides the seal between the well and the cover slip.

Alternatively, one can use sitting drop vapor diffusion where the drop is placed in a raised depression within the well. The sitting drop vapor diffusion experiment requires special plates or devices but set-up and retrieval of crystals is much simpler. Once the plate is prepared, it is then incubated at either 4 or 18°C for a period of time ranging from a few days to weeks. During incubation, each well of the plate is checked periodically under a microscope for the presence of crystals or precipitate. Conditions derived from the initial screen are usually further refined by adjusting pH, additives, and concentrations to grow suitable crystals for data collection and structure analysis. Crystal seeding (Stura et al., 1992) can also be explored to improve the crystal size and quality.

Several crystallization-screening protocols have been developed over the past decade to help identify initial crystallization conditions. These include footprint screening (Sutra et al., 1992), sparse matrix sampling (Jancarik and Kim, 1991), statistical methods (Carter and Carter, 1979), and MON-48 (Shieh et al., 1991). Reagent kits for some of these screens are now commercially available (Hampton Research [*www.hamptonresearch.com*], Emerald BioStructures [*www.emeraldbio-structures.com*]) and are commonly used in the laboratory.

In addition to the vapor diffusion method described previously, other techniques such as the batch and micro-batch methods, bulk and micro dialysis, free interface diffusion, liquid bridge, and concentration dialysis have also been developed to produce crystals for x-ray diffraction analysis (see McPherson, 1982 and McPherson, 1999).

Automated systems for vapor diffusion and micro-batch crystallization screening are now commercially available from Cyberlab, Inc. [*www.gilson.com/cyberlab.htm*] and Douglas Instruments Ltd. [*www.douglas.co.uk*]. These systems are particularly useful for rapid screening of new samples.

X-RAY SOURCES AND DETECTORS

X-Ray Sources

Today, most macromolecular x-ray diffraction experiments are carried out using x-radiation produced either by a rotating anode x-ray generator or by electrons or positrons losing angular momentum while orbiting in high energy synchrotron storage rings. The most common laboratory x-ray source for macromolecular crystallography is the rotating anode generator. In the rotating anode generator, electrons produced by a tungsten filament (cathode) are accelerated to high velocities by a potential difference toward a metal target (anode) where their impact produces x-rays of a wavelength characteristic of the target material (1.5418 for the CuKα characteristic line). The anode is usually a sealed copper cup that is water cooled and rotated at high speed (6500 rpm) to dissipate the heat generated during x-ray production. The removal of heat from the target material limits the x-ray intensity to 5 to 18 kW for most rotating anode generators. Commercial rotating anode generators are marketed in the U.S. by Rigaku [*www.msc.com*], Bruker AXS [*www.bruker-axs.com*], and Nonius [*www.nonius.com*].

Synchrotron rings are very large and expensive facilities occupying a full city block. X-ray ports (called beamlines) are placed every 10° around the ring, allowing for a total of 30 to 35 user stations. The large number of user stations makes the synchrotron cost-effective. The x-rays emitted by electrons and positrons moving at relativistic velocities as they orbit in high-energy synchrotron storage rings offers the crystallographer

1. High intensity, which allows micro crystals to be used.
2. The ability to select the wavelength of the X-radiation emitted. The high intensity of the synchrotron source (100 to 1,000,000 times that of a rotating anode generator) has allowed extension of the tractable resolution for an increasing number of protein structures to near atomic resolution, roughly 1.2 Å or better. This, in turn, has allowed the use of common techniques used in small molecule crystallography, such as the refinement using anisotropic temperature factors. The ultra-high resolution structures produced in this manner have improved the accuracy of structures and given researchers a deeper understanding of the biochemical behavior of the protein.

Another potential use of the intense x-ray beam and beam characteristics of 3rd generation synchrotron sources is micro-crystallography. X-ray diffraction analysis of macromolecular micro-crystals (dimensions <10 μm) is generally not successful using conventional x-ray sources because they diffract too weakly. However, by focusing the x-ray beam using special optics in conjunction with undulator beamlines at 3rd generation synchrotron storage rings, researchers are now able to collect data from micro-crystals previously thought to be unusable.

The tunability of synchrotron radiation allows for data collection at or near the x-ray absorption edge of anomalous scatterers present in the protein or crystal to provide experimental phase information. Using techniques such as multi-wavelength anomalous dispersion (MAD) and single-wavelength anomalous diffraction (SAD) researchers are now able to solve macromolecular structures in a matter of days or weeks, a process that required months, or even years, a decade ago.

Because synchrotron light sources are expensive to build and operate, most are funded at the national level. In the U.S., five light sources are available to researchers: the Advanced Photon Source (APS), Argonne National Laboratory [*www.aps.anl.gov*], the Advanced Light Source (ALS) Lawrence Berkley National Laboratory [*www.als.lbl.gov*], the National Synchrotron Light Source (NSLS) Brookhaven National Laboratory [*www.nsls.bnl.gov*], Stanford Synchrotron Research Laboratory (SSRL), Stanford University [*www.ssrl.slac.stanford.edu*], and the Macromolecular Diffraction Facility Cornell High Energy Synchrotron Source [*www.chess.cornell.edu*]. Researchers can apply for time at these facilities by submitting proposals to their General User Programs.

Detectors

The introduction of commercial multi-wire x-ray area detectors, such as the Nicolet X100 in the mid-1980s, marked a new chapter in macromolecular data collection. By recording many reflections simultaneously, these new detectors offered the ease of fast, automated data collection without the expense and effort associated with film data collection. Multi-wire detectors, although fast (with readout times under 10 s), were limited by their active area and dynamic range. This led to the development of storage-phosphor-based imaging plate (IP) systems such as the Rigaku, MarResearch [*www.marresearch.com*], and Mac Science systems in the early 1990s. These large, 30-cm IP systems allowed for the collection of higher resolution data during each exposure and were not as easily saturated compared to their multi-wire counterparts. A disadvantage of IP systems is the relatively slow readout time of 1 to 3 min, which makes IPs less efficient for data collection using the short exposure times commonly used for synchrotron data collection. However, IP systems remain the most popular means of collecting macromolecular x-ray diffraction data in the home laboratory.

The demand for a large detector with high dynamic range and fast readout for synchrotron applications led to the development of the charge-coupled device (CCD) based area detector system found on most synchrotron beam lines for macromole- cular crystallography. In the CCD detector, x-ray photons are converted to visible light by an x-ray-sensitive phosphor and focused onto the CCD chip either by a reducing fiber-optic taper or a lens system. The light photons induce a charge proportional to the number of x-ray photons striking the phosphor in the wells of the chip. CCD detectors are, however, lim*ited by the size and cost of the CCD chip and the light loss in the focusing system. Thus, most CCD systems for macro- molecular crystallography consist of four or more CCD modules placed in a tiled 2 × 2 (ADSC Quantum 4 and Quantum 210 [*www.adsc-xray.com*], and the MSC Jupiter 210 [*www.msc.com*]), or 3 × 3 (ADSC 310, and the SBC-2 [*www.sbc.anl.gov*]) arrays. However, three systems with larger CCD chips and a single CCD module, the MarResearch 165, the Bruker SMART 6000, and Bruker Proteum 300, are also commercially available.

Currently under development are the next generation x-ray detectors such as the pixel-array, amorphous silicon, and solid state detectors. These detectors offer a larger active area, lower background, faster readout, and higher dynamic range than the current CCD or IP systems.

DATA COLLECTION AND PROCESSING

Crystal Selection and Mounting

Once crystals are obtained, they must be placed in the x-ray beam in order to record the diffraction pattern. The experimenter should use care in choosing the crystal. First, it should have the same morphology or habit as the majority of crystals in the drop because crystals having similar habits should, in most cases, have the same unit cell and crystal symmetry. It also makes it easier to find isomorphous crystals for heavy atom derivatization, if needed. The crystal should be well formed with sharp faces and no re-entrant angles between its faces. The presence of re-entrant angles between crystal faces usually indicates that the crystal is twinned. Twinned crystals contain two of more independent crystalline components (present in varying amounts), each capable of producing a diffraction pattern. Thus, the diffraction pattern of a twinned crystal is a complicated combination of the diffraction patterns from the multiple crystalline components. Data from twinned crystals is usually difficult to process because the independent diffraction patterns from the various twin components must be identified and processed separately. Finally, it is always a good idea to survey the crystals using a polarizing microscope before choosing one for mounting. Single crystals will extinguish uniformly (every 90°) as they are rotated between the polarizer and analyzer of the polarizing microscope.

Macromolecular crystals contain an appreciable amount of solvent. To prevent the crystal from drying out, it must either be mounted and sealed in a glass capillary containing a small quantity of its mother liquor or flash frozen (Hope, 1988) into an amorphous glass and maintained at cryogenic temperatures during data collection. Although the latter technique using a small fiber loop (Teng, 1990) has become popular, capillary mounting (see McRee, 1999) is still employed for data collection at room temperature for crystals that cannot be flash frozen. It generally is good practice to first survey a new crystal at room temperature before trying to flash freeze it because the freezing process may affect or destroy the diffraction quality of the crystal. Flash freezing the crystal, however, offers certain advantages. The crystal is less susceptible to radiation damage, usually allowing for the collection of a complete data set from a single crystal. The amount of handling and the resulting mechanical stress on the crystal is minimized, which may lead to better diffraction quality. Finally, the crystal can be recovered and stored at cryogenic temperatures for future experiments.

The commonly used loop freezing technique is illustrated here. A small fiber or glass loop (0.05 to 0.5 mm diameter) is mounted at the end of a metal pin that is connected to a magnetic base for easy attachment to the goniometer head. Prior to mounting the crystal, the pin is centered in a cold (100 K) stream of nitrogen gas intersecting the position of the x-ray beam on the goniometer (this position is usually the center of the cross-hairs of the alignment microscope). The crystal is then scooped up with the loop directly from the solution and while the cold gas stream is momentary blocked, it is quickly transferred to the goniometer head. The cold stream obstruction, usually an index card or a microscope slide, is then removed allowing the cold gas stream to flash freeze the crystal. Once in place, the crystal is checked with the alignment microscope to ensure that it is centered in the cold stream and the x-ray beam.

One disadvantage of the flash-freezing technique is the increase in mosaicity that usually accompanies the process. Cryo-protectants (Garman, 1999) such as glycerol (13 to 25%), ethylene glycol (11 to 30%), PEG 400 (25 to 35%), 2-methyl-2,4-pentanediol (MPD) (to 28%), and glucose (to 25%) can be used to limit this increase in mosaicity. An easy method for determining cryo-protectant conditions is to simply flash freeze a loop full of cryo-protectant solution in the cold stream. The proper conditions have been determined if the solution freezes in the loop as a clear glass. If cryo-protection is needed, the crystal is transferred momentarily to a small drop of cryo-protectant solution prior to placing it into the cold stream.

Capillaries, cryo-loops, and other crystal mounting supplies are available from Charles Supper Company [*www.charles-supper.com*] and Hampton Research [*www.hamptonresearch.com*].

Data Collection

Before the actual data collection can begin, several important parameters need to be determined. These include the direct beam position (at $2\theta = 0$) on the face of the detector, the crystal-to-detector distance, $\Delta\phi$ or oscillation step, detector swing or 2θ angle, and exposure time. The direct beam position is usually determined by the site manager following realignment of the x-ray optics associated with changing the filament and recorded in a computer file or notebook. The remaining parameters usually can be obtained from a series of still photographs taken 90° apart. The crystal-to-detector distance is crystal dependent and should be chosen such that the reflections are well resolved (i.e., the reflection profile goes to background between reflections). At the distance where the reflections are resolved, if the observed diffraction pattern does not cover the entire detector surface, the detector may be moved further away from the crystal until 90 to 95% of the detector is covered. This will increase the signal-to-noise ratio of the data set. If, on the other hand, the diffraction pattern is observed to extend beyond the edge of the detector, the detector in some cases can be moved off center on its 2θ axis to allow collection of data at high resolution. The choice of the oscillation step size is both crystal and detector dependent and is usually chosen to match the crystal mosaicity and the readout speed of the detector. Image plate detector systems have slow readout times so data is usually collected using an oscillation step ranging from 0.5 to 1.5° while 0.1 to 0.5° oscillation steps are used for multi-wire and CCD detectors, which have faster readout speeds. The smaller oscillation step also increases the signal-to-noise ratio of the data set.

Once the crystal-to-detector distance, detector swing angle, and oscillation steps have been decided, the exposure time can be chosen by taking a series of images with increasing exposure times. It is generally a good rule of thumb to select an exposure time such that the average intensity of the high resolution data is at least three times the background on a single image. In addition, low-resolution reflections should also be checked for possible saturation of the detector.

Generally, one would want to determine the lattice parameters and orientation of the crystal (indexing) prior to collecting a data set. This is usually done by determining the positions (in X, Y, and ϕ) of a number of reflections on an image and utilizing auto-indexing algorithms. Indexing the crystal prior to collecting the data ensures that the data can be processed. Indexing also allows the estimation of the solvent content of the crystal and the number of molecules in the unit cell by Matthews' method (1968),

$$V_s = 1 / (1.23M / (V/Z))$$

where

V_s = the solvent content
M = the molecular weight of the protein

V = the volume of the unit cell
Z = the number of protein molecules in the unit cell

Crystals that give an unreasonable number of molecules in the unit cell or solvent content, an indication of potential problems can, therefore, be identified and removed before beginning data collection. Prior indexing also provides a means of generating a data collection strategy, thus ensuring that a complete data set is collected in the most efficient manner. Data collection strategies become increasingly important when beam time is limited or the detector is placed off axis.

Most commercial detectors provide their own software to collect and process data. Bruker, MarResearch, and Rigaku provide Proteum, MarFLM, and Crystal-Clear, respectively. In addition, programs such as HKL (Otwinski and Minor, 1997), d*TREK (Pflugrath, 1999), DPS (Steller et al., 1997, Bolotovsky et al., 1998, Rossmann and van Beek, 1999) and MOSFLM (Leslie, 1990) can also be used to process data from a variety of detectors. The details for each package can be found at [*www.ccp14.ac.uk/solution/ccd_software/*].

In general, the data reduction procedure consists of three or four steps: indexing, integration, scaling, and, in some cases, post refinement. First the data must be indexed. During indexing, the unit cell parameters and crystal orientation, are determined and a three-integer index is assigned to each reflection. Based on the unit cell, crystal orientation, and detector geometry, the reflection or peak positions for the entire data set can be predicted. Integration of the x-ray intensity at these positions then is carried out using either a two-dimensional summation or three-dimensional profiles. In both cases, model reflection profiles must be constructed to provide the program with information about peak size and shape so that the background measurements can be made correctly. Finally, the data must be merged and scaled together, based on space group symmetry, to provide a unique list of intensities. Indexing is usually judged by how well the predicted diffraction pattern fits the observed diffraction pattern or how close to integer values the calculated Miller indices are. Integration is usually monitored by the average rms deviation in X, Y, and ϕ of the predicted peak position compared to the intensity maxima found at or near that position. Before the data can be scaled, the space group of the crystal must be determined. This is done by looking for systematic absences in certain classes of reflections (see International Tables for Crystallography, Vol. A,) which indicate the presence of symmetry within the crystal lattice. The data then is merged and scaled according to the space group symmetry. Most scaling programs provide detailed statistics about the data set including R_{merge}, completeness, and $<I/\sigma>$ as a function of resolution. These values can be used to judge the quality of the data and to determine the effective resolution of the data set,

$$R_{\text{merge}(I)} = \frac{\sum_{hkl} \backslash \langle I \rangle - I_{hkl} \backslash}{\sum_{hkl} \langle I \rangle}$$

where I_{hkl} is the intensity of reflection *hkl*.

R_{merge} is a indication of the agreement between the scaled symmetry-related reflections in the data set and is usually in the range from 0.03 to 0.08 for well-diffracting crystals. The R_{merge} value as a function of resolution can also be used to estimate the effective resolution (the resolution at which R_{merge} exceeds 0.25) of the data set. Another quality indicator is the signal-to-noise ratio or I/σ of the data set. Larger $<I/\sigma>$ should provide a better estimate of the integrated intensities. The I/σ value as a function of resolution can also be used to estimate the effective resolution (the resolution at which the $< I/\sigma>$ falls below 2) of the data set.

THE PHASE PROBLEM

The data measured in an x-ray diffraction experiment are processed and eventually used as either intensities, I_{hkl}, or structure factor amplitudes, $|F_{hkl}|$. A structure factor amplitude, $|F_{hkl}|$, is simply the squareroot of the intensity, I_{hkl}, so that $|F_{hkl}| = \sqrt{I_{hkl}}$. A typical protein structure determination will measure thousands of intensities with each data point having a unique set of Miller indices, *hkl*. However, the diffraction experiment is unable to directly determine another vital piece of information, the phase of each intensity. This constitutes the so-called phase problem in x-ray diffraction. If the phase could be measured for each individual reflection, then an electron density map and, consequently atomic positions, could be calculated in a direct and straightforward way. To restate the phase problem: in order to reconstruct the protein image from its diffraction pattern both the intensity and phase angle, which can assume any value from 0 to 2π, of each of the thousands of measured reflections must be known. However, the phase information is lost in the diffraction experiment and a major effort must be made to assign correctly enough of these phases in order to produce a reasonable and chemically meaningful electron density map. Once a viable map is available, the completion of the structure determination is merely a matter of refining and improving the original solution.

The phase problem can be best understood from a simple mathematical construct. The structure factors (F_{hkl}) are treated in diffraction theory as complex quantities, i.e., they consist of a real part (A_{hkl}) and an imaginary part (B_{hkl}). If the phases, Φ_{hkl}, were available, the values of A_{hkl} and B_{hkl} could be calculated from very simple trigonometry,

$$A_{hkl} = |F_{hkl}|\cos(\Phi_{hkl})$$

$$B_{hkl} = |F_{hkl}|\sin(\Phi_{hkl})$$

which leads to the relationship:

$$(A_{hkl})^2 + (B_{hkl})^2 = |F_{hkl}|^2 = I_{hkl}$$

These relationships are often illustrated on an Argand diagram (see Figure 2.3). From the Argand diagram, it is obvious that A_{hkl} and B_{hkl} may be either positive or negative, depending on the value of the phase angle, Φ_{hkl}. Units of A_{hkl}, B_{hkl}, and F_{hkl} are in electrons. With this information, the electron density, ρ_{xyz}, at a given *xyz* position can be calculated using the equation below (V is equal to the volume of crystallographic unit cell),

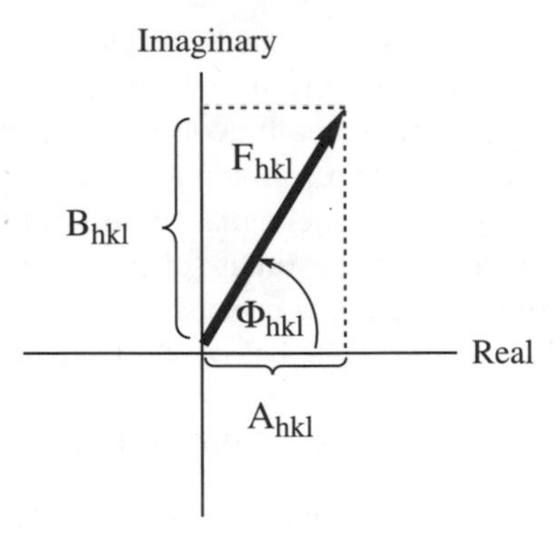

FIGURE 2.3 An Argand diagram of structure factor $\boldsymbol{F}_{hkl}$ with phase Φ_{hkl}. The real ($\boldsymbol{A}_{hkl}$) and imaginary ($\boldsymbol{B}_{hkl}$) components are also shown.

$$\rho = \frac{1}{V}\left\{\sum_{hkl} A_{hkl}\cos[2\pi(hx + ky + lz)] + \sum_{hkl} B_{hkl}\sin[2\pi(hx + ky + lz)]\right\}$$

Note that ρ_{xyz} is units of electrons/volume, i.e., electron density. In practice, the electron density for one unit cell of three-dimensional space is calculated by starting at $x, y, z = 0, 0, 0$ and stepping incrementally along each axis, summing the terms as shown in the equation for all *hkl* at each point in space. The electron density map will consist of areas of high values, corresponding to electron density around atomic positions. In proteins, large regions of low electron density generally correspond to areas containing mostly nonordered solvent (water) molecules. Electron density maps can be contoured at regular intervals to produce spheres or globular regions of high electron density. When viewed in three-dimensions, helical, and/or pleated sheet regions can be readily recognized.

The most common method for solving the phase problem in macromolecular crystallography these days is the multiple-wavelength anomalous diffraction (MAD). The MAD phasing technique (Hendrickson, 1991) became increasingly popular in the early 1990s and continues to find wide and successful application. MAD phasing exploits anomalous scattering information gained from diffraction experiments that are carried out at several different wavelengths near the x-ray absorption edge of an anomalous scatterer, such as selenium or iron. Because a tunable x-ray source is needed, MAD experiments must be conducted at a synchrotron source.

Recently, the requirement of data collected at several wavelengths has been shown to be unnecessary, in some cases, because anomalous scattering data collected at a single wavelength, if it is of sufficient accuracy, can yield interpretable electron density maps. Although this means of structure determination was first demonstrated in 1985 (Wang), using a process called iterative single-wavelength anomalous scattering (ISAS), and was successfully used by Chen et al. (1991) to solve the structure of a neurophysin-dipeptide complex, it has gained general acceptance only recently. Wide use of the ISAS process or the SAD (single-wavelength diffraction) method, as some call it, is expected to come into general use in the future using x-rays from home sources as well as from synchrotrons.

In addition to the MAD and SAD methods, there are the traditional isomorphous replacement methods that include multiple isomorphous replacement (MIR), which uses several heavy atom derivatives, and single isomorphous replacement (SIR), which uses only one heavy atom derivative. The underlying principle to all these methods is the phase–triangle relationship. To understand this relationship we shall begin our discussions with the isomorphous replacement method.

Isomorphous Replacement Methods

Until recently, the isomorphous replacement method (Green et al., 1954) has been the traditional technique for solving the phase problem. Here, a heavy atom (atom with a large number of electrons) must be introduced into the crystal lattice in such a way as not to disturb the conformation and packing of the protein in the crystal. Furthermore, it is necessary that the crystals of the heavy atom derivative be isomorphous with the native protein crystals, i.e., have similar cell dimensions and symmetry (space group). Use of data from a single heavy atom derivative, combined with native protein data, in the phasing process, is referred to as the single isomorphous replacement (SIR) method; use of native data and two or more heavy atom derivatives in phasing is called the multiple isomorphous replacement (MIR) method. These derivatives are usually obtained by soaking protein crystals in a solution containing an added heavy atom (usually as ionic salts) in the hope that the crystals will incorporate the heavy atom.

The large number of electrons associated with the heavy atom will produce large scattering amplitudes from this atom and, as a result, many of the intensities will be dominated by scattering from this heavy atom. In order to locate the position of the heavy atom, a Patterson function (Patterson, 1935) is calculated which represents a three-dimensional vector space map, also known as an $|F|^2$ map. The Patterson function is

$$P(uvw) = \frac{1}{V}\sum_{hkl} |F_{hkl}|^2 \cos[2\pi(hu + kv + lw)]$$

This function is similar to the electron density function given earlier. Here, $P(uvw)$ is the value of the Patterson function at Patterson coordinates u, v, w; these are the traditional coordinate symbols (instead of x, y, z) used for squared ($|F_{hkl}|^2$) space. All other symbols have their usual meaning. The Patterson function is a Fourier summation using the intensities as coefficients and setting all Φ_{hkl} equal to 0. The resulting contoured map will have peaks corresponding to vector differences between all atoms in the structure. A vector between an atom and itself is a zero vector; therefore, the Patterson functions always have a very large peak at $u, v, w = 0, 0, 0$.

In macromolecular crystallography, the Patterson coefficients used in the SIR and MIR techniques are the squared differences in the intensities between the native and heavy atom derivative data, $[\Delta F_{(hkl)}]^2$. If the native data set is represented by $F_{P(hkl)}$ and the heavy atom derivative by $F_{PH(hkl)}$, the differences are $|\Delta F_{(hkl)}| = ||F_{PH(hkl)}| - |F_{P(hkl)}||$; thus $|\Delta F_{(hkl)}|^2$ are the Patterson coefficients. Because the differences primarily are due to the heavy atoms, the resulting isomorphous difference Patterson map reveals the location of the heavy atoms. Programs, such as *SOLVE* (Terwilliger

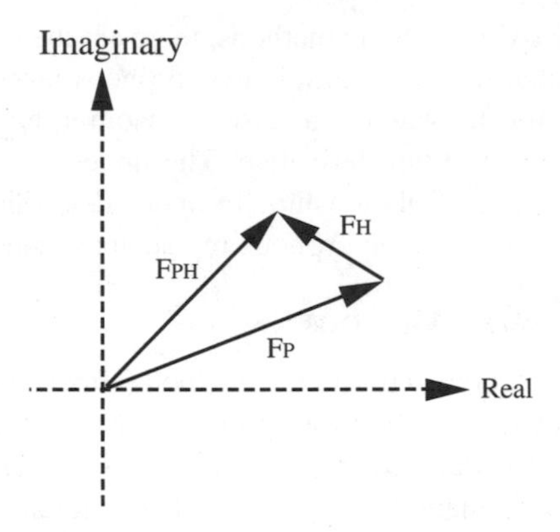

FIGURE 2.4. An Argand diagram showing the vector sum $\mathbf{F}_{PH} = \mathbf{F}_P + \mathbf{F}_H$.

and Berendzen, 1999, *www.solve.lanl.gov*), will automatically search Patterson space for heavy atom positions. Alternatively, *Shake-and-Bake* (Miller et al., 1993, *www.hwi.buffalo.edu/SnB*) uses direct statistical methods to locate heavy atoms.

The structure factor of the heavy-atom derivative may be expressed vectorially as $\mathbf{F}_{PH}$, where $\mathbf{F}_{PH} = \mathbf{F}_P + \mathbf{F}_H$ **:** $\mathbf{F}_P$ and $\mathbf{F}_H$ are the structure factors of native protein and the heavy atom, respectively, for the same reflection. Figure 2. 4 illustrates this phase triangle on an Argand diagram.

A triangle has three sides. For this triangle, its sides represent the magnitudes of F_{PH}, F_P, and F_H. If we know the orientation (phase or Φ) of one of the sides, then the orientation of the whole triangle may be fixed. If the orientation of the triangle is fixed, then the orientations of its three sides are also fixed. So, if we know the phase of one of the three quantities, F_{PH}, F_P, or F_H, then we will know the phases of the other two. F_{PH} and F_P are measured quantities, but F_H is calculated from the heavy atom parameters and the phase angle of F_H can also be calculated. This means if we know the heavy atom positions in the crystals, then the isomorphous replacement method should give us the phase angle of F_P for all the reflections. In simplest terms, isomorphous replacement finds the orientation of the phase triangle from the orientation of one of its sides. It turns out, however, that there are two possible ways to orient the triangle if we fix the orientation of one of its sides (see Figure 2.5). Even with this ambiguity, we have reduced the number of possibilities of orienting the phase triangle from an infinite number to just two possibilities. This is a tremendous improvement from no phase information to two possible values. The situation of two possible SIR phases is called the phase ambiguity problem because we obtain both a true and a false phase for each reflection. Both phase solutions are equally probable, i.e., the phase probability distribution is bimodal.

The phase ambiguity problem can also be understood from the mathematical expression shown next. We can express the protein phase angle α_P from the following equation,

$$|\mathbf{F}_{PH}|^2 = |\mathbf{F}_P|^2 + |\mathbf{F}_H|^2 + 2|\mathbf{F}_P||\mathbf{F}_H| \cos(\alpha_P - \alpha_H)$$

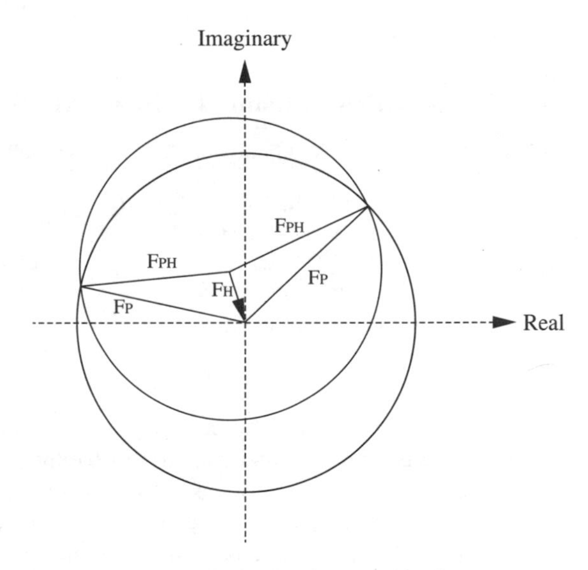

FIGURE 2.5 Harker construction of protein phase ambiguity.

which leads to

$$\alpha_P = \alpha_H + \cos^{-1}[(|\mathbf{F}_{PH}|^2 - |\mathbf{F}_P|^2 - |\mathbf{F}_H|^2) / 2|\mathbf{F}_P||\mathbf{F}_H|] = \alpha_H \pm \beta$$

Here we have two possible signs for β, because we have an arccosine function, which gives two possible solutions for a single argument. Thus, the phase ambiguity is actually a mathematical problem.

Because we get one true phase and one false phase, we do not know which one is the correct one to choose. Thus, the next step in the isomorphous replacement method is to resolve the phase ambiguity. Although the phase ambiguity is a mathematical problem, most crystallographers routinely solve it by collecting more experimental data.

For example, if we have another heavy atom isomorphous derivative available with heavy atom sites different from those found in the first derivative, when the preceding process is repeated, we will get two solutions, one true and one false for each reflection from the second derivative as well. The true solutions should be consistent between the two derivatives while the false solution should show a random variation. Thus, by comparing the solutions obtained from these two calculations, one (the computer) can establish which solution represents the true phase angle. This is the principle of the MIR method. One can also utilize the anomalous scattering (AS) data of the first derivative to resolve the phase ambiguity. In this case, the technique is called the SIRAS approach. If two derivatives and anomalous data are used, then it is called the MIRAS approach.

Usually, isomorphous replacement methods are coupled with a density modification process called solvent flattening. The solvent flattening process was made practical by the introduction of the ISIR/ISAS program suite (Wang, 1985) and other phasing programs such as DM (CCP4, 1994) and PHASES (Furey and Swaminathan,

TABLE 2.2
Protein Residues and Their Affinities for Heavy Metals

Residue	Affinity for	Conditions
Histidine	K_2PtCl_4, $NaAuCl_4$, $EtHgPO_4H_2$	pH > 6
Tryptophan	$Hg(OAc)_2$, $EtHgPO_4H_2$	
Glutamic, aspartic acids	$UO_2(NO_3)_2$, rare earth cations	pH > 5
Cysteine	Hg, Ir, Pt, Pd, Au cations	pH > 7
Methionine	$PtCl_4\backslash^{2-}$ anion	

1997) are based on this approach. Wang pointed out that the phase ambiguity in both the SIR or SAS data can be resolved by an iterative noise filtering process using mathematics, not additional experimental data, to resolve the phase ambiguity problem. Using solvent flattening SIR data alone is sufficient for the determination of protein crystal structure, provided that the derivative crystal is isomorphous to the native crystal. Likewise, SAD data alone should also be sufficient for the protein structure determination. This will be described in more detail in the next section.

Key to the MIR and SIR methods is the production of an isomorphous heavy atom derivative. This is traditionally done by soaking crystals in dilute (2 to 20 mm) solutions of heavy atom salts. Crystal cracking is generally a good indication that the heavy atom is interacting with the crystal lattice, and suggests that a good derivative can be obtained by soaking the crystal in a more dilute solution.

Table 2.2 lists some common heavy metals and their affinities for specific amino acid residues (Petsko, 1985). This list can serve as an initial guide in preparing heavy atom derivatives. Once data on a potential isomorphous heavy atom derivative have been collected, the merging *R* factor between the two data sets is another indication of both heavy atom incorporation and isomorphism between the native and derivative crystals. The merging *R* factor measures agreement of identical reflections between the two data sets. Typically, R_{merge} values for isomorphous derivatives range from 0.05 to 0.15, with values below 0.05 indicating that there is little heavy atom incorporation and values above 0.15 indicating a lack of isomorphism between the two crystals. Finally, an isomorphous difference Patterson should be calculated and inspected for consistent heavy atom peaks.

Anomalous Dispersion Methods

In the past 10 years, anomalous dispersion (AD) effects have been used more and more frequently to solve the phase problem. All elements display an AD effect in x-ray diffraction. However, the elements in the first and second row of the periodic table, for example, C, N, O, and so on, have negligible AD effects. For heavier elements, especially when the x-ray wavelength approaches an atomic absorption edge of the element, these AD effects can be very large. The scattering power of an atom exhibiting AD effects is

$$f_{AD} = f_n + \Delta f' + i\Delta f''$$

The f_n value is the normal scattering power of the atom in the absence of AD effects. The $\Delta f'$ value arises from the AD effect and is a real factor added to f_n; it may be positive or negative. The $\Delta f''$ is an imaginary term that also arises from the AD effect and is always positive and 90° ahead of $(f_n + \Delta f')$ in phase angle. The values of $\Delta f'$ and $\Delta f''$ are highly dependent on the wavelength of the radiation. An excellent discussion of AD is provided on the University of Washington Web site [*www.bmsc.washington.edu/scatter/*].

In the absence AD effects, the intensity of I_{hkl} is the same as the intensity of its inverse, I_{-h-k-l}. However, with AD effects, I_{hkl} and I_{-h-k-l} will be unequal. I_{hkl} and I_{-h-k-l} are referred to as Friedel pairs. For many crystallographic applications, it is assumed that Friedel pairs are, at least approximately, equal. However, accurate measurement of Friedel pair differences can be used to extract starting phases if the AD effect is large enough.

Figure 2.6 illustrates how AD from a single heavy atom can cause the intensities of a Friedel pair to be different. In the left-hand diagram (a), the vectors $\mathbf{F}_{n+}$ or $\mathbf{F}_{n-}$ represent the total scattering by "normal" atoms without AD effects. The vectors $\mathbf{F}_{A+}$ or $\mathbf{F}_{A-}$ represent the sum of the normal and real AD scattering values $(f_n + \Delta f')$.

The $\mathbf{F}_{A''+}$ or $\mathbf{F}_{A''-}$ vectors are the imaginary AD components and appear 90° (at a right angle) ahead of the $\mathbf{F}_{A+}$ or $\mathbf{F}_{A-}$ vectors. The total scattering vector in the top of the diagram is $\mathbf{F}_+$ and the bottom counterpart is the negative vector sum $\mathbf{F}_-$. The resultant vector, $\mathbf{F}_-$ in this case, is obviously shorter than the $\mathbf{F}_+$ vector. The $\mathbf{F}_-$ vector, however is not always shorter than the F_+ vector, most of the time it is longer. What is interesting is that for each reflection we can have two measurable quantities, $|\mathbf{F}_+|$ and $|\mathbf{F}_-|$, with different magnitudes.

In Figure 2.6(b), the $\mathbf{F}_-$ and associated vectors are shown with the imaginary axis inverted. Figure 2.6(c) uses $\mathbf{F}_+$ from Figure 2.6(a) and $\mathbf{F}_-$ from Figure 2.6(b) in order to construct the phase triangle relationship between $\mathbf{F}_+$, $\mathbf{F}_-$ and $2\mathbf{F}_{A''}$.

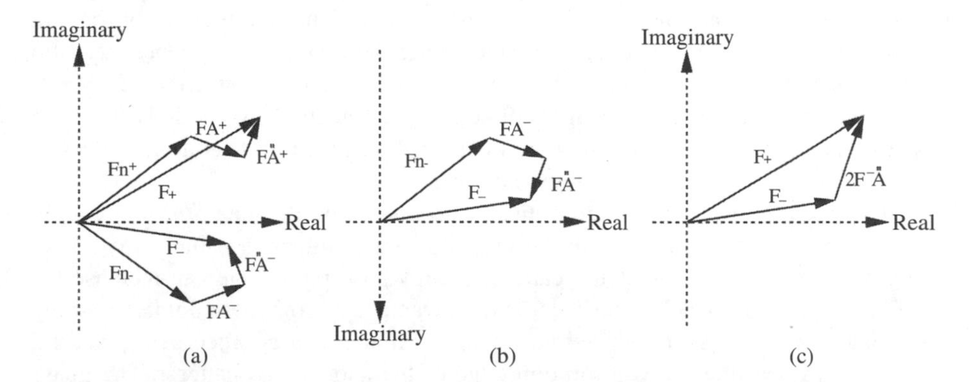

FIGURE 2.6 Illustration of the phase triangle in anomalous dispersion. (a) Diagram showing vector sums of $\mathbf{F}_{n+}$, $\mathbf{F}_{A+}$, and $\mathbf{F}_{A''+}$, which produce $\mathbf{F}_+$; corresponding vectors are shown for $\mathbf{F}_-$ at the bottom. (b) Diagram showing the $\mathbf{F}_-$ and associated vectors from (a) with the imaginary axis inverted. (c) Diagram composed of the $\mathbf{F}_+$ and $\mathbf{F}_-$ vectors from (a) and (b), respectively, showing the phase triangle relationship with $2\mathbf{F}_{A''}$.

The preceding observation is also significant in terms of designing our experiments for solving the phase problem. In the isomorphous replacement method described previously, heavy atoms were introduced into the native crystals to obtain $|F_{PH(hkl)}|$ so that for each reflection there would be two quantities, $|F_{PH(hkl)}|$, and $|F_{P(hkl)}|$ with different magnitudes, which can be experimentally measured. Using anomalous scattering data there is no need to add heavy atoms in the crystals because both $|F_{+++}|$ and $|F_{---}|$ are measurable quantities.

With this introduction, one can immediately see that the anomalous scattering data technique to obtain phase information can be done in a very similar manner using the isomorphous data technique. So, the statement made earlier about the principle for the isomorphous replacement, "to find the orientation of a phase triangle from one of its sides," should also be applicable to the use of anomalous scattering data. From this point of view, the MAD method is conceptually the counterpart of the MIR method and the SAD method is the counterpart of the SIR method. The major difference is only how the two measurable quantities in the experiments are obtained, i.e., are they measured in two separate experiments or are they obtained directly from a single experiment. Compared with the isomorphous method, the anomalous scattering technique should save considerable time and effort in obtaining phase information.

Much has been said about the advantage of using anomalous scattering data for gaining phase information, but little has been said about the problems that one has to overcome in order to obtain the accurate $|F_{+++}|$ and $|F_{---}|$ values. The main problem in collecting anomalous scattering data is that the difference between $|F_{+++}|$ and $|F_{---}|$, or anomalous signal, is generally about one order of magnitude smaller than that between $|F_{PH(hkl)}|$, and $|F_{P(hkl)}|$. Thus, the signal-to-noise (*S*/*n*) level in the data plays a critical role in the success of AD experiments, that is, the higher the *S*/*n*, the greater the probability of success in producing an interpretable electron density map. Thanks to recent advances in x-ray optics and detector technology, we can now measure diffraction data with much higher precision than ever before. This means that we can now consider anomalous scattering from elements having very weak anomalous scattering signals, such as sulfur, which were previously thought to be too weak for AD experiments. In fact, there have been several recent successful examples of direct determination of protein crystal structure from the native crystals using the anomalous scattering signal of sulfur (see Liu et al., 2000). This approach is still under development and its use is expected to increase as future advances in x-ray detector and optics technology continue.

Because AD experiments are critically dependent on the data *S*/*n*, introducing a strong anomalous scatterer into the crystal or data collection using x-ray wavelengths at or near an anomalous scatterer's x-ray absorption edge can increase the *S*/*n* of the data. The MAD experiment takes advantage of both these points. A strong anomalous scatterer is introduced into the crystal and data are recorded at several wavelengths near the x-ray absorption edge of the anomalous scatterer. The phase ambiguity in each of the SAD experiments then is resolved by the use of additional SAD data collected at other wavelengths in a manner similar to the use of multiple derivatives in the MIR technique. Thus, the MAD method is geometrically similar to the MIR method but uses anomalous scattering information from different wavelengths instead of different derivatives.

The MAD method requires that the proteins under investigation contain anomalous scatterers whose x-ray absorption edges are in the tunable range (0.8 to 2.0 Å) of the synchrotron. Metalloproteins or seleno-methionine proteins are good candidates for a MAD study (Hendrickson et al., 1990). However, the preparation of the protein samples is often the limitation of this method. Seleno-methione incorporation involves the expression of the protein in a bacterial cell line incapable of producing methionine *de novo* (see Hendrickson et al., 1990). Seleno-methionine is then added to the minimal growth media and is incorporated into the expressed protein. Crystallization of the selenium labeled proteins usually produces crystals that are isomorphous with the native protein. Therefore, it is generally a good idea to first crystallize and characterize the protein before going to the expense of producing the seleno-protein.

For those proteins containing metal ions or metal clusters, such as Fe, Ni, and Cu, the MAD experiment can be carried out directly on the native protein crystals without the need of seleno-derivatives. Crystals containing heavy atoms incorporated via chemical modification or by soaking (McPherson, 1982) can also be used for MAD experiments. Finally, Dauter et al. (2000) has shown that ordered bromine in the protein's solvation shell could provide MAD phasing information.

The SAD method, which combines the use of SAD data and solvent flattening to resolve phase ambiguity, was first introduced in the ISAS program (Wang, 1985). The ISAS process has been used to phase several structures including a neurophysin *para-iodo*-Phe-Try-amide complex [I-SAD, λ = 1.54 Å, 46 kDa, 5 iodine atoms] (Chen et al., 1991) and ferrochelatase [Fe-SAD, λ = 1.54 Å, 86 KDa, 2[2Fe-2S] clusters] (Wu et al., 2001). It is important to note that data for these structures were collected on a normal laboratory x-ray generator using Cu radiation. Thus, structures containing atoms such as iron, iodine or xenon, with sufficiently large $\Delta f''$ for Cu radiation, can be very effectively solved with SAD using home source data.

The ultimate goal of the SAD method is the use of S-SAD to phase protein data. Most proteins contain sulfur and, thus, it would be very convenient and efficient if data of sufficient accuracy could be measured to phase using the sulfur AD signal alone. Compare the anomalous scattering signal of sulfur ($\Delta f''$ = 0.56 electrons for Cu radiation) with that of iron ($\Delta f''$ = 3.20 electrons for Cu radiation) or xenon ($\Delta f''$ = 7.35 electrons for Cu radiation) and you begin to see the nature of the problem; the sulfur AD signal is very weak. Despite this, Hendrickson and Teeter (1981) were able to determine the structure of crambin (5 kDa, 6 sulfurs) using 1.5 Å resolution data collected in-house. The technique used, termed resolved anomalous phasing (RAP), estimates phases based on the sulfur substructure in the protein. Later, Wang showed that Rhe (12.5 kDa, 1 disulfide) could be solved at 3 Å resolution from simulated sulfur SAD data using the ISAS method. The ISAS method differs theoretically from the RAS method in that it uses the filtered image of the molecule to compute the protein phases. Accuracy in phase prediction is improved because the filtered image of the molecule is an approximate structure of the protein. Thus, iterative image filters, also known as solvent flattening, can be used to resolve the phase ambiguity from the anomalous scattering signal recorded using only a single wavelength.

To use sulfur as a phasing probe in S-SAD, several inherent problems must be addressed. First, because the sulfur K absorption edge is at 5.02 Å, well outside the tunable energy range of most current synchrotron facilities, the optimal wavelength must be identified in order to maximize the sulfur SAD signal and minimize the radiation damage for the crystal for data collection. A strategy collecting S-SAD data has been developed in Wang's laboratory and three protein structures have been solved using only low (3 to 4 Å) resolution data (Liu et al., 2000, submitted; Chen et al., manuscript in preparation). In these studies, the ISAS program has routinely resolved the phase ambiguity from S-SAD data collected using 1.54 or 1.74 Å X-radiation.

Molecular Replacement

Molecular replacement (MR) (Rossmann, 1972, 1990) has proven effective for solving macromolecular crystal structures based upon the knowledge of homologous structures. This method is straightforward and reduces the time and effort required for structure determination because there is no need to prepare heavy atom derivatives and collect their data. Model building is also simplified because little or no chain tracing is required. This method is gaining increasing importance as more structures become available. However, the three-dimensional structure of the search model must be very close to that of the unknown structure for the technique to work. Sequence homology between the model and unknown protein is helpful but not strictly required. Successful structure determinations have been carried out with search models with only 17% sequence similarity to the unknown protein. Several software programs such as AmoRe (Navaza, 1994) and X-PLOR/CNS (Brunger, 1990; Brunger et al., 1998) are available for MR calculations.

The MR method involves the following steps:

1. A rotational search to orient the homologous (known) model in the unit cell of the target (unknown) using a Patterson correlation search
2. A translational search where the newly orientated model is positioned in the unit cell as described later on

If M_1 represents the search model and M_2 represents the target structure, a simple relation can be described

$$M_2 = M_1[\mathbf{R}] + \mathbf{T}$$

where [**R**] is the rotation matrix and **T** is the translation vector. A suitable search model (M_1) can be obtained by sequence or function homology searches (such as BLAST) of known crystal and NMR structures. The rotation function [**R**] is first applied to M_1 to maximize the orientational agreement (largest Patterson correlation) between M_1 and M_2. Once the orientation of the model is known, a simple three-dimensional translation R-factor search over the asymmetric unit yields the solution. In a successful case, the Patterson correlation coefficient between M_1 and M_2 should be large and the R factor from the translational search should generally be lower than 50%. A rigid body refinement of the orientated and translated search model against the unknown data is usually carried out in order to refine the initial solution and to generate the initial phases of M_2 for model rebuilding and further refinement.

Because MR is dependent on a model structure, one must be careful not to introduce model bias into the unknown structure.

MODEL BUILDING

For structures not determined by molecular replacement, the chemical sequence of the protein must be fit into the experimental electron density map (Figure 2.7). This is called model building or chain tracing. As one would expect, the success or failure of chain tracing is dependent upon the quality of the electron density map. Thus, map quality evaluation is very important before one attempts to trace the chain. Good (traceable) electron maps should display most of the following features.

- There should be a large contrast difference between the protein and solvent regions, resulting in a clear boundary.
- The protein region of the map should have connected and clearly defined electron density. Density heights in the protein region should be consistent.
- Secondary structure elements, helices, and β-sheets should be clearly visible.
- The distribution of the electron density should have a characteristic histogram.

This histogram is independent of the fold and space group but is a function of resolution range and crystal solvent content. By comparing the histogram of the

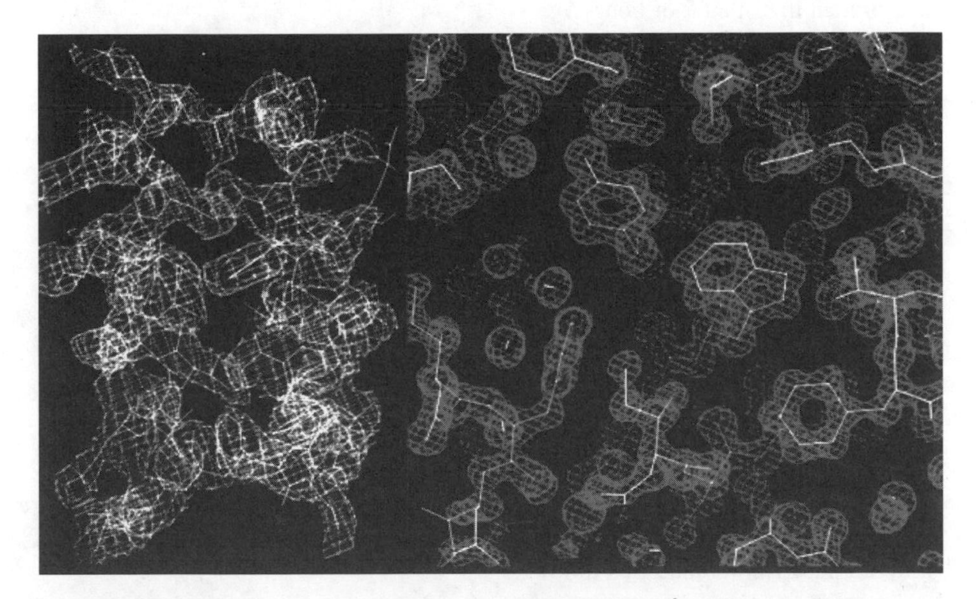

FIGURE 2.7. Left: a section of electron density calculated using 3 Å resolution data. The overall main chain fold, in this case two helices, can be observed. Side chain electron density is also apparent but lacks fine detail. Right: a section of electron density calculated using 1.1 Å resolution data. Near atomic resolution electron density maps show much more detail including fine detail in the side chain electron density as observed for the tyrosine, tryptophan, and phenylalanine residues. Also shown in the high resolution map are two solvent molecules (lower right).

unknown map with that of the correctly phased map, one will get an objective guide to the interpretability of the map.

If the map quality is good, chain tracing can begin. There are several steps in making the first trace of the electron density map. A molecular boundary needs to be defined and the asymmetric unit identified. If the asymmetric unit contains molecules related by noncrystallographic symmetry, only one molecule needs to be traced because the other molecules can be generated from the noncrystallographic symmetry. Noncrystallographic symmetry can be identified from a self-rotation function analysis, while the translation vector can be identified from the native Patterson or by analysis of the electron density map. Next, the initial C_α positions must be assigned either manually from a skeleton/bones map using programs such as *O* (Jones et al., 1991) and XtalView (McRee, 1999, *www.scripps.edu/pub/dem-web/toc.html*), or if the resolution of the data permits, automatically using wARP (Perrakis et al., 1999). Once the initial C_α positions have been placed, the protein sequence is then fitted into the electron density using the C_α positions as guides and the size and shape of the side chain electron density to define the amino acid type. In the case of Se-MAD or S-SAD experiments, the selenium and sulfur atoms can also serve as guide points during the fitting process. If noncrystallographic symmetry is present, this information should also aid in the interpretation of the election density map. Again, if the resolution of the data and phases is high enough, computer programs such as wARP can be used to fit the entire protein sequence of a moderate size protein in a matter of hours.

REFINEMENT

Once the initial model has been built, it must be refined against the x-ray diffraction data. Each nonhydrogen atom in the structure can be described by four to nine parameters, the *X*, *Y*, and *Z* positional plus one to six parameters, called thermal or temperature factors, defining the size, shape, and orientation of the electron density surrounding the atom. Based on these atomic parameters, one can calculate the structure factor (F_c) associated with each reflection (*hkl*) from the following equation:

$$F_{hkl} = \sum_j f_j e^{2\pi_i(hx_j + ky_j + lz_j)}$$

Refinement of the model then consists of minimizing the difference between the observed and calculated structure factors. The refinement is usually monitored by calculating the overall agreement (*R* factor) between the observed and calculated structure factor amplitudes, as follows:

$$r = \frac{\sum ||F_o| - |F_c||}{\sum |F_o|}$$

In order for the refinement to work (converge) properly, the number of observations (reflections) must exceed the number of parameters refined (usually four per atom) by a factor of six or more. The atomic positions for atoms in each residue must also obey the stereochemistry (bond lengths, bond angles, peptide planarity, etc.) constraints associated with the particular amino acid. These constraints can be viewed as additional observations (e.g., observed bond length vs. ideal bond length based on high-resolution structures (Engh and Huber, 1992) and the refinement program can include additional minimization terms to reflect this.

The x-ray and stereochemistry terms in the refinement are usually given different weighting factors that can be adjusted during refinement to obtain the best-refined model, that is, the model that agrees with both the x-ray data and the stereochemistry of the protein. The addition of stereochemical information to the refinement process is also important because it usually adds enough extra observations for least squares refinement to converge properly. Programs commonly used for model refinement include X-PLOR (Brunger et al., 1989, *xplor.csb.yale.edu*); CNS (Brunger et al., 1998, *cns.csb.yale.edu/v1.0*), REFMAC (Murshudov et al., 1999, *www.dl.ac.uk/ccp/ccp4*), and SHELX (Sheldrick, 1998, *shelx.uni-ac.gwdg.de/SHELX/*).

The refinement of the model generally begins with a positional refinement where the positions (x, y, and z) of every atom in the model are refined against the x-ray data and stereochemistry. The selection of the resolution limit used in the refinement is based on the data quality and individual refinement requirements. Generally, one should start to refine the structure using medium resolution (3.0 to 2.0 Å) data and gradually include all the higher resolution data. Free R factor (R_{free}) monitoring, introduced by Brunger (1992), throughout the refinement is recommended to assess the accuracy of the model and prevent overfitting of the data. The R_{free} calculation measures the agreement between observed and computed structure factor amplitudes for a test set of randomly selected reflections (usually 5 to 10%), which is set aside during the modeling and refinement process.

Once the positional refinement has reached convergence, manual revision of the model is carried out using residue deleted OMIT $2F_o - F_c$ electron density maps. OMIT maps remove model bias in the region under study because the investigated residues (usually 5% of the total) were omitted from refinement and phase calculation. The $2F_o - F_c$ coefficients used for the electron density map calculations further reduce model bias, because the observed data (F_o's) are given more weight in the calculation. At this point, missing residues, binding substrates, ions, or ligands can also be identified from difference ($F_o - F_c$) electron density maps and added to the model. Finally, the model should be adjusted such that the main chain torsion angles Φ and Ψ fall into the allowed regions of the Ramachandran plot (Ramakrishnan and Ramachandran, 1965).

Subsequently, simulated annealing (SA) refinement is carried out followed by manual revision as described previously. Simulated annealing refinement, introduced in 1989 by Brunger, serves to ensure that the refined model represents the true energy minimum of the system because conventional positional refinement sometimes converges to false local minima. SA refinement is essentially a standard crystallographic refinement linked with molecular dynamics. During SA, an energy term is added to the refinement targets to simulate a rise in the temperature of the system to the point where all the atoms dissociate and randomly arrange themselves in the liquid state.

As the system cools, the atoms can rearrange into their lowest energy state. Generally, SA refinement heats the system to 3000 to 5000 K and then slowly cools it in 25 K steps to 300 K. At each temperature step, 0.5 femto seconds of energy minimization is carried out and the total energy of the system is monitored.

Once the SA refinement has converged and if the resolution (2.5 Å or higher) of the data permits, solvent molecules identified from difference Fourier electron density maps can be included in the model. Generally, solvent molecules are accepted if

- The identified peak in the difference $F_o - F_c$ electron density correlates with a similar peak in the corresponding $2F_o - F_c$ electron density map
- The peak height is well above the noise level of the map
- The potential solvent atom forms one or more hydrogen bonds (2.6 to 3.3 Å) with the protein or with other solvent molecules

After the addition of solvent, the model is further refined (positional and group or individual thermal factors) until convergence. Again, it is a good idea to visually check the model after refinement and make manual revisions of the model if required.

Refinement can be monitored using both the R- and R_{free}- values, which should decrease as the refinement nears convergence. If this trend is not observed, it could indicate problems with correctness of the refined model or an overinterpretation of the data. In general, a good model at 2.5 Å resolution should refine to an R value in the neighborhood of 20% with an R_{free}- value below 30%. However, these values are dependent on data resolution and data quality.

VALIDATION

The quality of the resulting x-ray structure can be judged by how well the model fits the x-ray data (the R and R_{free} values) and its stereochemistry (the rms deviations from ideality in bond lengths and angles). In addition, the Lusatti plot (Lusatti, 1952) can be calculated to measure the general accuracy of the coordinate set as a function of resolution. Validation programs such as PROCHECK (Laskowski et al., 1993b, *http://www.biochem.ucl.ac.uk/~roman/procheck*) and WHAT_CHECK (Hooft et al, 1996, *www.cmbi.kun.nl/whatif*) have been developed to allow for a detailed analysis of the structure with regard to stereochemistry, hydrogen bonding, and other parameters. The Protein Data Bank (Berman et al., 2000, *www.rcsb.org/pdb*), validates all structures it receives and these reports are available online.

Typically, a well-refined structure at 2.5 Å resolution should meet or exceed most of the following parameters: R-value approximately 20%; R_{free}- value below 30%; rms deviations from ideality (Engh and Huber, 1992) in bond lengths and angles of 0.01 Å and 1.2°, respectively; have over 95% of the residues in the most favored, additionally allowed regions, or in generously allowed regions of the Ramachandran plot; and have reasonable thermal factors. Finally, the structure should in most cases make biochemical and biophysical sense.

FURTHER READING

Glusker, P. J., Lewis, M., and Rossi, M. (1994). *Crystal Structure Analysis for Chemists and Biologists*, VCH Publishers, New York.

Jones, C., Mulloy, B., and Sanderson, M. R., Eds. (1996). *Crystallographic Methods and Protocols*, Humana Press, Totawa, NJ.

McPherson, A. (1999). *Crystallization of Biological Macromolecules*, Cold Spring Harbor Press, Cold Spring Harbor, NY.

McRee, D. E. (1999). *Practical Protein Crystallography*, 2nd ed., Academic Press, San Diego.

REFERENCES

Ban, N., Nissen, P., Hansen, J., Moore, P. B., and Steitz, T. A. (2000). The complete atomic structure of the large ribosomal subunit at 2.4 angstrom resolution, *Science*, 289, 905–920.

Berman, H. M., Westbrook, J., Feng, Z., Gilliland, G., Bhat, T. N., Weissig, H., Shindyalov, I. N., and Bourne, P. E. (2000). The protein data bank, *Nucl. Acids Res.*, 28, 235–242.

Bolotovsky, R., Steller, I., and Rossmann, M. G. (1998). The use of partial reflections for scaling and averaging x-ray area-detector data, *J. Appl. Crystallogr.*, 31, 708–717.

Bragg, W. L. (1913). The structure of some crystals as indicated by their diffraction of x-rays, *Proc. Roy. Soc. (London)*, A89, 248–277.

Bravis, A. (1849). Memorie sur les polyedres forme symetrique, *J. Math. (Louville)*, 14, 137–180.

Brunger, A. T. (1990). Extension of molecular replacement: a new search strategy based on Patterson correlation refinement., *Acta Crystallogr.*, A46, 546–593.

Brunger, A. T., Karplus, M., and Petsko, G. A. (1989). Crystallographic refinement by simulated annealing: application to crambin, *Acta Crystallogr.*, A45, 50–61.

Brunger, A. T. (1992). Free R value: a novel statistical quantity for assessing the accuracy of crystal structures, *Nature*, 355, 472–475.

Brunger, A. T., Karplus, M., and Petsko, G. A. (1989). Crystallographic refinement by simulated annealing: application to crambin, *Acta Crystallogr.*, A45, 50.

Brunger, A. T., Adams, P. D., Clore, G. M., DeLano, W. L., Gros, P., Grosse-Kunstleve, R. W., Jiang, J. S., Kuszewski, J., Nilges, M., Pannu, N. S., Read, R. J., Rice, L. M., Simonson, T., and Warren, G.L. (1998). Crystallography and NMR system: a new software suite for macromolecular structure determination, *Acta Crystallogr.*, D54, 905–921.

Carter, C. W., Jr. and Carter, C. W. (1979). Protein crystallization using incomplete factorial experiments, *J. Biol. Chem.*, 254, 12219–12223.

CCP4 (1994). The Crystallographic Computing Project4 (CCP4) suite: programs for protein crystallography, *Acta Crystallogr.*, D50, 760–763.

Chen, L., Rose, J. P., Breslow, E., Yang, D., Chang, W. R., Furey, W. F., Sax, M., and Wang, B.C. (1991). Crystal structure of a bovine neurophysin II dipeptide complex at 2.8 Å determined from the single-wavelength anomalous scattering signal of an incorporated iodine atom., *Proc. Natl. Acad. Sci. U.S.A.*, 88, 4240–4244.

Chen, C.-J., Liu, Z.-J., Rose, J. P., Rosenbaum, G., and Wang, B. C. (2001). Low resolution sulfur super atom phasing for proteins containing disulfide bonds. *Acta Crystallogr.* Section D, submitted.

Dauter, Z., Dauter, M., and Rajashankar, K. R. (2000). Novel approach to phasing proteins: derivatization by short cryo-soaking with halides, *Acta Crystallogr.*, D56, 232–237.

Edmonds, C. G. and Smith, R. D. (1990). Electrospray ionization mass spectrometry, *Meth. Enzymol.*, 193, 412–431.

Engh, R. A. and R. Huber. (1992). Accurate bond and angle parameters for x-ray protein structure refinement, *Acta Crystallogr.* A47, 392–400.

Furey, W. F. J. and Swaminathan, S. (1997). Phases95: a program package for the processing and analysis of diffraction data from macromolecules, *Meth. Enzymol.*, 277, 590–620.

Garman, E. (1999). Cool data: quantity and quality, *Acta Crystallogr.*, D55, 1641–1653.

Georgalis, Y. and Saenger, W. (1999). Light scattering studies on supersaturated protein solutions, *Sci. Prog.*, 2, 271–294.

Green, D. W, Ingram, V. M., and Perutz, M. F. (1954). The structure of haemoglobin IV, Sign determination by the isomorphous replacement method, *Proc. Roy. Soc.*, A225, 287–307.

Hahn, T., Ed. (1987). *International Tables of Crystallography*, Vol. A., *Space Group Symmetry*, 2nd ed., rev., D. Reidel, Dordrecht.

Hendrickson, W. A. (1991). Determination of macromolecular structures from anomalous diffraction of synchrotron radiation, *Science*, 51–58, 254.

Hendrickson, W. A. and Teeter, M. M. (1981). Structure of the hydrophobic protein crambin determined directly from the anomalous scattering of sulphur, *Nature* (London), 107–113, 290.

Hendrickson, W. A., Horton, J. R., and LeMaster, D. M. (1990). Selenomethionine proteins produced for analysis by multiwavelength anomalous diffraction (MAD): a vehicle for direct determination of three-dimensional structure, *EMBO J.*, 9, 1665–1672,.

Hessel, J. F. C. (1830). Krystallometric oder Kystallonomic und Krystallographie, in *Gehlers Physikalisches Wörterbuch*, Vol. 5, Schwickert, Leipzig, 1023–1360.

Hitscherich, C., Kaplan, J., Allaman, M., Wiencek, J., and Loll, P. J. (2000). Static light scattering studies of OmpF porin: implications for integral membrane protein crystallization, *Protein*, 9, 1559–1566.

Hooft, R. W. W., Vriend, G., Sander, C., and Abola, E. E. (1996). What check Errors in protein structures, *Nature*, 381, 272.

Hope, H. (1988). Cryocrystallography of biological macromolecules: a generally applicable method, *Acta Crystallogr.*, B44, 22–26.

Jancarik, J. and Kim, S.-H. (1991). Sparse matrix sampling: a screening method for crystallization of proteins, *J. Appl. Crystallogr.*, 24, 409–411.

Jardine, I. (1990). Molecular weight analysis of proteins, *Meth. Enzymol.*, 193, 441–455.

Jones, T. A., Zou, J.-Y., Cowan, S. W., and Kjeldegaard, M. (1991). Improved methods for building protein models in electron density maps and the location of errors in these models, *Acta Crystallogr.*, A47, 110–119.

Laskowski, R. A., MacArthur, M. W., Moss, D. S., and Thornton, J. M. (1993). PROCHECK: a program to check the stereochemical quality of protein structures, *J. Appl. Crystallogr.*, 26, 283–291.

Leslie, A. G. W. (1990). Recent changes to the MOSFLM package for processing film and image plate data, in *Crystallographic Computing*, Oxford University Press, New York.

Liu, Z. J., Vysotski, E. S., Chen, C. J., Rose, J. P., Lee, J., and Wang, B. C. (2000). Structure of Ca+2–regulated photoprotein obelin at 1.7 Å resolution determined directly from its sulfur substructure, *Protein Sci.*, 9, 2085–2093.

Lusatti, V. (1952). Traitement statistique des erreurs dans la determination des structures crystallines, *Acta Crystallogr.*, 5, 802–810.

Malkin, A. J. and McPherson, A. (1994). Light-scattering investigations of nucleation processes and kinetics of crystallization in macromolecular systems, *Acta Crystallogr.*, D50, 385–395.

Matthews, B. W. (1968). Solvent content of protein crystals, *J. Mol. Biol.*, 33, 491–497.

McPherson, A. J. (1982). *Preparation and Analysis of Protein Crystals*, John Wiley & Sons, New York.

McPherson, A. J. (1999). *Crystallization of Biological Macromolecules*, Cold Spring Harbor Press, Cold Spring Harbor, NY.

McRee, D. E. (1999). XtalView/Xfit—A versatile program for manipulating atomic coordinates and electron density, *J. Struct. Biol.*, 25, 156–165.

Merril, C. R. (1990). Gel-staining techniques, *Meth. Enzymol.*, 182, 477–488.

Miller, R., DeTitta, G. T., Jones, R., Langs, D. A., Weeks, C. M., and Hauptman, J. A. (1993). On the application of the minimal principle to solve unknown structures, *Science*, 259, 1430–1433.

Murshudov, G. N., Vagin, A. A., Lebedev, A., Wilson, K. S., and Dodson, E. J. (1999). Efficient anisotropic refinement of macromolecular structures using FFT, *Acta Crystallogr.*, D55, 247–255.

Navaza, J. (1994). AMORE: an automated package for molecular replacement, *Acta Crystallogr.*, A50, 157–163.

Otwinowski, Z. and Minor, W. (1997). Processing of x-ray diffraction data collected in oscillation mode, *Meth. Enzymol.*, 276, 307–326.

Patterson, A. L. (1935). A direct method for the determination of the components of interatomic distances in crystals, *Z. Krist.*, 90, 517–542.

Perrakis, A., Morris, R., and Lamzin, V. S. (1999). Automated protein model building combined with iterative structure refinement, *Natl. Struct. Biol.*, 6, 458–463.

Petsko, G. A. (1985). Preparation of isomorphous heavy atom derivatives, *Meth. Enzymol.*, 114, 147–156.

Pflugrath, J. W. (1999). The finer things in x-ray diffraction data collection, *Acta Crystallogr.*, D55, 1718–1725.

Ramakrishnan, C. and Ramachandran, G. N. (1965). Stereochemical criteria for polypeptide and protein chain conformations. II. Allowed conformations for a pair of peptide units, *Biophys. J.*, 5, 909–933.

Rossmann, M. G., Ed. (1972). *The Molecular Replacement Method*, Gordon and Breach, New York.

Rossmann, M. J. (1990). The molecular replacement method, *Acta Crystallogr.*, A46, 73–82.

Rossmann, M. G. and van Beek, C. G. (1999). Data processing, *Acta Crystallogr.*, D55, 1631–1640.

Sambrook, J., Fritsch, E. F., and Maniatis, T. (1989). *Molecular Cloning: a Laboratory Manual*, 2nd ed., Cold Spring Harbor Laboratory, Cold Spring Harbor, NY.

Scoble, H. A. and Martin, S. A. (1990). Characterization of recombinant proteins, *Meth. Enzymol.*, 193, 519–536.

Sheldrick, G. M. (1998). *Direct Methods for Solving Macromolecular Structures*, Kluwer Academic Publishers, Dordrecht, pp. 401–411.

Shieh, H. S., Stevens, A. M., and Stegeman, R. A. (1991). MON48: a screening procedure for protein crystallization at Monsanto, Monsanto Corporate Research, St. Louis, MO.

Steller, I., Bolotovsky, R., and Rossmann, M. G. (1997). An algorithm for automatic indexing of oscillation images using Fourier analysis, *J. Appl. Crystallogr.*, 30, 1036–1040.

Sutra, E. A. and Wilson, I. A. (1992). Seeding techniques, in *Crystallization of Nucleic Acids and Proteins, a Practical Approach*, Ducruix, A., and Giege, R., Eds., IRL, Oxford, UK, pp. 99–126.

Teng, T. Y. (1990). Mounting crystals for macromolecular crystallography in a free standing thin film, *J. Appl. Crystallogr.*, 23, 387–391.

Terwilliger, T. C. and Berendzen, J. (1999). Automated MAD and MIR structure solution, *Acta Crystallogr.*, D55, 849–861.

Wang, B.-C. (1985). Resolution of phase ambiguity in macromolecular crystallography, *Meth. Enzymol.*, 115, 90–112.

Wu, C. K., Dailey, H. A., Rose, J. P., Burden, A., Sellers, V. M., and Wang, B.C. (2001). The 2 Å structure of human ferrochelatase, the terminal enzyme of heme biosynthesis, *Nat. Struct. Biol.*, 8, 156–160.

3 Protein NMR Spectroscopy

Gordon S. Rule and T. Kevin Hitchens

CONTENTS

Introduction 37
Fundamentals of Magnetic Resonance Spectroscopy 39
Chemical Shift 40
Relaxation of Nuclear Spins 41
One-, Two-, and *N*-Dimensional NMR 46
Effect of Chemical Exchanges on the Shape of NMR Resonance Lines 49
Applications and Practical Advice 51
Assignment of Resonances 52
General Approach to Assignments 53
Assignment of Small Unlabeled Proteins (<8 kDa) 54
Assignment of Moderate-Sized Labeled Proteins (10 to 15 kDa) 56
Assignment of Large Proteins (15 to 25 kDa) 57
Assignment of Very Large Proteins (25 to 60 kDa) 59
Commonly Performed Experiments 60
Measurement of Kinetic On-Rates and Equilibrium Constants 60
Topological Mapping of Ligand Binding Sites 61
Structure Determination by NMR 62
Relaxation Measurements 65
References 66
Additional Reading 68
Glossary 68

INTRODUCTION

Nuclear magnetic resonance (NMR) spectroscopy detects the transitions of nuclear spin states. Although both spin 1/2 (i.e., ^{1}H, ^{15}N, ^{13}C, ^{19}F) and spin 1 (i.e., ^{2}H, ^{14}N) nuclei can be detected with NMR, this article will focus solely on the use of spin 1/2 nuclei. Spin 1 nuclei have large quadrupole moments because of the asymmetrical charge distribution at the nucleus. This large quadrupole moment results in efficient relaxation of spin 1 nuclei, leading to broad resonance lines. Types of nuclear spins that are most commonly observed in proteins are ^{1}H, ^{15}N, ^{13}C, and ^{19}F. Of these, only the proton (^{1}H) is naturally abundant.

0-8493-9453-8/02/$0.00+$1.50

Consequently, it is necessary to actively incorporate ^{15}N, ^{13}C, or ^{19}F to high isotopic enrichment to observe signals from these spins. This usually can be accomplished readily by biosynthesis.

NMR is a form of spectroscopy that is unique in its ability to provide answers to a broad range of biochemical questions. The type of information that can be obtained from NMR spectroscopic studies of biological macromolecules can be placed broadly into the following categories.

- Information on the chemical kinetics of protein–ligand interactions
- Delineation of protein folding pathways
- Detection of changes in protein structure due to ligand binding
- Determination of protein/nucleic acid structure
- Detailed characterization of protein/nucleic acid dynamics

Other biophysical techniques can, to varying degrees, also provide similar information. For example, stop-flow kinetics can be used to investigate chemical kinetics. Changes in protein structure due to ligand binding can be detected by a number of spectroscopic techniques, such as circular dichroism, UV-VIS spectroscopy, IR spectroscopy, and fluorescence spectroscopy. However, all of these techniques either report on an average change in structure (e.g., CD, IR), or report a change in the environment of a small number of residues in the protein (e.g., UV-VIS, fluorescence). In contrast, with NMR it is possible to detect the effect of ligand binding on virtually every single nuclear spin in the protein. The richness and detail of this information is unparalleled by any spectroscopic technique.

Structure determination of proteins and protein–ligand complexes by NMR is rivaled only by x-ray crystallography. In general, it is preferable to determine protein structures by x-ray methods because the resolution of the structure is usually higher and the time required to obtain the structure is usually shorter. However, in many cases it is difficult to obtain suitable crystals for diffraction and NMR may be the only method available for structure determination. In addition, critical questions regarding the effects of lattice packing on the conformation of proteins in the crystalline environment can be addressed using NMR. Finally, NMR is the premier method of investigating the dynamic properties of proteins and nucleic acids. There are two other common methods of investigating the dynamical properties of proteins: fluorescence depolarization and thermal factors in x-ray studies. Although fluorescence depolarization experiments can provide similar information on the dynamics of proteins as NMR, they suffer from the drawback that only a small number of fluorescent sites can be probed. In contrast, NMR provides the opportunity to investigate the dynamic properties of almost any atom in the protein. High thermal factors (B factors) of x-ray derived structures can indicate disorder; however, it can be difficult to separate static disorder from true dynamics without additional information, such as the temperature dependence of the B-factor. In addition, it is important to note that the dynamic properties of proteins in crystalline lattices may be perturbed by lattice contacts.

In summary, NMR is a particularly useful tool to study a wide variety of biochemical questions. The preceding discussion may have implied that NMR is sufficiently powerful that it can stand alone as a biochemical tool. However, its utility is greatly enhanced when combined with other biochemical and biophysical techniques. A clear example of this is the synergy between x-ray and NMR studies. The former can be used to obtain detailed information about the structure of a protein and a protein–ligand complex while the latter can be used to investigate the chemical kinetics of ligand binding as well as the dynamics of the protein and protein–ligand complex.

The purpose of this chapter is to introduce the reader to the utility of NMR spectroscopy for probing structural, dynamic, and biophysical properties of proteins and protein–ligand complexes. The scope of this chapter is limited; for additional information we offer several texts at the end of the chapter where more detailed information may be found. An understanding of the applications and limitations of NMR spectroscopy requires knowledge of basic magnetic resonance theory. Therefore, the chapter begins by introducing some of the important principles of magnetic resonance, including nuclear-spin relaxation, effects of chemical exchange, dipolar coupling, and scalar coupling. Then the chapter is directed toward the application of NMR spectroscopy to specific biochemical problems and addresses sample requirements, instrumentation requirements, and provides some information regarding the assignment of resonances, measurement of chemical kinetics, topology mapping, structure determination, and relaxation measurements. A brief glossary at the end of the chapter defines most of the common terms used in the discipline.

FUNDAMENTALS OF MAGNETIC RESONANCE SPECTROSCOPY

The degree of information that can be obtained from a biochemical system using NMR depends on the system under investigation. To fully understand this relationship, it is necessary to understand some of the fundamentals of NMR spectroscopy.

The interaction between a nuclear spin and the magnetic field component of light is a consequence of the fact that the spin angular momentum gives rise to a magnetic dipole,

$$\vec{u} = \gamma_X \vec{I} \tag{3.1}$$

In this equation, $\vec{u}$ equals the magnetic dipole, γ_x equals the gyromagnetic ratio for spin X, and I equals the spin angular momentum.

The energy of interaction between the nuclear spin and the magnetic dipole is given by the dot product between u and the external magnetic field (B),

$$E = \vec{u} \bullet \vec{B} \tag{3.2}$$

Usually, the direction of the magnetic field is defined to be along the z-axis so this equation becomes

$$\begin{aligned} E &= u_Z B_Z \\ &= \gamma_X m_Z B_Z \end{aligned} \tag{3.3}$$

where m_Z is the z component of the spin angular momentum.

From quantum mechanics it is known that spin 1/2 nuclei can have only two possible values of m_Z, + 1/2 and −1/2, thus defining the ground ($m_Z = +1/2$) and the excited states ($m_Z = -1/2$). The energy difference between the ground and excited states is

$$\Delta E = \gamma B_Z \tag{3.4}$$

There are two important consequences of this equation. First, because the population difference polarization between the ground state and the excited state is proportional to the energy difference between the two states (Boltzmann equation), the sensitivity of the NMR experiment depends on both γ and B. The gyromagnetic ratio is an inherent property of the nuclear spin. The gyromagnetic ratio of the ^{1}H is larger than that of ^{19}F, ^{13}C, or ^{15}N. Therefore the sensitivities of ^{19}F, ^{13}C, or ^{15}N spins relative to the ^{1}H, at full isotopic enrichment are 0.97, 0.25, and 0.10, respectively. The population difference between the ground and excited states is on the order of 10^{-5} for ^{1}H at typical magnetic field strengths (11.7 Tesla), revealing that NMR spectroscopy is the least sensitive of all spectroscopic methods. Application of higher magnetic field strengths only provides modest gains in sensitivity, so signal averaging is generally employed to increase the signal to noise in the spectrum.

The second important consequence of Equation (3.4) is that the absorption frequency ($\omega = \Delta E/\eta$) of the nuclear spin is directly proportional to the magnetic field strength. Thus, the separation of spectral lines increases as the magnetic field strength increases. With rare exceptions, increasing the magnetic field strength provides better sensitivity and increased resolution.

Chemical Shift

The field strength of an NMR spectrometer is conventionally described by the absorption frequency of protons for that instrument. Commercially available instruments have field strengths that range from 11.7 to 21.06 Tesla, corresponding to proton frequencies between 500 and 900 MHz. Because NMR experiments are performed at different field strengths, it is necessary to normalize the absorption frequency by the field strength. Consequently, the actual absorption frequencies in NMR are not reported in absolute frequencies (or wavelengths). Rather, the observed frequencies (ν) are reported as chemical shifts (δ, units are ppm) which are referenced to an absorption frequency of a defined standard (ν_0) and normalized to account for the effect of the magnetic field strength on the absorption energy.

$$\delta = \frac{(\nu - \nu_0)}{\nu_0} \times 10^6 \tag{3.5}$$

The chemical shift of a nuclear spin depends on the strength of the magnetic field present at the nucleus. This is the fundamental reason why spins of the same type (i.e., ^{1}H) can have different chemical shifts and why it is possible to observe signals from virtually every nuclear spin within a biological polymer. The strength of the magnetic field at the nucleus ($B_{Effective}$) differs from the applied magnetic field ($B_{Applied}$). This difference can be attributed to several factors, the most important being the electron density surrounding the nucleus and nearby magnetic dipoles (both from nuclear as well as electron spins). Thus, the absorption frequency of a spin is shifted from that of an isolated nuclear spin according to

$$B_{Effective} = (1 - \sigma)B_{Applied} \quad (3.6)$$

Sigma (σ) is the shielding factor that accounts for the change in the effective field or the chemical shift. The largest determinant of σ is the chemical bonding environment. Consequently, each type of proton in a protein has a reasonably well-defined range over which its chemical shifts usually fall. For example, the amide protons have chemical shifts around 8 ppm, whereas methyl protons have chemical shifts at about 0 ppm. The behavior of other nuclear spins is similar. In particular, the ^{13}C chemical shifts are very predictable and can be quite diagnostic for the type of residue to which they belong.

In addition to chemical bonding effects, the proximity of charged groups, the spatial location of other magnetic dipoles, and the distribution of aromatic ring systems close to the observed spin also cause chemical shift changes. The varied environment that is generated in a folded protein modifies all of these factors, causing chemical shift changes that depend on the location of a nuclear spin within the tertiary structure of the protein. Thus, chemically equivalent spins (e.g., the H_α proton of alanines) can have different absorption frequencies. For example, a methyl group next to a positively charged group (e.g., lysine) may show a chemical shift that is several ppm higher than in the absence of the charged group. Alternatively, a methyl group placed above an aromatic ring will experience a large negative change in its chemical shift.

Ideally, the varied environment within the folded protein will generate unique chemical shifts for all of the nuclear spins. In practice, many of the protons (as well as other types of spins) in a protein show differences in chemical shift that are less than the spectral line width. These resonance lines are termed degenerate because it is not possible to resolve their individual spectral lines. In some cases, this degeneracy can be removed by increasing the dimensionality of the spectra. Multi-dimensional NMR experiments will be discussed in more detail later.

Relaxation of Nuclear Spins

A very important observable in NMR spectroscopy is the effect of the chemical and magnetic environment on the relaxation properties of the nuclear spins. In particular, changes in the environment resulting from molecular motion can be readily detected from the relaxation rates. The small energy difference between the ground and excited state results in a very slow rate of spontaneous emission of energy from the excited state (the rate of spontaneous emission from an excited state is proportional to the cube of the energy difference between the ground and excited state). Because spontaneous emission is far too slow to account for the observed rates of nuclear spin relaxation, the relaxation of nuclear spins must depend on stimulated emission. Stimulated emission is caused by time dependent fluctuations of the effective magnetic field at the nucleus. These magnetic field fluctuations are largely generated by two mechanisms: an anisotropic chemical shift or dipolar coupling. If the chemical shift of a spin is not isotropic, then the orientation of the molecule will change the effective field at the nucleus (chemical shift anisotropy). As the molecule (or part of molecule) changes its orientation, the apparent field at the nucleus become time dependent. The magnitude of this effect depends on the square of the magnetic field strength. This relaxation mechanism is important for ^{15}N, ^{13}C, and ^{19}F nuclei, especially at high fields.

Dipolar coupling between nearby spins can also generate field fluctuations. Dipolar coupling occurs when one spin interacts with the magnetic field generated by another spin. This effect occurs through space and does not require that the spins be connected by chemical bonds. In fact, the two spins need not be on the same molecule. The strength of the interaction with this field depends on three factors: the types of interacting spins, the distance between the spins, and the angle between the vector that joins the spins and the external magnetic field. These effects are summarized by the following equation,

$$B_{LOCAL} \propto \gamma_X \gamma_Y \frac{1}{r^3}[1 - 3\cos(\theta(t))] \tag{3.7}$$

B_{LOCAL} is the local magnetic field at the position of one spin (X) resulting from the presence of another spin (Y). γ_X and γ_Y are the gyromagnetic ratios of the two spins; r is the distance between the two spins; and θ is the angle between the direction of the external magnetic field (along z) and the line joining the two spins.

Because the strength of the dipolar coupling depends on the relative orientation of the two spins with respect to the external magnetic field, any reorientation of the two coupled spins will generate a fluctuating magnetic field (note the explicit time dependence of the angle, θ). The rapid isotropic tumbling of protein molecules in solution effectively average this field, thus no change in the chemical shift is usually observed. However, the tumbling of the molecule causes fluctuations in the local field in a manner similar to that observed with chemical shift anisotropy.

The field fluctuations caused by dipolar coupling or chemical shift anisotropy can be effective in causing stimulated emission. For these fluctuations to cause stimulated emission, they must occur at a frequency that matches the energy difference between the various quantum states of the system. For two coupled spins, there are six possible transitions that link the four possible states of the two spins. These transitions can be grouped into one zero quantum transition, four single quantum transitions (two for each spin), and one double quantum transition (see Figure 3.1). For example, efficient relaxation of the spins by single quantum transitions would require the presence of magnetic field fluctuations at the frequency of the single quantum transition.

The stimulated transitions between the four energy levels of two coupled spins result in two distinct types of relaxation. The first is called spin-lattice relaxation. This form of relaxation involves the net loss of energy from the excited state and is analogous

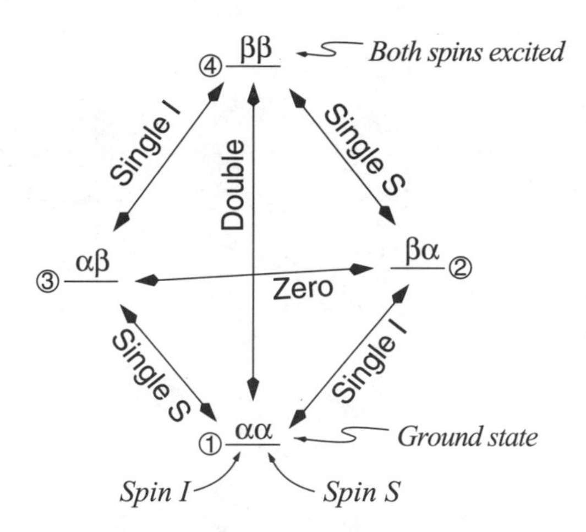

FIGURE 3.1 Possible energy states for two coupled spins. The quantum energy levels for two coupled spins (spin *I* and spin *S*) are shown. Each isolated spin has two possible states, α and β. The ground state of the coupled spins is represented by the αα state (1), the two single quantum transitions link the ground state to the two singly excited states (2, 3). Of these transitions, one causes excitation of the *I* spin while the other transition causes excitation of the *S* spin. There are two additional single quantum transitions that connect these intermediate states to the doubly excited state (4). The two singly excited intermediate states, being of nearly equal energy, can be interchanged by a zero quantum transition. The double quantum state can be converted directly to the ground state via a double quantum transition.

to the decay of excited fluorescence states. The rate constant for this decay is commonly denoted as R_1 and the corresponding relaxation time is called T_1 ($R_1 = 1/T_1$). A net loss of energy from the excited spins to the surrounding environment (lattice) occurs with either a single or a double quantum transition. In contrast, the zero quantum transition does not cause a net loss of energy to the lattice because it only transfers the energy of one excited spin to another. The spin-lattice relaxation rate dictates the repetition rate of the experiment because it is the mechanism that re-establishes the population difference between the ground and excited states. Typical T_1 times are on the order of 1 s and, thus, it is not favorable to acquire spectra much faster than 1/s.

Although zero-quantum transitions cannot provide a mechanism for the net loss of energy from the system in larger molecules, they do provide a means by which inter-proton distances can be measured (see Figure 3.2). If the population difference between the ground and excited state of one spin is perturbed, then dipolar coupling between the spins will also cause the population difference of the coupled spin to change as well. This population change will change the intensity of the peak belonging to the coupled spin. The rate at which this change occurs is proportional to $1/r^6$ (the square of the fluctuating magnetic field determines the relaxation rate). For example, in the case of two coupled spins, suppose that the ground and excited states of the first spin (*I*) were saturated. This implies that the population difference between levels 1 (αα) and 2 (βα) as well as the difference between levels 3 (αβ) and 4 (ββ) will become equal (Figure 3.2, middle section). The population difference of the coupled spin (spin *S*) is given by the difference between level 4 (ββ) and 2 (βα) as

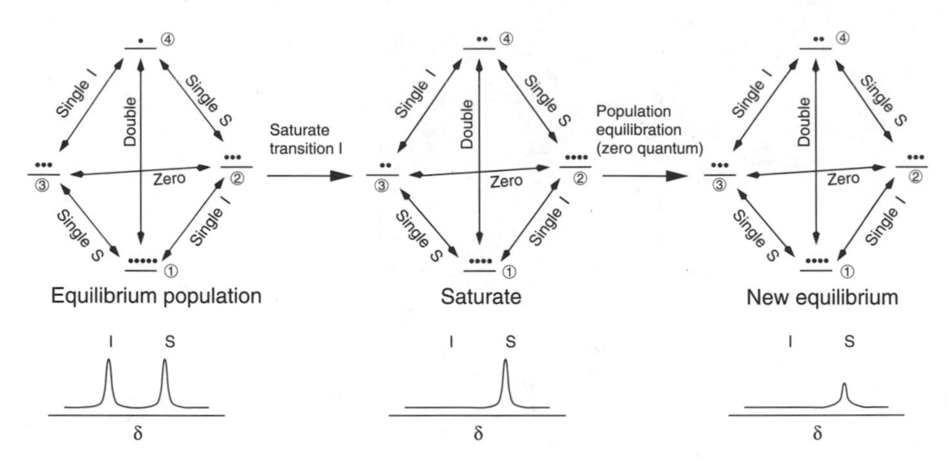

FIGURE 3.2 Perturbation of spin populations by dipolar coupling. The four-level energy diagram for two coupled spins is shown and the expected NMR spectrum is drawn underneath. The dots drawn on each level represent the number of spins, or population, of that level. At equilibrium (left section), the ground state contains 5 spins, each of the singly excited states contain 3 spins, while the double quantum state contains 1 spin. The intensity of the NMR absorptionline is calculated from the population difference between levels that are connected by single quantum transitions. At equilibrium, the net difference is two units. The total intensity of the absorptionline of both the *I* and the *S* spin is 4. Selective irradiation of spin *I* (middle section) will result in an equalization of the population difference between levels which are connected by a single quantum transition of the *I* spins. Consequently, the absorptionline for the *I* spin will disappear. If spin *S* is not dipolar coupled to spin *I*, then its intensity will not change. However, if magnetic field fluctuations exist to promote efficient zero quantum transitions, then the populations of levels 2 and 3 become equal (right section). This causes the population difference between levels connected by single quantum *S* transitions to decrease. Consequently, the intensity of absorptionline for spin *S* decreases (right section, bottom).

well as the difference between levels 3 (αβ) and 1 (αα). The population difference is 4 at equilibrium (Figure 3.2). However, if there is efficient exchange of magnetization between levels 2 and 3 (zero quantum transition) owing to dipolar coupling, then the population of level 3 will become equal to that of level 2. This will cause a reduction in the population difference of the coupled (*S*) spin from 4 to 2. Consequently, the intensity of the absorption peak of spin *S* will decrease by a factor of two because of the saturation of the *I* spin single quantum transition.

The change in population difference of one spin owing to a change in population of the coupled spin is called the nuclear overhauser effect, or NOE. The magnitude of this change is easily measured in a NOESY (nuclear overhauser effect spectroscopy) experiment (Neuhaus and Williamson, 1989). Because the dipolar coupling depends on the distance between the two couple spins, the NOE can be used to measure the distance between two protons in a protein. Because the size of the NOE also depends on the gyromagnetic ratio of the coupled spins, it is usually only feasible to measure proton–proton NOEs couplings. An important exception is the dipolar coupling between a proton and its attached heteronuclear (^{13}C or ^{15}N) spin. These two nuclei are sufficiently close that dipolar coupling between them can be readily detected. Because the field fluctuations depend on motion and the proton–nitrogen or proton–carbon distance is fixed, measurement of the heteronuclear NOE is useful in characterizing the dynamics of the main-chain of a protein.

The second form of relaxation is called spin–spin relaxation. This form involves any change in the quantum state of the spin. Thus, any of the transitions shown in Figure 3.1 can cause spin–spin relaxation. In particular, the exchange of magnetization between spins via a zero quantum transition is a very effective mechanism for spin–spin relaxation. Thus, spin–spin relaxation is analogous to fluorescence energy transfer. Because spin–spin relaxation limits the lifetime of the excited state, it affects the line width of the observed resonance lines due to the uncertainty principle; short-lived states have ill-defined frequencies. The actual relationship between the spin–spin relaxation rate and the line width ($\Delta\nu$) is given by R_2, the rate of spin–spin relaxation; T_2 is the time constant for spin–spin relaxation,

$$\Delta\nu = \frac{1}{\pi}R_2 = \frac{1}{\pi T_2} \tag{3.8}$$

The relationship between molecular motion and the magnetic field fluctuations that cause relaxation are well established. The R_1 and R_2 relaxation rates are directly related to the magnitude of the magnetic field fluctuations at zero, single, and double quantum frequencies. The magnitude of the magnetic field fluctuations at any given frequency is described by the spectral density function, $J(\omega)$. For a molecule that is undergoing random isotropic motion, the spectral density function is

$$J(\omega) = \frac{2}{5}\frac{\tau_C}{(1 + \omega^2\tau_C^2)} \tag{3.9}$$

where ω is the frequency and τ_C is the rotational correlation time for the molecule.

The rotational correlation time can easily be thought of as the time constant for loss of memory regarding the previous orientation of the molecule. It is roughly equal to the time it takes a molecule to rotate 1 radian while undergoing random rotational motion. Small molecules have short τ_C values while large molecules have long τ_C values.

The shape of the spectral density function depends on the overall rotational correlation time of the molecule. The latter is proportional to the molecular weight. The theoretical spectral density functions for a small protein and a large protein are shown in Figure 3.3.

The key point is the effect of molecular weight on the spectral density function. As the molecular size increases, the intensity of fluctuations with a frequency close to the zero quantum transitions also increases. Hence, the spin–spin relaxation rate increases as the molecular weight increases. This has two very important consequences. First, the spectral line width will increase as the molecular size increases. Consequently, the NMR spectra of larger proteins show increased degeneracy because of the increased number of resonances and the increased line width. The second consequence of the shortened lifetime of the excited state is a reduction in the efficiency by which magnetization can be passed from one nucleus to another

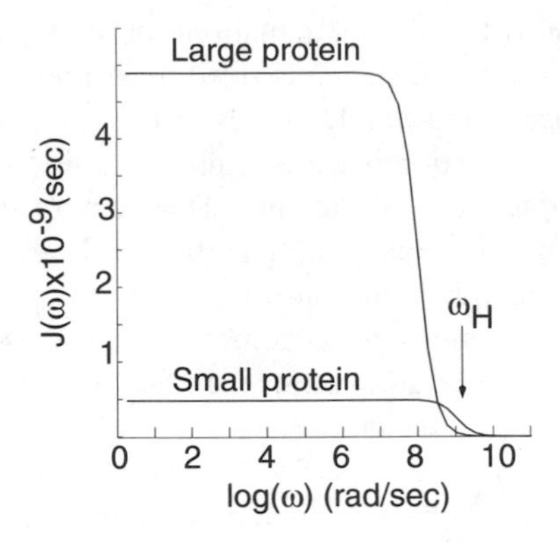

FIGURE 3.3 Effect of molecular weight on the spectral density function. The spectral density function, J(ω), is plotted vs. the frequency of the magnetic field fluctuations. The spectral density functions for a large protein (25 kDa) and a small protein (2.5 kDa) are shown. The frequency for single quantum transition of the ^1H spins is indicated by the arrow.

in multi-dimensional experiments (see below). Thus, the sensitivity of these experiments decreases for larger molecules.

One-, Two-, and *N*-Dimensional NMR

NMR spectra are not generally acquired using the traditional energy scanning methods utilized in other forms of spectroscopy, such as UV-VIS absorption spectroscopy. Rather, all of the nuclear spins are excited simultaneously by a short electromagnetic pulse in the radio frequency range (RF pulse). The response of the nuclear spins to this excitation pulse is acquired as a function of time and the NMR spectrum of the sample is calculated from the time-domain data by Fourier transform methods (see Figure 3.4). There are two significant advantages to this method. First, an increase in the rate by which spectra can be acquired for signal averaging purposes is realized. Second, the excited state generated by the first RF pulse can be manipulated by subsequent pulses to yield multi-dimensional spectra (Jeener et al., 1979; Bax and Lerner, 1986; Clore and Gronenborn, 1991).

In multi-dimensional experiments, either scalar or dipolar coupling between spins is utilized to pass magnetization from one spin to another. Scalar coupling involves the perturbation of the magnetic field at one spin by the spin state (α or β) of another spin. Scalar coupling is propagated by the intervening electrons between the two coupled spins. Thus, scalar coupling can be observed generally only between spins that are within one, two, or three chemical bonds of each other. While the magnetization exists on a particular spin during a multi-dimensional experiment, it is possible to encode the magnetization with the chemical shift of that spin in much the same way that a passport might be stamped at a border crossing.

One of the simplest and most utilized two-dimensional NMR experiments is the proton–nitrogen heteronuclear single quantum correlation experiment (^1H–^{15}N HSQC, see Figure 3.5) (Bax et al., 1983). This experiment has four distinct intervals. First, the more intense magnetization of the amide proton is transferred to the amide

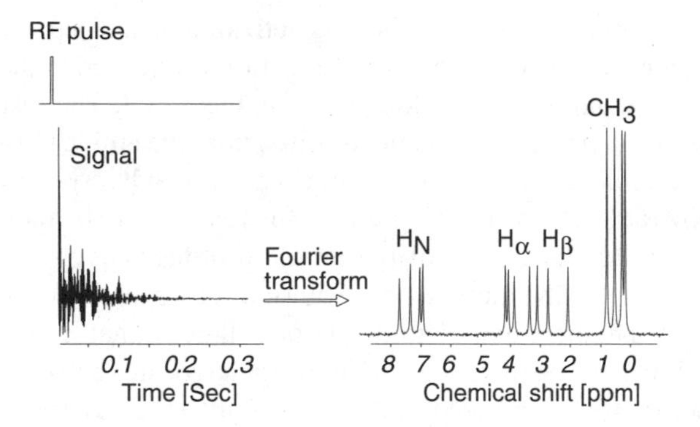

FIGURE 3.4 Illustration of the process of acquiring a one-dimensional NMR spectrum. The steps involved in obtaining an NMR spectrum are shown. The sample is a tetra-peptide (Val-Ala-Ser-Ala). A short (10 μsec) intense RF pulse is applied to the sample. This pulse excites all of the nuclei and they emit energy at their characteristic absorption frequencies. This signal is called the free induction decay (FID) and is collected as a function of time. This time domain signal is converted to spectrum by Fourier transformation. Note the characteristic chemical shifts for amide protons (H_N), α-protons, β-protons, and methyl protons. Also note that the two alanine residues, although chemically equivalent, have different chemical shifts because they experience different local environments.

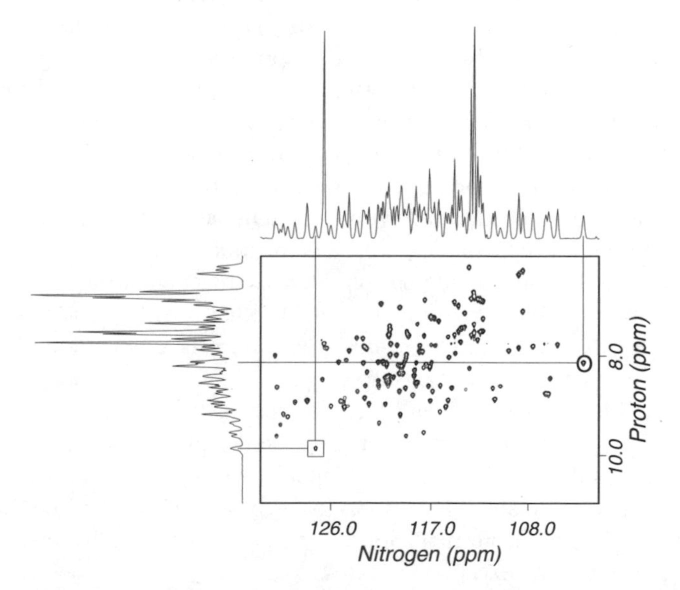

FIGURE 3.5 One and two-dimensional NMR. This figure shows a one-dimensional proton (top spectrum) and nitrogen spectrum (right side) of a 130-residue protein. The two-dimensional HSQC spectrum is shown in the middle. This is a contour plot (similar to a topographical map) in which lines are drawn to connect points of equal intensity. Each amide group gives a peak that is located at its proton and nitrogen chemical shifts. The boxed resonance is from Asn120 and it is resolved in both the one- and two-dimensional spectra. In contrast, the resonance from Gly53 (circled) is resolved in a one-dimensional nitrogen spectrum but is degenerate with several other amides in a one-dimensional proton spectrum.

nitrogen via scalar coupling. Second, the magnetization is labeled with the chemical shift of the amide nitrogen. Third, the magnetization is returned to the amide proton. Fourth, the chemical shift of the amide proton is measured. The resultant magnetization is now associated with two chemical shifts: both the amide nitrogen and amide proton. The principle advantage to acquiring an HSQC spectrum vs. a one-dimensional NMR spectrum is the increase in resolution. In most proteins, the one-dimensional proton spectrum of the amide protons is very overlapped.

Figure 3.5 illustrates the increase in resolution that is afforded by increasing the dimensionality of the spectrum. Neither the one-dimensional proton spectrum nor the one-dimensional nitrogen spectrum shows a large number of resolved peaks. Most of the peaks are overlapped and are difficult to analyze. However, in the two-dimensional HSQC spectrum most of the resonances are resolved; thus, information can be obtained from virtually all of the amide groups within the protein.

This technique can be easily extended into three or higher dimensions. For example, the three-dimensional NOESY-HSQC experiment records the following path of magnetization:

$$H_X \rightarrow H_N \rightarrow N \rightarrow H_N$$

In the first part of this experiment the magnetization is labeled with the first frequency (that of H_X, where X means any proton). This magnetization is then transferred to the nearby amide protons (H_N) by dipolar coupling. Because the degree of dipolar coupling depends on the distance between the two spins, the magnetization at this stage contains information on both the frequency of the H_X proton as well as the distance between H_X and the amide proton [c.f. Equation 3.7]. This magnetization is then passed, via scalar coupling, to the bound amide nitrogen. While on this spin, the magnetization becomes labeled with the chemical shift of the amide nitrogen. In the final phase of the experiment, the magnetization is transferred, again by scalar coupling, back to the same amide proton. Here, the last chemical shift is recorded, that of the amide proton. The result is that the final signal is encoded with three frequencies: H_X, N_H, and H_N. The resultant spectrum is a three-dimensional cube. Within this cube are regions of intensity or "peaks." The location of a peak is defined by the chemical shift of H_X, N, and H_N. The intensity of a peak in this particular experiment is proportional to $1/r^6$, where r is the distance between H_X and H_N. The range of distances that can be measured depends on the signal to noise of the data. It is very common to be able to measure distances as long as 4 Å. Distances as far as 5 Å can be obtained from most spectra. Distances that are larger than 5 Å are usually only observed in exceptional cases.

The resolution that is gained by the additional frequency dimension (N_H in this case) is illustrated in Figure 3.6. The two-dimensional NOESY spectrum of a 50 residue α-helical protein is shown in the left part of this figure. In this experiment, the magnetization is transferred from proton H_X to proton H_Y by dipolar coupling. The first frequency dimension corresponds to the chemical shift of H_X and the second frequency dimension corresponds to the chemical shift of H_Y. As with the NOESY-HSQC described previously, the intensity of the peak is related to the distance between H_X and H_Y. Note the large number of unresolved overlapping peaks in the

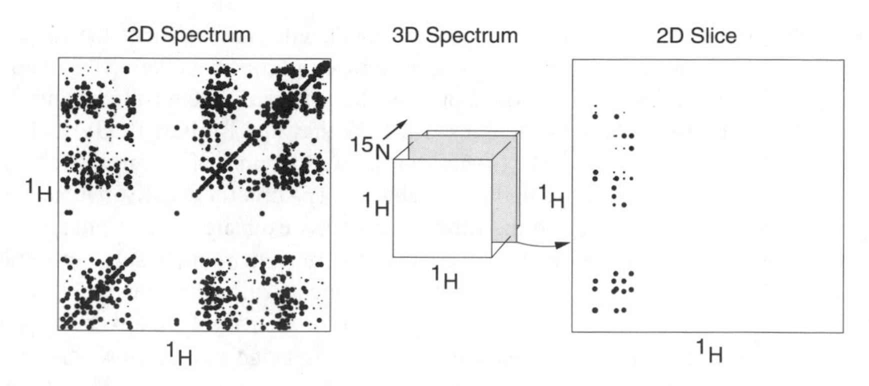

FIGURE 3.6 Increased resolution owing to higher dimensionality. The spectrum on the left is a representation of a two-dimensional NOESY spectrum obtained from a 50-residue 3-helix bundle protein. The size of each dot (peak) is proportional to the inverse of the sixth power of the distance between the two protons. Note that this spectrum is symmetrical because magnetization is transferred to and from a spin at the same time (i.e., the two paths of magnetization transfer are $H_X \rightarrow H_Y$ and $H_Y \rightarrow H_X$. The middle portion of the figure shows a three-dimensional NOESY-HSQC spectrum in the form of a cube. The additional frequency, the amide nitrogen chemical shift, extends out of the plane of the page. The right-hand portion of the figure shows a slice from the cube at a single ^{15}N frequency. Note the reduction in the number of peaks. The spectrum shown in this slice is not symmetrical because of the unidirectional nature of the magnetization transfer path in this experiment ($H_X \rightarrow H_N \rightarrow N \rightarrow H_N$).

two-dimensional spectrum. The addition of the nitrogen frequency gives a three-dimensional spectrum, shown as a cube in the middle of Figure 3.6. A single plane in this cube will only contain signals from residues whose amide nitrogen shifts are almost identical. All other residues will be found on different slices in this cube. One such slice (shaded in the center diagram) has been extracted and is shown in the right part of Figure 3.6. This plane provides the same information as the two-dimensional spectrum on the right, but only contains signals from five residues because they all have the same N_H shift. Consequently, most of the peaks can be resolved and studied.

The increased resolution that is gained by the addition of another frequency dimension is not without cost. Acquiring the three-dimensional cube is equivalent to acquiring 32 to 128 two-dimensional spectra. If the two-dimensional spectra required 8 h, to obtain, the three-dimensional spectrum will require 24 to 64 h, depending on how many different slices are required and how much time is spent acquiring each slice. However, it is clear that the two-dimensional spectra cannot be analyzed owing to severe overlap of peaks, while the three-dimensional spectra can be analyzed.

EFFECT OF CHEMICAL EXCHANGE ON THE SHAPE OF NMR RESONANCE LINES

The exchange of a nuclear spin between two distinct environments can have a profound effect on the shape and position of the resonance line. If the two environments result in a different chemical shift, it is possible to obtain information about the rates of chemical exchange as well as to determine the equilibrium constant relating the relative population of two states. A detailed account of the effects of chemical exchange on the NMR spectra can be found in Jenkins, 1991; Feeney and Birdsall 1993; Lian and Roberts, 1993; and Lian et al., 1994. In this review, two extremes of chemical exchange will be discussed: slow exchange and fast exchange. The criteria that distinguish slow exchange from fast exchange are the time scale of the exchange rate relative to the frequency separation that is generated by the two different environments. For example, if the chemical shifts in the two different environments differ by 100 Hz, then an exchange rate of 10 s^{-1} would result in slow exchange. Conversely, an exchange rate of 1000 s^{-1} would result in fast exchange. If the exchange rate is slow, then two resonance lines are observed and the exchange rate affects the line width. The lifetime of the excited state is shortened when the exchange event occurs. Thus, slow exchange results in an increase in natural line width by the exchange rate,

$$\Delta\nu = \frac{1}{\pi}R_2 + k_{EXCHANGE} \tag{3.10}$$

There are two common applications of exchange measurements. The effect of temperature on the exchange rate can be easily determined by line width measurements at different temperatures. The kinetic on-rate for the binding of a ligand can also be determined by measuring the effect of ligand concentration on the width of the resonance line for the free ligand. In this case, the observed exchange rate, $k^{observed}$, is given by the product of the kinetic on-rate and the concentration of ligand-free protein $[P]$,

$$k_{ON}^{OBSERVED} = k_{ON}[P] \tag{3.11}$$

Note that conditions can be arranged (i.e., low ligand concentration) such that slow exchange is always observed. It is also possible to measure the kinetic on-rate by detecting broadening of the protein resonances. In this case,

$$k_{ON}^{OBSERVED} = k_{ON}[L] \tag{3.12}$$

Figure 3.7 illustrates the effects of slow to intermediate exchange on the line width and position of the resonance lines. Spectrum “A” shows the NMR line from the protein in the absence of any ligand, while spectrum “F” is the NMR line of the fully saturated protein. As ligand is added (spectra $B \rightarrow E$), the line initially broadens (spectra B and C) due to slow exchange. As the ligand concentration reaches half saturation (spectrum D), intermediate exchange occurs. As the ligand concentration increases further, the fraction of free protein decreases, and the line moves to the position observed for the fully bound protein.

If the rate of exchange is faster than the frequency difference between the two states, then it is not possible to obtain any information on the microscopic rate constants because both the line width and the position of the resonance line are averaged owing to rapid exchange. The observed chemical shift is

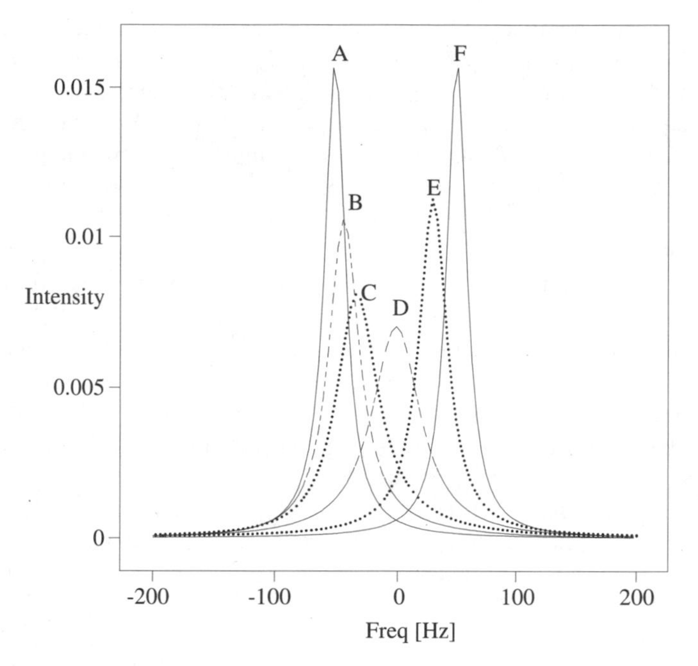

FIGURE 3.7 Effect of chemical exchange on the NMR spectrum. Spectra were simulated assuming a kinetic on-rate of 100 M^{-1} s^{-1} and a kinetic off-rate of $10^7 s^{-1}$ (overall affinity constant of 10^5 M^{-1}). The fractional saturations for each simulation were 0 (*A*), 0.05 (*B*), 0.20 (*C*), 0.50 (*D*), 0.70 (*E*), and 1.0 (*F*).

$$\delta_{OBS} = f_A \delta_A + f_B \delta_B \tag{3.13}$$

where f_A and f_B are the fractional occupancy of the two states. The equilibrium constant can be obtained directly from the fractional occupancies. This is a particularly good method of measuring the pK_a of an ionizable group. In the case of ligand binding, equilibrium constants can be obtained from measurements of this type, but the high concentrations of protein required for NMR studies reduces the accuracy of the measurement, especially for high binding constants.

APPLICATIONS AND PRACTICAL ADVICE

In general, NMR experiments will require protein concentrations that range from 0.5 to 1.5 m*M* in a sample volume of approximately 0.4 ml. It is generally not advisable to work much above 1.5 m*M* due to increases in viscosity. The increase in viscosity causes the apparent molecular weight of the protein to increase, causing an increase in linewidth. It is usually advantageous to perform the experiment at the highest magnetic field available (see the exceptions on page 61).

The experimental limitations, resources required, and time to completion of a NMR study depend entirely on the biochemical question to be addressed. There are many questions that can be answered with little material and a small investment of time. Others place significant demands on resources as well as time. The principle difference between simple experiments and costly experiments is the level of resonance assignments required. In simple experiments, resonance assignments can be made on the basis of the spectral properties of the sample alone. For example, the binding kinetics of a ligand can be measured using ligand containing a single ^{19}F spin while amide exchange kinetics of a protein can be obtained by integrating the spectrum over the region known to contain amide protons.

Information at the atomic level is only possible with detailed resonance assignments. In general, the level of detail of the information is directly proportional to number of assignments. For a large number of experiments, it is only necessary to obtain the assignments of the amide proton and nitrogen. These experiments include the amide exchange properties of individual amides, the relaxation properties of the amide nitrogen, and chemical shift changes due to ligand binding. However, structure determination by NMR requires assignment of virtually all of the protons in the protein. In many cases, strategies directed at obtaining amide assignments also produce the assignments for the H_α, C_α, H_β, and C_β spins. Therefore, only a relatively small number of experiments need to be performed to complete the resonance assignments. Strategies for obtaining protein chemical shift assignments are presented below.

Assignment of Resonances

Most, but not all, of the applications of NMR discussed here require the assignment of NMR lines to specific nuclear spins within the protein. Clearly, such assignments are essential if the properties of the spectral line (i.e., dipolar coupling, relaxation, etc.) are to be associated with the properties of a specific atom within the protein. There are two distinct steps in the assignment process. Although these steps can be conceptually separated, they are often pursued simultaneously. First, resonance lines are assigned to spins within an amino acid residue without the distinction of the location of the amino acid within the primary sequence. Second, these assignments are placed into the correct location within the known primary sequence of the protein. The data required to make these assignments are obtained from a number of different NMR experiments. The experiments required, the most suitable field strength, the amount of instrument time required, and the amount of analysis time required all depend on the size of the protein under study. These items are summarized in Table 3.1.

The time from initial studies to assignment of most of the resonance lines is fairly constant, about 1 to 4 months depending on the size of the protein. This time requirement will certainly decrease as software becomes available for more automated assignments. The consistency in time is due to the fact that as the protein gets larger it is absolutely necessary to utilize isotopic labeling to obtain the assignments. The use of isotopic labeling simplifies the assignment task such that more complex problems (i.e., more residues) can be solved in a proportionally shorter time. For example, it is generally not possible to assign proteins with molecular weights between 10 to 15 kDa without ^{15}N labeling, nor is it generally possible to

TABLE 3.1
Summary of Experiments Requirements for Assignments

Protein Size	Isotopic Labeling	Types of NMR Experiments	Field Strength	Instrument Time	Analysis Time
< 10 K	None	2-D 1H–1H NOESY	300–	1 week	1 month
		2-D TOSCY	500 MHZ		
10–15 K	^{15}N	3-D HSQC-TOCSY	500–	1–2 weeks	1 month
		3-D HSQC-TOCSY	600 MHz		
15–30 K	^{15}N	3-D-HSQC-NOESY	500–	1 month	1–2 months
	^{13}C	HNCA			
		HN(CO)CA	750 MHz		
		HN(CACB)HAHB			
		HN(CO)CACB			
		HN(COCACB)HAHB			
		Carbon TOCSY, etc.			
30–60 K	^{15}N	3-D/4-D-HSQC-NOESY	600–	2 months	2 months
	^{13}C	HNCA	900 MHz		
	^{2}H	HN(CO)CA			
		HNCB			
		HN(CO)CB			
		HN(COCACB)CG			

assign proteins with molecular weights between 15 to 20 kDa without both ^{15}N and ^{13}C labeling. Proteins with molecular weights above 25 kDa usually require replacement of protons with deuterons in order to decrease the linewidths and to reduce signal losses due to relaxation during the experiment. However, the time required to obtain assignments of smaller proteins can be reduced greatly by the use of ^{13}C labeling. Therefore, if expression is efficient, it would be highly advantageous to label the protein with ^{13}C.

General Approach to Assignments

The approach that would be taken to assign the NMR spectrum depends on two critical factors: the size of the protein and whether isotopic labeling with ^{15}N and ^{13}C is possible. If the protein is approximately 10 kDa or less, then it will be possible to obtain assignments without isotopic labeling. If the size is in the range of 10 to 15 kDa labeling with ^{15}N is required. For proteins in the size range of 15 to 30 kDa, it will be necessary to label with both ^{15}N and ^{13}C. Proteins that are larger than 30 kDa will require labeling with ^{15}N, ^{13}C, and ^{2}H.

To illustrate the assignment process consider the following tetrapeptide sequence that is part of a protein:

$$-Val_{10}-Ala_{11}-Ser_{12}-Ala_{13}-$$

The assignment strategies under different experimental conditions (molecular weight, isotopic labeling) are discussed below.

Assignment of Small Unlabeled Proteins (<8kDa)

Without the benefit of isotopic labeling, it is only possible to acquire two-dimensional ^1H–^1H spectra. The main advantage of using unlabeled material is that proteins from natural sources can be studied. Furthermore, the instrument time required to complete the assignments is small. The major disadvantage in performing ^1H spectroscopy alone is that the ^1H proton spectrum is only resolved for small proteins. Furthermore, the sequential assignments have to be obtained solely via dipolar coupling. Thus, a relatively long time is required to complete the assignments because the dipolar coupling between protons on nonadjacent residues can lead to confusion.

The first step in the assignment scheme would be to determine the chemical shifts of the protons contained in each residue. This collection of intra-residue chemical shifts is called a spin system. Spin systems are usually defined using a TOCSY experiment (total correlation spectroscopy) (Bax and Davis, 1985). In the TOCSY experiment magnetization is passed from one spin to another by scalar coupling in much the same way a baton is passed in a relay race (see Figure 3.8, top). While the magnetization exists on a spin it is possible to record the chemical shift of that particular spin. Under favorable conditions, the chemical shifts of all of the spins in an amino acid can be obtained.

The pathways of magnetization transfer in the TOCSY experiments are illustrated in Figure 3.9 for a serine. A simplified description of the proton–proton TOCSY involves decomposing the experiment into three parts and ignoring magnetization that originates on the amide proton. In the first part, the frequency of the α- and β-protons is recorded. In the second, this frequency information is passed to the amide proton. In the case of the β-protons, this involves a two-step process via the α-proton. In the third, the magnetization is labeled with the chemical shift of the amide proton. The resultant spectrum gives the chemical shifts of all protons in the amino acid residue. For most proteins of 8 kDa or less, this experiment can provide most of the side-chain proton assignments. However, spectral overlaps may result in 10 to 20% missing assignments at the initial stages. However, a number of the missing assignments can become apparent after analysis of the NOESY spectrum.

As the molecular weight increases, the sensitivity of the ^1H TOCSY decreases owing to the loss of magnetization by efficient spin–spin relaxation. Once the molecular weight exceeds 12 to 15 kDa, the ^1H TOCSY will provide incomplete information because many of the peaks are of weak intensity and are buried in the noise.

The TOCSY experiments provide the chemical shifts of the aliphatic protons. Depending on the residue type, it is usually possible to assign a spin system to a class of amino acids. In favorable cases, it may be possible to uniquely identify the residue type of a spin system. For example, residues like Asn, Asp, Glu, and Gln are easily distinguished from hydrophobic residues based on characteristic proton chemical shifts. A number of residues, such as Ser, Ala, and Thr, can be uniquely assigned from their characteristic proton shifts. Note that if any of these residues

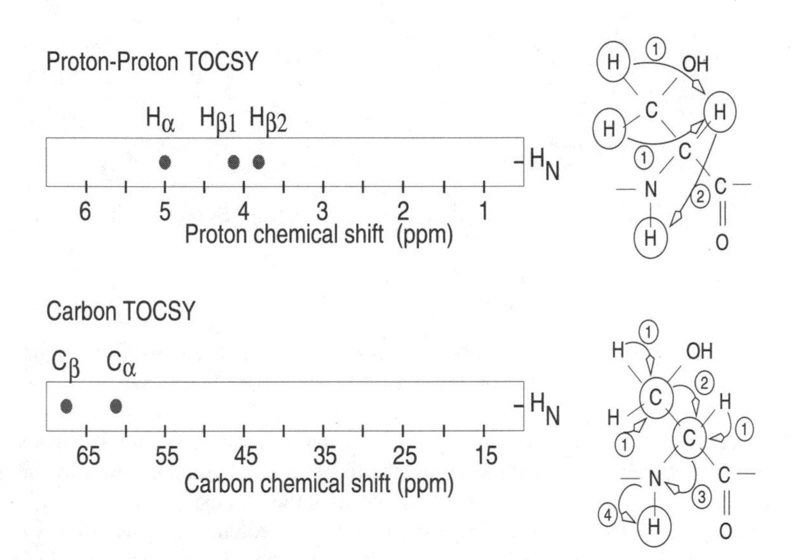

FIGURE 3.8 TOCSY experiments for intra-residue assignments. The structure of serine is shown on the right. Nuclei whose chemical shifts are measured in the experiment are circled. The pathway of magnetization transfer is indicated by the arrows. Top: In a proton–proton TOCSY magnetization is first labeled with the frequency of the aliphatic protons. This magnetization is transferred from the β-protons to the α-proton (step 1) and then from the α-proton to the amide proton (step 2). The magnetization that is labeled with the chemical shift of the α-proton is transferred directly to the amide proton. In the second part of the experiment, the magnetization is labeled with the chemical shift of the amide proton. A portion of the resultant two-dimensional spectrum is shown on the left. The chemical shift of the amide proton defines the *y* position of the peaks and the chemical shifts of the aliphatic protons define the *x*-position of the peaks. In this particular case, the chemical shifts of the H_α, $H_{\beta 1}$, $H_{\beta 2}$, are 5.0, 4.2, and 3.8 ppm, respectively. Bottom: In the carbon TOCSY, the magnetization of the carbons is transferred via carbon–carbon and then carbon–nitrogen coupling, to the amide group. The resultant spectra give the carbon chemical shifts of the aliphatic carbons. In this particular case, the C_α and C_β carbons resonant at 61 and 67 ppm, respectively.

only occurs once in the protein, then it is possible to assign the spin system to a particular amino acid based on its characteristic chemical shifts alone. However, if the residue occurs more than once, it cannot be assigned by this method. In the example presented here, the two alanines easily could be distinguished from the serine on the basis of chemical shifts. However, it would not be possible to distinguish Ala_{11} from Ala_{13}.

The next step in the assignment process is to assign each spin system to a particular residue in the protein. This can be accomplished in one of two ways. The first method utilizes dipolar coupling between protons on adjacent amino acids (observed in a NOESY spectra). In a protein, one would expect that some of the protons on one amino acid are within 5 Å of the protons on the adjacent amino acid. In this particular example, dipolar coupling between adjacent residues would be used to infer that Val_{10} is adjacent to Ala_{11}, that Ala_{11} is adjacent to Ser_{12}, and so on. The difficulty with this method is that a proton will show dipolar coupling to a large

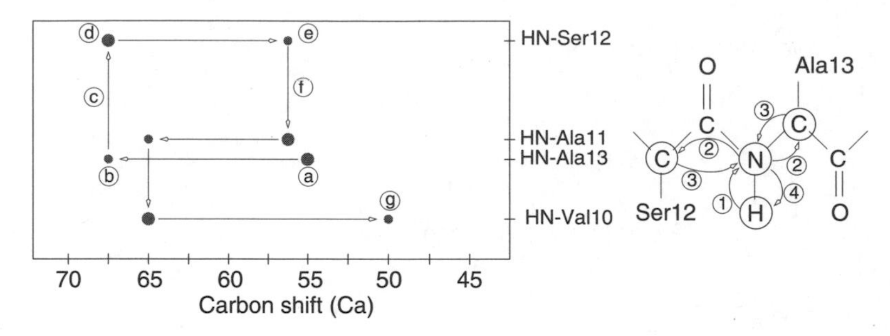

FIGURE 3.9 The use of the HNCA experiment in assignments. The path of magnetization transfer in the experiment is shown on the right. The steps are (1) transfer of magnetization from the amide proton to the nitrogen; (2) transfer of magnetization from the amide nitrogen to both α-carbons, followed by recording of their chemical shifts; (3) transfer back to the amide nitrogen followed by recording of its chemical shift; (4) transfer back to the amide proton and recording of its chemical shift. Note that all of these transfers utilize scalar coupling and are, thus, independent of the tertiary structure of the protein. The use of the spectra in the assignment of the tetra-peptide sequence is shown on the left. This is a two-dimensional slice from the three-dimensional matrix. Note that in this example all four residues have the same amide nitrogen shift. In practice, these residues would be found on different planes owing to the fact that their nitrogen chemical shifts are likely to be different. The steps in assignment are as follows: (a) marks the intra-residue C_α peak for Ala_{13}; (b) marks the inter-residue C_α peak; (c) shows that the inter-residue peak for Ala_{13} matches the intra-residue peak for Ser_{12}; (d) marks the intra-residue peak for Ser_{12}; (e) marks the inter-residue peak for Ser_{12}; (f) shows that the inter-residue peak for Ser_{12} has the same chemical shift as Ala_{11}; (g) shows the position of the C_α carbon of residue 9. The pathway of arrows shows the direction of the sequential assignments from the c-terminal to amino-terminal.

number of other protons, some of which will be on distant parts of the polypeptide chain. The dipolar coupling between nonsequential residues causes considerable confusion in the early stages of assignment.

Assignment of Moderate-Sized Labeled Proteins (10 to 15 kDa)

For these proteins, the ^1H–^1H spectrum is insufficiently resolved in two-dimensional experiments such that it is necessary to increase the resolution by a third frequency dimension. Although ^{13}C could also be used, ^{15}N is less expensive and the amide proton and nitrogen assignments are useful in other experiments, such as relaxation measurements and binding kinetics. The incorporation of ^{15}N can be easily accomplished by replacing the ^{14}N ammonia in minimal media with ^{15}N ammonia. This is quite inexpensive, usually less than $50 per liter of media. For those cell lines that grow poorly on minimal media, it is possible to obtain, at higher cost, rich media that is ^{15}N labeled. For the most part, the experiments that are applied to ^{15}N labeled proteins are simply three-dimensional versions of the same two-dimensional ^1H–^1H experiments that are used for smaller, unlabeled proteins. The actual assignment process still utilizes dipolar coupling between protons on adjacent residues. The time required for acquisition of the data is similar to that required for small proteins.

The reduced complexity of the data, owing to the addition of the third dimension, facilitates the assignment process such that a more complicated problem (i.e., more residues) can be solved in about the same period of time similar to a smaller protein.

The assignment of side-chain assignments using scalar coupling (i.e., TOCSY experiment) can be problematic toward the upper end of this molecular weight range because the larger molecular weight causes faster spin–spin relaxation rates. This, in turn, reduces the sensitivity of the TOCSY experiment because of signal loss during the transfer steps. Although the degeneracy in the three-dimensional HSQC-TOCSY spectrum is reduced relative to the two-dimensional TOCSY spectrum, fewer resonances from side-chain spins may be assigned using scalar coupling in moderately large proteins. However, a number of these side-chain assignments can be obtained in a tedious manner by detecting intra-residue proton–proton dipolar coupling in the relatively well-resolved three-dimensional HSQC-NOESY spectrum.

Although the addition of the third dimension increases the resolution of the NOESY spectra, the assignment of sequential residues is more difficult due to the larger number of possible assignments. In many cases, the assignment process can be greatly facilitated by obtaining some starting points for assignment by determining the type of residue for a number of the amide groups. This is accomplished by producing a small number of samples that are selectively labeled in one ^{15}N amino acid. A number of ^{15}N-labeled amino acids are inexpensive and can be incorporated into the protein by growth of the cells in a more complete media. Usually the ^{15}N-labeled amino acid is provided with all of the other (^{14}N) amino acids. The resultant HSQC spectrum of these selectively labeled proteins will only show signals from the labeled residues, thus indicating the type of residue for these peaks. Unfortunately, *E. coli* has a number of transaminases that catalyze the exchange of the amide nitrogen with the nitrogen pool. This event will dilute the supplied ^{15}N label as well as cause the incorporation of the added ^{15}N into other types of amino acids. Therefore, it is preferable to use a transaminase deficient strain for this type of labeling (McIntosh and Dahlquist, 1990). The strain, DL39 (Le Master and Richards, 1988) lacks most of the known transaminase activities. The regulatory elements for a number of expression systems (i.e., lambda, T7) can be incorporated into this strain. In the event a suitable transaminase strain cannot be obtained, it is possible to use short induction times and higher levels of the labeled amino acids to prevent dilution and mixing. Alternatively, some of the more expensive ^{15}N-labeled amino acids (e.g., Lys, Met, Thr, Trp) do not appear to be affected by transaminase activity and can be used for labeling under normal growth conditions with the normal host strain.

Assignment of Large Proteins (15 to 25 kDa)

In this case, it is absolutely necessary to label the protein with both ^{13}C and ^{15}N. The larger number of residues increase the complexity of ^{15}N separated three-dimensional spectra such that it is difficult to obtain the assignments by use of dipolar coupling between protons on adjacent residues because of the potentially misleading dipolar coupling to nonadjacent residues. Consequently, it is necessary to employ ^{15}N–^{13}C scalar coupling across the peptide bond to obtain the assignments.

This greatly simplifies the assignment task because scalar coupling only occurs through chemical bonds.

Doubly labeled protein is obtained in a similar manner to ^{15}N-labeled protein. The bacterial cell line expressing the protein are simply grown in minimal media containing ^{15}N ammonia and ^{13}C glucose. The cost of this media is considerably more than the ^{15}N-labeled media, approximately $1000 per liter. It is often advantageous to have a separate sample that is labeled with ^{15}N alone because the ^{15}N–^{13}C coupling increases the line width of the ^{15}N resonances, decreasing the resolution of the ^{15}N separated NOESY and TOCSY spectra. Thus both a ^{15}N-labeled and a ^{13}C–^{15}N-labeled sample are desirable.

The development of NMR techniques for double-labeled material began in the late 1980s, and there are dozens of different types of experiments that exploit scalar couplings for assignment purposes. These experiments are called triple resonance experiments because they involve three different nuclear spins, ^{1}H, ^{13}C, and ^{15}N (Ikura et al., 1990; Bax and Grzesiek, 1993). To perform experiments of this type, it is usually necessary to isotopically enrich the protein to 99% for the ^{13}C and ^{15}N atoms. The goal of these experiments is to correlate intra- and inter-residue chemical shifts with the amide proton and nitrogen chemical shifts.

The path of magnetization transfer in triple resonance experiments is often used to name the experiment. An example is the HN(CO)CA experiment. Nuclei not enclosed within brackets will have their chemical shift recorded in the experiment. Nuclei enclosed in brackets are only utilized to relay the magnetization between spins. In the HN(CO)CA experiment, magnetization is transferred from the amide group, via the carbonyl carbon, to the α-carbon of the preceding residue. The chemical shifts of the amide proton, amide nitrogen, and the α-carbon of the preceding residue are recorded, giving a three-dimensional spectrum. Another example of a triple resonance experiment is the HNCA experiment. The name of the HNCA experiment indicates that the chemical shift of the amide proton, amide nitrogen, and CA carbon would be recorded. Because the amide nitrogen is directly scalar coupled to both its own α-carbon and the α-carbon of the preceding residue, two peaks are observed for each amide group. Conversely, in the HN(CO)CA experiment the magnetization is relayed to the CA atom by the intervening CO spin. Thus, the only carbon shift recorded is the α-carbon of the preceding residue.

The HNCA and HN(CO)CA experiments are clearly complementary and are used to determine the intra- and inter-residue shifts of the α-carbon for each amide group. The inter-residue α-carbon shift is used to determine which spin system precedes a spin system. Once the amide of the preceding spin system is identified, its inter-residue α-carbon shift can be used to find the next preceding residue. By matching the intra- and inter-residue alpha carbon shifts for all of the amides, it is possible to sequentially link one residue to the next residue (see Figure 3.9).

There are several problems with this approach. First, because prolines lack an amide proton, they do not appear in the spectrum. Consequently, they break the sequential linking of the α-carbon chemical shifts. Second, if two or more residues have the *same* α-carbon shift (i.e., they are degenerate), it is not possible to unambiguously determine which residue should be the preceding one. Consequently, it is necessary to perform additional experiments that associate other inter- and intra-residues shifts with the amide proton. It would be typical to acquire four to six different triple resonance experiments [i.e., HNCA, HN(CO)CA, HNCACB, CBCA(CO)NH, HNHAHB, HBHA(CBCACO)NH] over a 1- to 2-week period. Usually, the combination of the four pairs of intra- and inter-residue shifts (C_α, H_α, C_β, H_β) is sufficient to arrive at an unambiguous sequential assignment. As with the intra-residue assignment experiment (TOCSY), the sensitivity of the triple resonance experiments is adversely affected by the size of the protein. As the spin–spin relaxation rates increase due to molecular size, increasingly more magnetization is lost during the transfer steps. Although the HNCA and HN(CO)CA experiments can be used for proteins up to 25 kDa, the HNCACB and HN(CACB)HAHB experiments give poor spectra for this size of protein. A more complete discussion regarding the relative sensitivities of triple resonance experiments can be found in Bax and Grzesiek, 1993.

If it is desired to determine the structure of the protein, then it will also be necessary to obtain the side-chain assignments. For triple-labeled proteins, the ^{13}C present in the protein can be utilized to relay magnetization in TOCSY experiments. Furthermore, the ^{13}C can be used to generate an additional dimension to further decrease the degeneracy. In the carbon TOCSY experiment, magnetization is transferred by the stronger carbon–carbon couplings. In the example shown in Figure 3.8, magnetization is first passed from the aliphatic protons to their attached carbon via scalar coupling. This step utilizes the stronger polarization of the protons to enhance the polarization of the carbons. The next step is recording the various carbon chemical shifts. This information is then passed from carbon to carbon until it reaches the α-carbon. At this point, it is passed on to the nitrogen and, finally, to the amide proton. The chemical shift of the amide proton is recorded and the resultant three-dimensional spectrum contains the amide proton shift as well as all of the aliphatic carbon shifts. Four or five different experiments are typically performed to obtain the spectra required for side-chain assignments. Each of these experiments requires 2 to 3 days of instrument time.

Most TOCSY experiments that are applied to large ^{13}C-labeled proteins utilize the carbon–carbon couplings to relay magnetization from one spin to its scalar-coupled partner. Because the carbon–carbon couplings are approximately 4 to 6 times larger than the ^1H–^1H coupling, the rate of transfer is faster. Consequently, the loss due to spin–spin relaxation is reduced and near complete side-chain (both ^1H, and ^{13}C) assignments can be obtained from proteins in the lower to middle of this size range. As the upper limit of this size range is reached, the loss of magnetization due to spin–spin relaxation can be significant, again reducing the completeness of the assignments that can be obtained by the use of scalar coupling.

Assignment of Very Large Proteins (25 to 60 kDa)

Obtaining the assignments for this group of proteins is very challenging. The key problem is that the spin–spin relaxation rate may become so fast that very little magnetization can be transferred from one spin to another via scalar coupling and most of the experiments described previously fail for reasons of poor signal to noise. The loss of magnetization during the transfer steps in both the TOCSY and triple resonance experiments can be reduced if the spin–spin relaxation rate of aliphatic carbon nuclei can be decreased. The principal relaxation mechanism for these carbons is dipolar coupling to the attached proton. Replacement of protons with deuterons decreases the rate of spin–spin relaxation by a factor of 5 to 10. The theoretical decrease in the carbon spin–spin relaxation rate should be 49, $\gamma_H^2/\gamma_D^2 = 7^2$. The maximum decrease in the spin–spin relaxation rate is usually not realized because other relaxation mechanisms, such as CSA and carbon–carbon dipolar couplings, are not affected by deuteration. Although deuteration is essential for the application of NMR experiments to larger proteins (greater than 40 kDa), it can also be useful in the study of smaller proteins. The properties of the amide proton resonances are also favorably improved by deuteration. In protonated samples, the amide proton resonance lines are usually broad, due to efficient dipolar coupling to aliphatic protons. In deuterated proteins, the relaxation rate of the amide protons is decreased, leading to sharper lines.

The generation of suitably labeled ^{15}N, ^{13}C, and ^{2}H samples is challenging (Le Master, 1990; Gardner and Kay, 1998; Goto and Kay, 2000). The most inexpensive method of incorporating ^{2}H is to simply grow the bacterial host in D_2O solvent with a protonated carbon source (e.g., ^{13}C glucose). Most strains can be adapted to grow in 95 to 100% D_2O by successively increasing the D_2O content of the media. Under these conditions, almost all of the H_α protons are readily replaced with deuterons during amino acid biosynthesis and approximately 75% of the remaining aliphatic protons are deuterated. Because the amide protons are also deuterated by this procedure, it is necessary to incubate the protein in H_2O buffer for several days (weeks) to back exchange the amide protons. Most amide protons will exchange with solvent in this time frame. In some cases, it may be necessary to denature the protein to obtain complete replacement of amide deuterons with protons.

^{1}H–^{15}N HSQC spectra of these samples show improved signal to noise as well as resolution owing to a decrease in the amide protonline width. Because the α-carbon is now deuterated, experiments that utilize this carbon to transfer magnetization [e.g., HNCA, HN(CO)CA] become much more sensitive. Note that the H_α chemical shift can no longer be used for the sequential assignments because these protons no longer exist. Additional assignment information can be obtained from the HN(CA)CB and HN(COCA)CB experiments. Protein obtained from a culture grown in D_2O is only partially deuterated at the C_β carbon. For larger protein (e.g., 40 to 60 kDa), the enhancement of the spin–spin relaxation rate by the residual H_β protons is sufficient to render the experiment useless. Consequently, to obtain information about the C_β chemical shifts, it is necessary to utilize protein that is fully deuterated. This can be accomplished by growth on ^{2}H- and ^{13}C-labeled glucose. Alternatively, several forms of completely deuterated rich media are commercially available.

COMMONLY PERFORMED EXPERIMENTS

MEASUREMENT OF KINETIC ON-RATES AND EQUILIBRIUM CONSTANTS

The measurement of kinetic rates and equilibrium constants can provide important information about the mechanism of ligand binding. Ligand binding rates that are significantly slower than diffusion suggest the existence of nonproductive conformations of the ligand, protein, or both. NMR experiments designed to provide this type of information usually do not require extensive assignment of resonance lines or large amounts of material.

The kinetic on-rate is obtained by measuring the broadening of resonance lines for the free ligand. Owing to the relatively small size of most ligands, these lines are likely to be sharp(er) and, therefore, easy to distinguish from that of the protein. The kinetic on-rate is simply obtained from the additional broadening observed for the ligand lines in the presence of the protein. If the proton NMR spectrum of the ligand is too complicated, it is quite easy to introduce ^{19}F into various sites on the ligand. For example, it is straightforward to introduce a single fluorine into a DNA or RNA ligand and obtain ^{19}F spectra. Alternatively, the ligand could be labeled with either ^{13}C or ^{15}N and an HSQC experiment could be used to observe the lines from the labeled ligand. An alternative way of obtaining the kinetic on-rates by NMR is to observe the resonance lines from the free protein. These lines would experience broadening due to the shortening of the excited state lifetime by the binding of the ligand. To observe well-resolved resonances from the protein, it may be necessary to label the protein with either ^{15}N or ^{13}C. Note that it is not necessary to assign the peaks in the HSQC spectra to extract meaningful kinetic information.

Equilibrium constants can be easily measured if the exchange of the ligand and protein are fast compared to the chemical shift difference (in Hz) between the bound and free ligand. It is usually more convenient to observe resonances from the protein. The chemical shift of a line is recorded as a function of the ligand concentration and the fraction of protein saturated is obtained from this chemical shift. This experiment also requires the observation of well-resolved resonances from the protein, likely necessitating isotopic labeling. Again, it is not necessary to assign the resonances from the protein to establish the equilibrium constant.

Because the ligand lines are usually sharp, this experiment could be performed with a ligand concentration of 50 μM (0.4-ml sample), or a total of 60 μgs. The ligand may have to be labeled with ^{19}F, ^{13}C, or ^{15}N. Labeling with ^{19}F is relatively inexpensive, whereas labeling with ^{13}C and ^{15}N is costly. The amount of protein required would depend on the binding affinity; however, less than 1 equivalent would be required, or 50 μm (600 μg of a 30 kDa protein). Note that if resonances from the ligand are being observed, then the protein need not be labeled, nor is there a limitation on the size of the protein. If the protein resonances are being observed, then the limitations on the size of the protein will be similar to those discussed below for structure determination.

Sufficiently high field strength (i.e., 400 to 800 MHz) to resolve the ligand resonances is required. Note that if ^{19}F is used, it is advantageous to work at a lower field strength (i.e., 300 MHz or less) to reduce the broadening of lines that results for the CSA of ^{19}F. The measurements of kinetic on-rates and/or binding constants should only require a few days of instrument time.

Topological Mapping of Ligand Binding Sites

The binding site of ligands on proteins can be mapped by either monitoring ligand induced changes in chemical shift of the amide groups or in their solvent exchange kinetics (Hajduk et al., 1999; Moore, 1999; Roberts, 1999). Note that the ligand can be a small molecule, nucleic acid, or another protein. For these experiments, it is not necessary to have labeled protein unless the protein size exceeds 8 to 10 kDa. ^{15}N labeling would be advantageous for proteins in the size range of 10 to 25 kDa, while ^{15}N and ^{2}H labeling would be useful/essential for proteins in the size range of 25 to 60 kDa. Because these experiments require knowledge of the amide resonance assignments, it would be necessary to obtain a ^{13}C-labeled sample for assignment purposes. However, the mapping experiment itself can be performed without labeling with ^{13}C. If the exchange kinetics of the ligand binding are fast, then the position and line width of the protein resonance line will be the weighted average of the liganded and unliganded states. In this case, the assignments for the unliganded protein can be used to obtain the assignments of the liganded protein. Unfortunately, if fast exchange is not observed, it is necessary to obtain the assignments of the protein–ligand complex using the methods discussed previously.

The actual experiments are quite simple. Chemical shift changes can be measured by acquiring two-dimensional spectra (TOCSY for unlabeled protein, ^{1}H–^{15}N HSQC for ^{15}N labeled protein) in the presence of the ligand. Resonances that show the largest change in the chemical shift arises from residues that show the largest change in the chemical environment owing to ligand binding. These results should be interpreted with some care. Conformational changes distant from the binding site can cause chemical shift changes. In addition, residues in the binding site can experience compensatory changes and show little chemical shift change. However, if a number of residues in the same spatial location in the structure show relatively large changes in chemical shift, it is likely that this location is the site of ligand binding.

Ligand–induced changes in the rate of amide exchange can also be used to determine the site of ligand binding. These experiments are also quite simple. The exchange rate of amide protons in the protein and the protein–ligand complex are quantified by measuring the intensity of the amide resonances as a function of time after the protein, or the protein-ligand complex, is placed into D_2O. Amide groups on or near the surface of proteins readily exchange their protons with deuterons when the protein is placed in D_2O. The formation of the protein–ligand complex may shield some of these amides from the solvent, thus reducing their exchange rate. Similar to chemical shift changes, these results should also be interpreted with some care. Although the largest reduction in amide-exchange rates appear to be at the site of ligand binding, it is possible to observe reduced exchange rates for residues quite distant from the ligand-binding site (Saito and Paterson, 1996).

Structure Determination by NMR

The determination of structure by NMR involves the measurement of experimental constraints, followed by model building with the goal to arrive at a structure that shows good agreement with both the experimental constraints and the covalent geometry (Clore and Gronenborn, 1988; Dotsch and Wagner, 1998; Guntert, 1998; Wider, 2000; Ferentz and Wagner, 2000). For structure determination, the protein need not be monomeric, but the aggregation state should be well-defined (i.e., dimer, tetramer, octamer). However, the structure of monomeric forms are usually the most straightforward because of the smaller size.

The upper size limitation for complete structure determination is not well defined. Structures of proteins that are less than 30 kDa should be obtainable by NMR methods. However, it is certainly possible to obtain some degree of structural information on much larger proteins (e.g., 30–80 kDa). The five different types of constraints that are commonly used in structure determination by NMR spectroscopy are listed below. The first two are the most common forms of constraints for the generation of structures from the NMR data.

- Inter-proton distance measurements from dipolar coupling
- The orientation of chemical bonds with the respect to the external magnetic field
- The presence of hydrogen bonds
- Torsional angles
- Chemical shifts

Inter-proton distances are obtained from the intensity of peaks in NOESY spectra. In many cases (i.e., unlabeled or ^{15}N-labeled material), these spectra were acquired for assignment purposes. Thus, no additional data need to be obtained. For ^{13}C- and ^{15}N-labeled proteins, it would be necessary to collect several three- and four-dimensional NOESY spectra to obtain these constraints. Depending on the concentration of the sample, this could require an additional 1 to 2 weeks of instrument time. Because dipolar coupling is generally only observable between protons separated by 5 Å or less, inter-proton dipolar coupling defines the local structure well and the global structure poorly. However, structures of globular proteins to a moderate resolution (equivalent to about 2 Å x-ray resolution) can be obtained by using a large number of these constraints. The structures of highly elongated molecules (i.e., DNA, fiberous proteins) are ill defined by NOE constraints.

In the case of partially deuterated proteins, the number of available distance constraints is reduced due to the presence of the deuterons. Consequently, it is usually only possible to measure inter-proton distances between amide protons as well as the distance between amides and methyl protons. In the case of fully deuterated proteins, it is only possible to measure inter-proton distances between amides. Although these distances are more accurate in deuterated proteins they are insufficient in number to produce a reasonably high-resolution structure.

The accuracy of the structures, especially from fully deuterated samples, can be markedly improved by measuring the angle between the external magnetic field and the bonds in the protein. When protein molecules are partially oriented, then small changes in the chemical shift arise from dipolar coupling between heteronuclear spins. These small changes can be used to obtain information on the orientation of the bond with respect to the external magnetic field (i.e., the $\cos(\theta)$ term in Equation (3.7) no longer averages to zero) (Tjandra and Bax, 1997; Prestegard, 1998; Tjandra, 1999). There are several methods of orienting proteins in a magnetic field. In the case of heme-containing proteins, the interaction of the heme with the magnetic field is sufficient to partially orient the protein (Tolman et al., 1997). For other proteins, it is necessary to induce partial orientation of the solvent by the inclusion of lipid bicelles (Tjandra and Bax, 1997), filamentous phage (Hansen et al., 2000), or purple membrane (Sass et al., 1999). All three of these particles orient in the magnetic field and this order is imparted onto the proteins by a variety of mechanisms. Typical applications of this technique would involve the measurement of the direction of the N_H–H_N bond vector with respect to the magnetic field. The inclusion of bond orientation information from dipolar coupling provides information on the global orientation of segments of the molecule with respect to the external magnetic field. The information content of these constraints is very high such that their inclusion produces NMR structures that are equal in resolution to good quality x-ray structures (i.e., about 1.5 Å resolution).

Hydrogen bonds can also be used as constraints in model building. Hydrogen bond donors (usually HN protons) are identified because they exchange slowly with the solvent when the protein is placed in D_2O. Because it is possible only to identify the hydrogen bond donors from amide-exchange kinetics, hydrogen bonds cannot be used as a constraint until the acceptor can be inferred from structures of moderate resolution. Regardless of this difficulty, NMR is one of the few techniques that can explicitly identify protons involved in hydrogen bonds.

Torsional angles are obtained from measurements of scalar coupling. For three-bond couplings, the size of the coupling depends on the torsional angle of the bonds connecting the atoms. In general, the phi angle is routinely measured. It is also possible to measure the chi angles at a number of positions on the side chain. Because these measurements utilize scalar coupling, the sensitivity of the experiment drops as the molecular size increases. It is generally difficult to obtain this information for proteins that are larger than 20 to 25 kDa.

All of the above constraints are automatically employed in various refinement programs. As with x-ray crystallography, simulated annealing is extensively used in an effort to avoid obtaining structures whose energy is not at the global minimum. In contrast to x-ray crystallography, the final NMR structure that is deposited in the protein database is usually a collection of 5 to 50 of the lowest energy structures that have been aligned to either the lowest energy structure or the average structure. Inspection of this collection of structures shows regions of the protein whose conformation is well described (possessing a low root-mean-square deviation) vs. those regions whose conformation is not well described (high RMSD). There are several possible reasons for high RMSD for certain residues. The conformation of these residues may be ill defined due to an absence of constraints. Alternatively, the molecular motion affects the strength of dipolar coupling between protons, thus affecting the accuracy of the distance measurement. The existence of such motion can be readily detected by relaxation measurements and it is becoming common practice to report both the structure of a protein and the preliminary relaxation data. The inclusion of the latter is important because it allows the reader to determine which regions of the structure are poorly determined due to the presence of molecular motion.

The amount of time required to complete a structure by NMR spectroscopy is difficult to estimate. The assignment of peaks in the various NOESY spectra may require 1 to 3 months, depending on the degree of automation utilized. The actual refinement of the structure requires the completion of 2 to 20 cycles of building structures and assigning additional distance constraints based on the newest structures. Each of these cycles can consume 1 day to 1 week, depending on the degeneracy in the data and the size of the structure. A time period of 2 to 3 months to refine the structure is optimistic. Again, recent progress in the automation of this procedure should reduce the refinement time.

Relaxation Measurements

The spin-lattice relaxation rate, the spin–spin relaxation rate, and heteronuclear NOE are easily measured and all are sensitive to molecular motion. The intent of relaxation measurements is to infer something about the extent and range of internal motions in the protein (Lane and Lefevre, 1994; Dayie et al., 1996; Kay, 1998; Palmer, 2001; Stone, 2001). Of considerable interest are ligand-induced changes in these motions. In general, the relaxation properties of the ^{15}N spin are observed because the amide nitrogen is largely relaxed by its interaction with the amide proton (thereby greatly simplifying the theory required to analyze the data). Furthermore, the $^{1}H–^{15}N$ HSQC spectra of rather large proteins are sufficiently resolved to provide information on a large number of residues within the protein.

To quantify local (internal) motion, the spectral density function (Equation 3.9) is modified to include the effects of the local motion of the intensity of magnetic field fluctuations at the frequencies of zero, single, and double quantum transitions. The goal of these studies is to define both the extent and frequency of the internal motions. Most approaches assume that the motion of the atoms can be represented as restricted diffusion on the surface of a sphere. The degree of restriction is represented by the order parameter, S. If S is close to one, the atom diffuses in a highly restricted manner over a small region of the sphere. If S approaches zero, the atom is free to diffuse over most of the sphere in an unrestricted manner. The order parameter, S, is analogous to the thermal factor (B) in x-ray crystallography. Residues with high thermal factors often show low values for S (Powers et al., 1993). The rate of the internal motion also can be obtained from this analysis; however, it is difficult to characterize motions on a time scale that is slower than the overall correlation time of the protein.

A complete characterization of molecular motion requires the measurement of spin-lattice relaxation, spin–spin relaxation, and the heteronuclear NOE at several magnetic field strengths. This comprehensive approach requires a considerable amount of instrument time as well as analysis time. A more qualitative description of internal motions can be obtained from measurements of the proton-nitrogen NOE. For amide groups that do not show any internal motion, the NOE is close to 1.0. Amide groups that are undergoing very rapid motion show a negative NOE (see Figure 3.10).

Measurements of amide hydrogen exchange rates can characterize structural fluctuations on a much slower time scale, ranging from minutes to days (Woodward et al., 1982; Clarke and Itzhaki, 1998; Raschke and Marqusee, 1998). Absolute rates of amide exchange are somewhat difficult to interpret. However, ligand-induced changes in these rates can provide valuable information on local and global changes in protein dynamics due to the binding of the ligand.

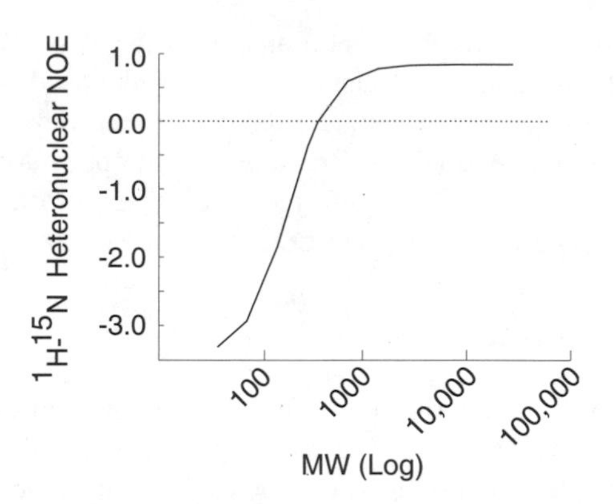

FIGURE 3.10 Effect of motion on the heteronuclear NOE. The ^{1}H–^{15}N heteronuclear NOE is shown as a function of molecular weight. A residue that is immobile in a protein of size 10 kDa or larger would show a heteronuclear NOE of 0.84. If the internal motion of this residue were similar to the isotropic motion of a 1 kDa peptide (octamer), then the NOE would be approximately 0.7. If the motion of this residue were similar to the isotropic motion of a tetramer (500 Da), the heteronuclear NOE would be approximately 0.4. If the motion of this residue were similar to the motion of a free amino acid in solution, the heteronuclear NOE would be approximately –2.0.

REFERENCES

Bax, A., Griffey, R. H., and Hawkins, B. L. (1983). Correlation of proton and nitrogen-15 chemical shifts by multiple quantum NMR, *J. Mag. Res.*, 55, 301–315.

Bax, A. and Davis, D. G. (1985). Practical aspects of two-dimensional transverse NOE spectroscopy, *J. Magn. Resonance*, 63, 207–213.

Bax, A. and Grzesiek, S. (1993). Methodological advances in protein NMR, *Acc. Chem. Res.*, 26, 131–138.

Bax, A. and Lerner, L. (1986). Two-dimensional nuclear magnetic resonance spectroscopy, *Science*, 232, 960–967.

Clarke, J. and Itzhaki, L. (1998). Hydrogen exchange and protein folding, *Curr. Opin. Struct. Biol.*, 8, 112–118.

Clore, G. M. and Gronenborn, A. M. (1988). NMR structure determination of proteins and protein complexes larger than 20 kDa, *Curr. Opin. Chem. Biol.*, 2, 564–570.

Clore, G. M. and Gronenborn, A. M. (1991). 2-Dimensional, 3-dimensional, and 4-dimensional NMR methods for obtaining larger and more precise 3-dimensional structures of proteins in solution, *Annu. Rev. Biophys. Biomol. Struct.*, 20, 29–63.

Dayie, K. T., Wagner, G., and Lefevre J. F. (1996). Theory and practice of nuclear spin relaxation in proteins, *Annu. Rev. Phys. Chem.*, 47, 243–282.

Dotsch, V. and Wagner, G. (1998). New approaches to structure determination by NMR spectroscopy, *Curr. Opin. Struct. Biol.*, 8, 619–623.

Feeney, J. and Birdsall, B. (1993). NMR studies of protein-ligand interactions, in *NMR of Macromolecules: A Practical Approach* (Roberts, G. C. K., Ed.). Oxford University Press, Oxford.

Ferentz, A. E. and Wagner G. (2000). NMR spectroscopy: a multifaceted approach to macromolecular structure, *Q. Rev. Biophys.*, 33, 29–65.

Gardner, K. H. and Kay, L. E. (1998). The use of ^{2}H, ^{13}C, ^{15}N multidimensional NMR to study the structure and dynamics of proteins, *Annu. Rev. Biomol. Struct.*, 27, 357–406.

Goto, N. K. and Kay, L. E. (2000). New developments in isotope labeling strategies for protein solution NMR spectroscopy, *Curr. Opin. Struct. Biol.*, 10, 585–592.

Guntert, P. (1998). Structure calculation of biological macromolecules from NMR data, *Q. Rev. Biophys.*, 31, 145–237.

Hajduk, P. J., Meadows, R. P., and Fesik, S. W. (1999). NMR-based screening in drug discovery, *Q. Rev. Biophys.*, 32, 211–240.

Hansen, M. R., Hanson, P., and Pardi, A. (2000). Filamentous bacteriophage for aligning RNA, DNA, and proteins for measurement of nuclear magnetic resonance dipolar coupling interactions, *Methods Enzymol.*, 317, 220–240.

Ikura, M., Kay, L. E., and Bax, A. (1990). A novel approach for sequential assignment of ^{1}H, ^{13}C, and ^{15}N spectra of proteins: heteronuclear triple-resonance three-dimensional NMR spectroscopy. Application to calmodulin, *Biochemistry*, 29, 4659–4667.

Jeener, J., Meier, B. H., Bachmann, P., and Ernst, R. R. (1979). Investigation of exchange processes by two-dimensional NMR spectroscopy, *J. Chem. Phys.*, 71, 4546–4553.

Jenkins, B. G. (1991). Detection of site-specific binding and co-binding of ligands to macromolecules using ^{19}F NMR, *Life Sci.*, 48, 1227–1240.

Kay, L. E. (1998). Protein dynamics from NMR, *Natl. Struct. Biol.*, 5, 513–517.

Lane, A. N. and Lefevre, J. F. (1994). Nuclear magnetic resonance measurements of slow conformational dynamics in macromolecules, *Methods Enzymol.*, 239, 596–619.

LeMaster, D. M. (1990). Uniform and selective deuteration in 2-dimensional NMR of proteins, *Annu. Rev. Biophys. Biomol. Struct.*, 19, 243–266.

LeMaster, D. M. and Richards, F. M. (1988). NMR sequential assignments of *E. coli* thioredoxin utilizing random fractional deuteration, *Biochemistry*, 27, 142–150.

Lian, L. Y. and Roberts, G. C. K. (1993). Effects of chemical exchange on NMR spectra, in *NMR of Macromolecules: a Practical Approach* (Roberts, G. C. K., Ed.). Oxford University Press, Oxford.

Lian, L. Y., Baruskov, I. L., Sutcliffe, M. J., Sze, K. H., and Roberts, G. C. K. (1994). Protein-ligand interactions: exchange processes and determination of ligand conformation and protein-ligand contacts, *Methods Enzymo.*, 239, 657–700.

McIntosh, L. P. and Dahlquist, F. W. (1990). Biosynthetic incorporation of ^{15}N and ^{13}C for assignment and interpretation of Nuclear Magnetic Resonance spectra of proteins, *Q. Rev. Biophys.*, 23, 1–38.

Moore, J. M. (1999). NMR techniques for characterization of ligand binding: utility for lead generation and optimization in drug discovery, *Biopolymers*, 51, 221–243.

Neuhaus, D. and Williamson, M. P. (1989). *The Nuclear Overhauser Effect in Structural and Conformational Analysis*. VCH Publishers, New York.

Palmer, A. G., III (2001). NMR probes of molecular dynamics: overview and comparison with other techniques, *Annu. Rev. Biophys. Biomol. Struct.*, 30, 129–155.

Powers, R., Clore, G. M., Garrett, D. S., and Gronenborn, A. M. (1993). Relationships between the precision of high-resolution protein NMR structures, solution-order parameters, and crystallographic B factors, *J. Mag. Res.*, B101, 325–327.

Prestegard, J. H. (1998). New techniques in structural NMR—anisotropic interactions, *Natl. Struct. Biol.*, 5, 517–552.

Raschke, T. M. and Marqusee, S. (1998). Hydrogen exchange studies of protein structure, *Curr. Opin. Biotechnol.*, 9, 80–86.

Roberts, G. C. K. (1999). NMR spectroscopy in structure-based drug design, *Curr. Opin. Biotechnol.*, 10, 42–47.

Saito, N. G. and Paterson, Y. (1996). Nuclear Magnetic Resonance spectroscopy for the study of B-Cell epitopes. *Methods*, 9, 516–524.

Saas, J., Cordier, F., Hoffmann, A., Rogowski, M., Cousin, A., Omichinski, J. G., Löwen, H., and Grzesiek, S. (1999). Purple membrane induced alignment of biological macromolecules in the magnetic field, *J. Am. Chem. Society*, 121, 2047–2055.

Stone, M. J. (2001). NMR relaxation studies of the role of conformational entropy in protein stability and ligand binding, *Acc. Chem. Res.*, 34, 379–388.

Tjandra, N. (1999). Establishing a degree of order: obtaining high-resolution NMR structures from molecular alignment, *Struc. Fold. Des.*, 7, R205–211.

Tjandra, N. and Bax, A. (1997). Direct measurement of distances and angles bio-molecules by NMR in a dilute liquid crystalline medium, *Science*, 278, 1111–1114.

Tolman, J. R., Flanagan, J. M., Kennedy M. A., and Prestegard, J. H. (1997). NMR evidence for slow collective motions in cyanometmyoglobin, *Natl. Struc. Biol.*, 4, 292–297.

Wider, G. (2000). Structure determination of biological macromolecules in solution using nuclear magnetic resonance spectroscopy, *Biotechniques*, 6, 1278–1282.

Woodward, C., Simon, I., and Tuchsen, E. (1982). Hydrogen exchange and the dynamic structure of proteins, *Mol. Cell. Biochem.*, 48, 135–160.

ADDITIONAL READING

Cavanagh, J., Fairbrother, W. J., Palmer, A. G., III, and Skelton, N. J. (1996). *Protein NMR Spectro-scopy Principles and Practice,* Academic Press, New York.

Clore, G. M. and Gronenborn, A. M. (1993). *NMR of Proteins,* CRC Press, Ann Arbor, MI.

Roberts, G. C. K., Ed. (1993). *NMR of Macromolecules: a Practical Approach*, Oxford University Press, New York.

GLOSSARY

Chemical shift—a method used for reference NMR spectra that are independent of the magnetic field strength. Units of chemical shift are parts per million (ppm).

Chemical shift anisotropy (CSA)—the dependence of the chemical shift on the orientation of the nuclear spin with respect to the magnetic field; a nuclear spin relaxation mechanism is important for ^{13}C, ^{15}N, and ^{19}F.

Degenerate—two or more resonance lines that cannot be distinguished from each other because the chemical shift difference between them is smaller than the line width of the resonance.

Dipolar coupling—coupling between spins that occurs through space; the strength of the interaction is dependent on the distance between the nuclei, their gyromagnetic ratios, and the molecular dynamics of the coupled pair.

Double quantum transition—a transition between nuclear spin states in which two spins are simultaneously excited from the ground state or relaxed to the ground state.

FID (free induction decay)—this is the time domain signal acquired after excitation of the nuclear spins; decay in amplitude is owing to relaxation effects.

Field strength—the strength of the external field applied to nuclear spins; in spectroscopy, the units are based on the absorption frequency of the proton at that field strength.

Fourier transform—a mathematical procedure that extracts the amplitudes of each frequency component from the time domain data; generates an NMR spectrum from the time domain data.

Frequency domain—the normal NMR spectrum; obtained from time domain data by Fourier transformation.

Gyromagnetic ratio (γ)—fundamental constant for a given type of nuclear spin; the magnitude of γ determines the sensitivity and absorption frequency of the spin; see Equation (3.4).

Heteronuclear NOE—dipolar coupling between two unlike spins (i.e., ^{1}H and ^{15}N) cause a perturbation of the population of one spin to be transferred to another.

HSQC (heteronuclear correlation spectroscopy)—a two-dimensional NMR experiment commonly used to show the correlation of the chemical shifts of ^{15}N or ^{13}C spins with their bound proton resonances.

Magnetization transfer—the transfer of magnetization from one spin to another; this can occur via scalar coupling or via dipolar coupling; commonly used to label magnetization with more than one frequency in multi-dimensional NMR experiments (scalar coupling) or to measure the intensity of the nuclear overhauser effect (dipolar coupling).

NOE (nuclear overhauser effect)—perturbation in the population levels of one spin owing to dipolar coupling to another spin whose population levels are not at thermal equilibrium.

NOESY (nuclear overhauser effect spectroscopy)—a two-dimensional NMR experiment commonly used to measure the strength of the NOE; principally used to measure the inter-nuclear distance between two dipolar coupled spins.

Nuclear relaxation—change in the state of nuclear magnetization due to stimulated emission.

Polarization — population difference between ground and excited states.

Rotational correlation time—typical time for a rotating body to tumble through an angle of one radian; it is also the time required for the orientation of a rotationing body to lose memory of its previous position.

Scalar coupling—coupling of nuclear spins that is mediated by chemical bonds; commonly utilized to transfer magnetization in triple resonance experiments.

Shielding factor—the effect of electrons contained in chemical bonds on the effective magnetic field felt by the nuclear spin; causes different chemical shifts for nuclei in different environments.

Single quantum transition—change in the energy level or quantum state of a nuclear spin owing to the absorption or emission of an amount of energy that is equal to the energy difference between the ground and excited state.

Spectral density function—characterizes time dependent magnetic field fluctuations; gives the intensity of these fluctuations at any given frequency.

Spin system—a group of nuclear spins related by scalar coupling to each other; traditionally restricted to those spins that show measurable ^{1}H–^{1}H couplings; a single amino-acid residue is an example of a spin system. A dipeptide would contain two spin systems.

Spin–spin relaxation—a change in the state of excited nuclear spins owing to spin–spin flipping; a nuclear spin in the ground state becomes excited by absorbing energy from a spin in the excited state. spin–spin relaxation does not cause a net loss of energy from the system, but does limit the lifetime of the excited state.

Spin-lattice relaxation—a change in the state of excited nuclear spins owing to loss of energy to the surroundings (lattice); this also limits the lifetime of the excited state (see spin–spin relaxation).

T1—spin-lattice relaxation time constant.

T2—spin–spin relaxation time constant.

Time domain—see free induction decay.

TOCSY—Total correlation spectroscopy; a NMR experiment that utilizes scalar coupling to transfer magnetization between spins; the magnetization that is initiated on one spin is transferred to all the other spins in the spin system. Therefore, the chemical shift of every nuclear spin within the spin system becomes correlated with all other chemical shifts in the spin system.

Zero quantum transition—change in the population levels that does not result in the net change of the quantum state.

4 Analysis of Proteins by Mass Spectrometry

Mark E. Bier

CONTENTS

Abstract 72
Introduction 72
Mass Spectrometry Instrumentation 73
 Ion Sources 73
  Electrospray Ionization 73
  Matrix-Assisted Laser Desorption Ionization 75
 Mass Analyzers 77
  Triple-Stage Quadrupole (QqQ) Mass Spectrometer 78
  Paul* Three-Dimensional Quadrupole Ion Trap (QIT) Mass Spectrometer 79
  Time-of-Flight (TOF) Mass Spectrometer 79
  Quadrupole Time-of-Flight (Qq-TOF) Mass Spectrometer 80
  Fourier Transform Ion Cyclotron (FT) Mass Spectrometer 80
Sample Preparation for MALDI and ESI 81
 Working at the Femtomole Level 81
 Sample Volumes Required for ESI and MALDI-MS Analysis 81
 Analyte Concentrations Required for MALDI and ESI-MS 81
 Sample Quality for ESI or MALDI 82
 Pipet Tip Packing Bed Purification Method 83
 Capillary-Liquid Chromatography MS 83
 Capillary-Electrophoresis ESI-MS 84
Molecular Weight Determinations of Peptides and Proteins 84
 Multiply Charged Molecular Ions and the Determination of Charge State 84
 Average vs. Exact Monoisotopic Molecular Weights 86
 Resolution 87
 Mass Accuracy 87
Protein Sequencing 88
 Peptide Sequencing by Mass Spectrometry–Mass Spectrometry 88
 Peptide Ladder Sequencing 92
Protein Identification 92
 Preparation of the Protein Sample for Identification 93

0-8493-9453-8/02/$0.00+$1.50

Protein Identification by Peptide Mapping 95
Sequence Tag Approach to Protein Identification
Using MS–MS Analysis 95
SEQUEST™ Approach to Protein Identification
Using MS–MS Analysis 98
Data Dependent Scanning 98
Protein Modifications, Folding, and Tissue Analysis 98
A Final Thought 99
Acknowledgments 99
References 100

ABSTRACT

The analysis of proteins by mass spectrometry has emerged as the technique of choice for obtaining high performance results from small amounts of analyte. Interest in this analytical technique has increased primarily as the result of the introduction of the electrospray ionization (ESI) and matrix-assisted laser desorption ionization (MALDI) sources. These innovations permit nonvolatile molecules to be readily introduced into the gas phase.

A number of mass analyzers in use today have been coupled to these sources. These include two- and three-dimensional quadrupole field, time-of-flight (TOF), quadrupole-TOF hybrids, magnetic sector, and Fourier transform mass spectrometers. Paramount to the mass spectrometer analyzer used in the analysis is proper sample preparation. With proper preparation of proteins and peptides, their molecular weights can be determined with high mass accuracy. Conversely, a poorly prepared sample will lead to poor or no mass spectrometer results. For peptides and proteins, the mass accuracy is typically better than 0.01%.

Sequence information at the femtomole and, in some cases, attomole levels can be obtained using product MS–MS scans of peptide molecular ions from protein digests. Mass spectrometry is now widely used to identify proteins by acquiring peptide molecular weight maps and/or MS–MS spectra and entering this data into a protein database search engine. Modifications such as phosphorylations or glycosylations on proteins can also be identified with the use of MS, and some researchers are also studying protein conformation using H/D exchange reactions.

INTRODUCTION

Early in the history of mass spectrometry (MS), large biomolecules were not analyzed because efficient methods to transport these molecules into the gas phase were unknown. Degradation typically occurred during vaporization of these nonvolatile molecules so that electron ionization of the intact molecular ion was not possible. Ionization by fast atom bombardment[1, 2] (FAB), field desorption (FD),[3] secondary ionization mass spectrometry (SIMS), and plasma desorption (PD)[4] from the radioactive decay of ^{252}Cf finally made the ionization and analysis of peptides possible. These latter techniques, although still used today, are not as popular as electrospray ionization[5–9] (ESI) and matrix-assisted laser desorption ionization (MALDI).[10–12] With the advent of MALDI and ESI, mass spectrometry has become the preferred analytical method of choice for the analysis of proteins and peptides. The increased use of MALDI and ESI has occurred because of the technical ease and the quality of the MS data produced. ESI and MALDI allow the routine analysis of proteins and peptides by MS, and since the early 1990s, many more biological researchers have taken advantage of these MS techniques. Anyone studying proteins or peptides should be aware of the analytical power of mass spectrometry!

This chapter is not intended for the seasoned mass spectrometrist who currently does proteomics research by MS, but rather, it was written for scientists who are inexperienced in the use of MS for protein research. This chapter includes a brief description of ESI, MALDI, current mass spectrometers, the basics in sample preparation, and examples of MS analyses of proteins and peptides.

MASS SPECTROMETRY INSTRUMENTATION

All mass spectrometers have three components: an ion source, a mass analyzer, and an ion detector. We will cover the two most common ion sources and five mass analyzers used today to analyze proteins.

Ion Sources

This section will cover the two most popular ionization techniques for analyzing proteins by mass spectrometry: electrospray ionization (ESI) and matrix-assisted laser desorption ionization (MALDI). A brief description of five common mass spectrometers that have been coupled to these ion sources will follow.

Electrospray Ionization

Remarkably, Malcolm Dole and his colleagues first studied electrospray ionization of molecules in 1968.[13] Dole's group electrosprayed the proteins lysozyme (MW = 14,000 amu) and zein (MW = 38,000 amu),[14–16] but did not introduce these ions into a mass spectrometer. It was not until the 1980s that John Fenn's group in the U.S.[17] and Maxim Alexanderov's group in Russia[18] coupled this ionization technique to a modern mass spectrometer. Their experiments soon revolutionized how biomolecular ions could be introduced into the gas phase. Fenn's group[19] showed that electrosprayed proteins carry multiple charges ($[M + H]^{+}$, $[M + 2H]^{+2}$, $[M + 3H]^{+3}$, . . ., $[M + nH]^{+n}$) and that a distribution of these multiply charged ions could be used to calculate accurate protein molecular weights. Soon after it was realized that electrospray could be used to ionize proteins and peptides, the rapid development and use of this ion source ensued.

The basic ESI source consists of a high voltage capillary sprayer. Many different research groups have made various electrospray devices, but all must apply a high voltage to the solution. A nanoflow sprayer similar to the one developed by Dohmeier, Jorgenson, and co-workers[20, 21] and Wilm and Mann[22, 23] is shown in Figure 4.1. In this figure the nanoflow electrospray ion source is shown with a gold-coated glass capillary and electrical connection. The first step in electrospray ionization is to dissolve the analyte in a liquid solution of an organic : aqueous solvent containing an acid or base. The analyte is then introduced into the electrospray tip as a result

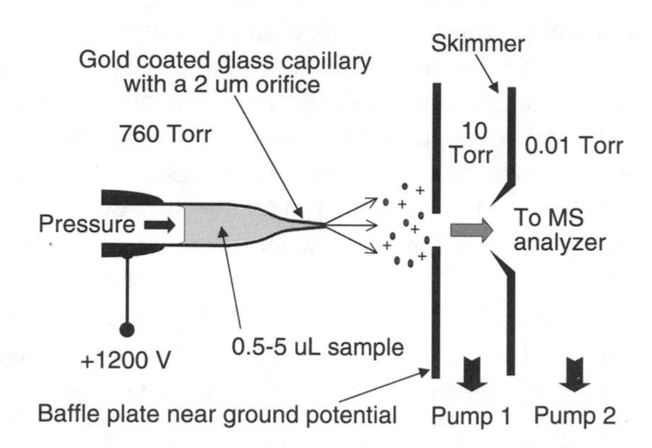

FIGURE 4.1 Nano flow (10 to 100 nl/min) electrospray ion source showing a gold coated glass capillary filled with a solution containing a peptide or protein. The high voltage applied to the solution causes a spray of micrometer-sized droplets from which ions emerge and enter the low pressure region of the mass analyzer.

of capillary action, the ESI process, and pneumatic, or hydraulic pressure. Depending on the solution, liquid flow rate, the electrospray tip geometry and its tip distance to the electrode ground plate, a potential of ±0.5 to 6 kV [(+) for positive or (–) for negative ions] is used to initiate and maintain the electrospray process. For nanoliter per minute flow rates (5 to1000 nL/min), in tips with 1 to 10 μm orifice, the electrolyte liquid is typically set to a voltage between 500 to 2000 V. For higher electrospray flow rates (1 to 1000 μL/min) in tips with a 100 μm orifice, the liquid is charged to a potential between 4 to 6 kV. Approximately 1 mL/min is considered the maximum flow rate for most high-flow rate ESI sources. The electrospray voltage causes a highly charged mist of micrometer-sized droplets to be sprayed from the tip. A properly functioning electrospray needle tip forms a *Taylor cone* at the tip where the micro droplets are formed. In addition to the applied voltage, a concentric flow of nitrogen or other suitable gas is often used at higher flow rates (greater than 1 μL/min) to help nebulize the solution and desolvate the analyte ions. Ions enter the gas phase and are sampled at an orifice/skimmer assembly that leads to the first low pressure region of the mass spectrometer (1 to 10 Torr). After several stages of differential pumping, a fraction of these ions enter the mass analyzer section of the mass spectrometer. Figure 4.2 show an ESI mass spectrum of multiply charged ions from the tetradecapeptide renin substrate (rat), MW_{ave} = 1823.09, $MW_{monoiso}$ = 1821.9200 amu ($MW_{monoiso}$, refers to the molecular weight of the monoisotopic ^{12}C peak), from a solution of 50 : 50, acetonitrile : water and 0.1 *M* acetic acid. The spectrum shows the $[M + H]^+$, $[M + 2H]^{+2}$, $[M + 3H]^{+3}$ ions and the sodiated ions. The inset shows an expanded view of the $[M + H]^+$ and $[M + Na]^+$ ions.

The ESI technique is an online ionization technique because the liquid flowing from a syringe or from a liquid chromatography column is coupled directly to the ESI source. ESI sources have been coupled to many different types of mass analyzers and these will be discussed after the next section on MALDI.

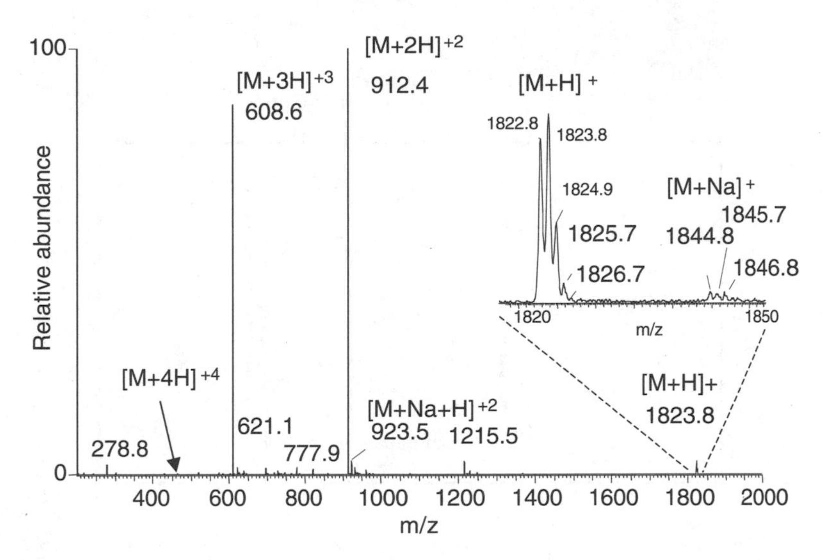

FIGURE 4.2 Electrospray ion trap mass spectrum of the tetradecapeptide renin substrate (rat) (MW_{ave} = 1823.09, $MW_{monoiso}$ = 1821.9200, $[M + H]^+$ = 1822.928) showing the $[M + H]^+$, $[M + 2H]^{+2}$, $[M + 3H]^{+3}$, and sodiated ions. The inset shows an expanded view of the $[M + H]^+$ and $[M + Na]^+$ ions.

Matrix-Assisted Laser Desorption Ionization

The MALDI source developed by Hillenkamp, Karas, and Giessmann of Germany[24] consists of a solid mixture of analyte and matrix on a sample plate, along with a laser, and light and ion optics, as shown in Figure 4.3. The matrix for positive ions consists of an organic chromaphore molecule such as sinapinic acid (3,5-dimethoxy-4-hydroxycinnamic acid), alpha-cyano-4-dihydroxycinnamic acid, 2,5-dihydroxy-benzoic acid, or nicotinic acid at a concentration of 10 mg/ml in a 50 : 50, organic : aqueous solution with 0.1 *M* acetic acid (AA) or trifluoroacetic acid (TFA). The peptide or protein is dissolved in an organic : aqueous solution with 0.1 *M* AA or TFA at 10 fmole/μl to 20 pmole/μL and mixed with the matrix solution. The final matrix-sample mixture is placed on a sample plate in a 0.5- to 1-μL aliquot and is allowed to dry to form crystals. Once the sample plate is inserted into the mass spectrometer through a vacuum interlock, it is positioned so that the pulsed laser light strikes a small portion of the sample. Use of photons at 337 nm from a nitrogen laser is common, but some researchers have used light at 266 nm from a Nd-YAG laser. Some MALDI plates can accommodate hundreds of samples. When the chromaphore absorbs the photons, it vaporizes and lifts the analyte ions from the surface and into a gas phase plume directly above the target plate. The gaseous plume is believed to consist of neutrals, metastable ions, positive ions,

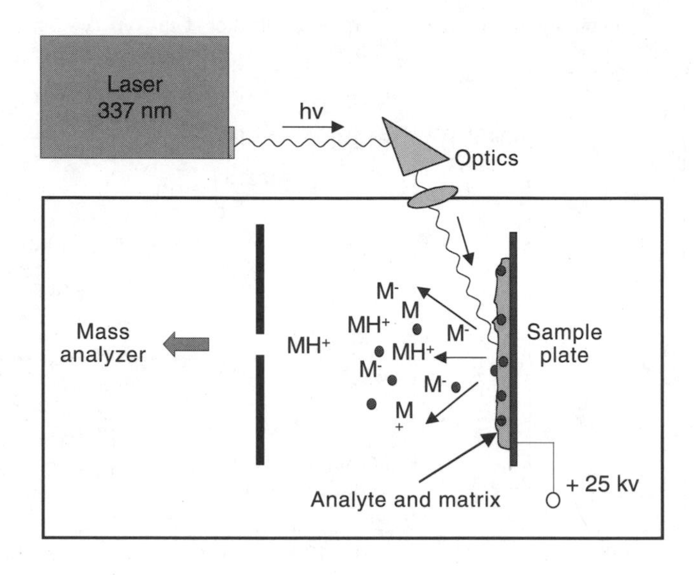

b. Three common matrix molecules for peptides and proteins

CH=CHCOOH; H_3CO; OCH_3; OH
Sinapinic acid
MW 224.07 amu

CH=C(CN)COOH; OH
Alpha-cyano-4-hydroxy cinnamic acid
MW 189.04 amu

COOH; OH; HO
2,5-dihydroxy benzoic acid
MW 154.03 amu

FIGURE 4.3 Matrix-assisted laser desorption ionization source. (a) Gas phase ions are formed when the nitrogen laser produces photons at 337 nm that strike the solid matrix-analyte sample. The ions above the sample plate are accelerated to 25,000 eV of kinetic energy and fly through a stainless steel tube (not shown) for TOF *m/z* analysis. (b) Three commonly used matrix molecules for MALDI of peptides and proteins.

negative ions, ion clusters, and fragments. After the laser fires, the sample plate is charged to 20 to 30 kV for positive ions. The positive ions are pushed from the region above the plate and into the analyzer. Many laser shots are usually collected and averaged for improved signal-to-noise. The time-of-flight (TOF) mass analyzer is the most common analyzer coupled to a MALDI source because it is ideal for this pulsing ion source. Current TOF analyzers are capable of femtomole detection limits and resolution values of 30,000 ($R = m/\Delta m$ FWHM). MALDI sources have also been coupled to ion trap mass spectrometers,[25, 26] but these have not been developed commercially. Figure 4.4 shows a MALDI-TOF

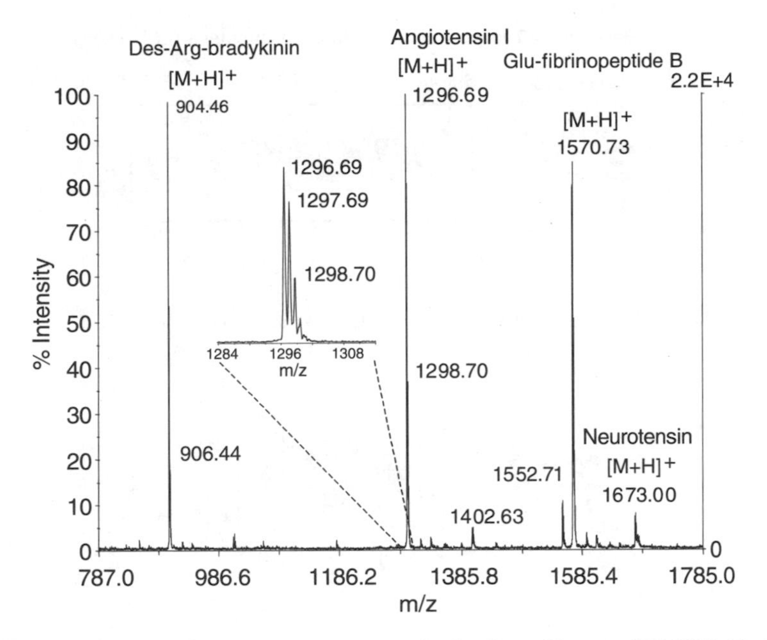

FIGURE 4.4 MALDI-TOF mass spectrum acquired using a Voyager DE-STR (PerSeptive Biosystems) in linear mode of four peptides: the mixture consisted of 1 pmole/µL each of des-Arg[1]-bradykinin, $MW_{monoiso}$ = 903.47; angiotensin I, $MW_{monoiso}$ = 1295.68; Glu-fibrinopeptide B, $MW_{monoiso}$ = 1569.68 and 50 fmole/µL of neurotensin, $MW_{monoiso}$ = 1672.92. The sample was prepared in α-cyano-4-hydroxycinnamic acid in 50 : 50, CH_3CN : H_2O and 0.3% TFA. The $[M + H]^+$ ion dominates over multiply charged peptide ions in MALDI mass spectrometry. The inset shows the resolved isotope peaks for angiotensin I ($MW_{monoiso}$ = ^{12}C isotope peak).

mass spectrum acquired in linear mode of 4 peptides in 50 : 50, acetonitrile : water and 0.3 *M* trifluoro-acetic acid (TFA): Glu-des-Arg[1]-bradykinin $MW_{monoiso}$ = 903.47; angiotensin I, $MW_{monoiso}$ = 1295.68; Glu-fibrinopeptide B, $MW_{monoiso}$ = 1569.68 and neurotensin, $MW_{monoiso}$ = 1672.92 ($MW_{monoiso}$ = ^{12}C isotope peak). The $[M + H]^+$ ion dominates over the multiply charged peptide ions in MALDI mass spectrometry. Sodiated $[M + Na]^+$ ions are also observed in the spectrum.

The MALDI technique is considered an offline ionization technique because the sample is purified, deposited, and dried on the sample plate before the analysis.

Mass Analyzers

ESI and MALDI sources have been coupled to many different mass analyzer types. Of these mass analyzers, five will be discussed next and are shown in Figure 4.5. To achieve the highest performance from these analyzers, an experienced operator will tune the mass spectrometer to improve the resolution, sensitivity, and signal-to-noise (*S/N*) level. Tuning is essential to achieve optimum performance, but it differs for each mass spectrometer and will not be covered in this chapter.

a. Electrospray Triple Quadrupole Mass Spectrometer (ESI-QqQ)

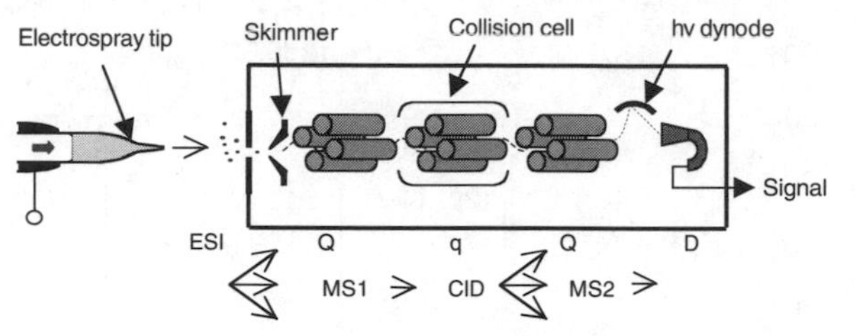

b. Electrospray Quadrupole Ion Trap Mass Spectrometer (ESI-QIT)

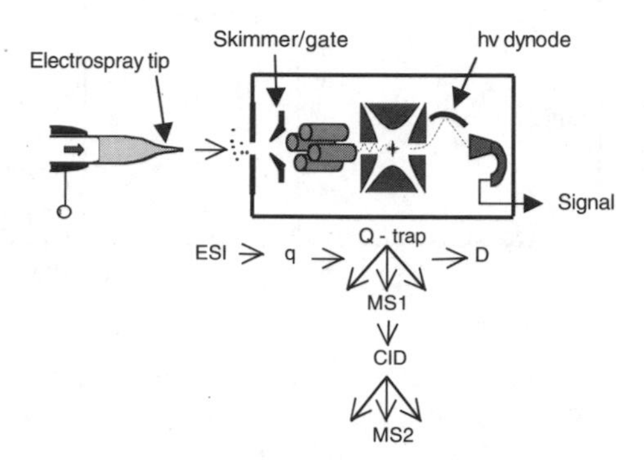

c. MALDI Time-of-Flight Mass Spectrometer (MALDI-TOF-MS)

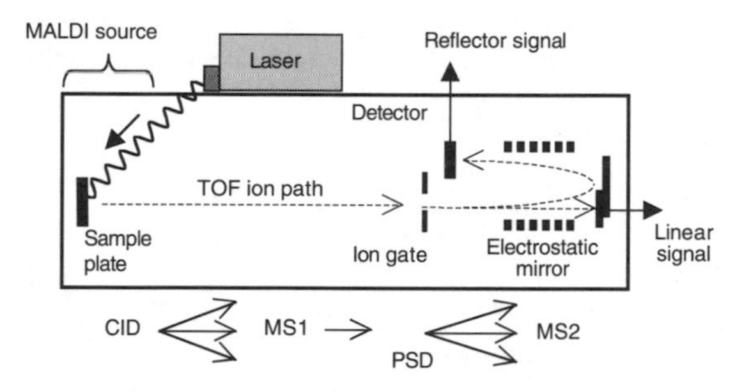

FIGURE 4.5 The five mass spectrometers commonly used for proteomic research. (a) ESI triple-stage quadrupole mass spectrometer; (b) ESI quadrupole ion trap (QIT) mass spectrometer; (c) MALDI time-of-flight mass spectrometer.

Triple-Stage Quadrupole (QqQ) Mass Spectrometer

The triple quadrupole (QqQ) mass spectrometer is comprised of two linear mass filters, Q1 and Q3, and a radiofrequency (RF) only quadrupole, q2, as shown in Figure 4.5(a). The QqQ analyzer is capable of MS and MS–MS scans and is commonly coupled to an ESI source for proteomic analysis. Each of the three quadrupoles consists of four linear rods that have a high voltage RF potential (1 to 3 MHz) and a DC potential applied to them. A triple quadrupole instrument performs three types of MS–MS scan modes: precursor, product, and neutral loss. The RF/DC ratio of Q1 or Q3, or Q1 and Q3 together, is scanned for each of the respective scan modes mentioned previously. For a product scan, ions from the ESI source are accelerated into Q1 where only one *m/z* is selected (fixed RF/DC). Next, the selected *m/z* ions enter the RF collision quadrupole where they undergo collision-induced dissociation (CID). This dissociation typically leads to the production of many fragments. The resulting product ions are then mass filtered by scanning the RF/DC of the third quadruple, Q3, to produce the product mass spectrum. For a precursor scan, a product ion is selected by Q3 (fixed RF/DC) and Q1 is scanned. The intensity of all the precursor ions that undergo CID in q2 to form the selected product ion is measured at the detector to record a precursor scan. Finally, a neutral loss scan determines all the molecular ions that fragment with the same loss, $\Delta m/z$, of a neutral. In this scan, both Q1 and Q3 are synchronized to scan together at the known $\Delta m/z$. Triple quadrupole mass spectrometers are now capable of high resolution with R = 10,000 FWHM at *m/z* 700.[27]

Paul* Three-Dimension Quadrupole Ion Trap (QIT) Mass Spectrometer

The three-dimensional quadupole field ion trap[28, 29] or Paul trap is a three-electrode device [see Figure 4.5(b)]. Ions are injected into the device and collected in packets from an ESI or MALDI source. The ion trap analyzer is capable of MS, MS^n (MS^3 = MS–MS–MS) and high-resolution scans (R = 20,000). The ion packets enter through an entrance-end cap and are analyzed by scanning the RF amplitude of the ring electrode. The ions are resonated sequentially from low to high *m/z* and are ejected from the ion trap through the exit-end cap electrode to a detector. Unlike the triple quadrupole (QqQ) mass spectrometer discussed previously, the ion trap performs tandem mass spectrometry (MS–MS) scan modes in the same analyzer.

Time-of-Flight (TOF) Mass Spectrometer

The time-of-flight (TOF) mass spectrometer is shown schematically in Figure 4.5(c). It is the simplest of all mass analyzers, but it has truly come to prominence in the mass spectrometry arena within the last 10 years. Both ESI and MALDI sources have been coupled to TOF mass analyzers. The TOF analyzer is capable of MS and MS–MS scan modes at high resolution (R = 30,000). The analyzer consists simply of a metal flight tube and the *m/z* ratios of the ions are determined by accurately and precisely measuring the time it takes the ions to travel from the MALDI source to the detector. Given that all ions of different *m/z* receive the same kinetic energy [$KE = (1/2)mv^2$], low *m/z* ions will reach the detector sooner than high *m/z* ions. High performance TOF instruments can be operated in linear or reflectron mode. High resolution is obtained by using a TOF-MS that has a reflectron (electrostatic mirror), delayed extraction, and a long flight path. Together these latter devices allow for energy and direction focusing. MS–MS is possible and mass accuracy errors of 5 ppm are possible for peptides with the MALDI-TOF mass spectrometer operating in reflectron mode.

---

* The ion trap is often called the "Paul" trap after the Nobel Prize winner, Dr. Wolfgang Paul, who invented the device in the 1950s.

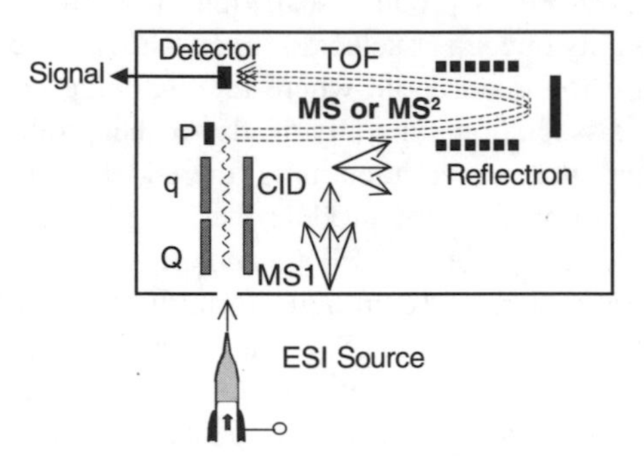

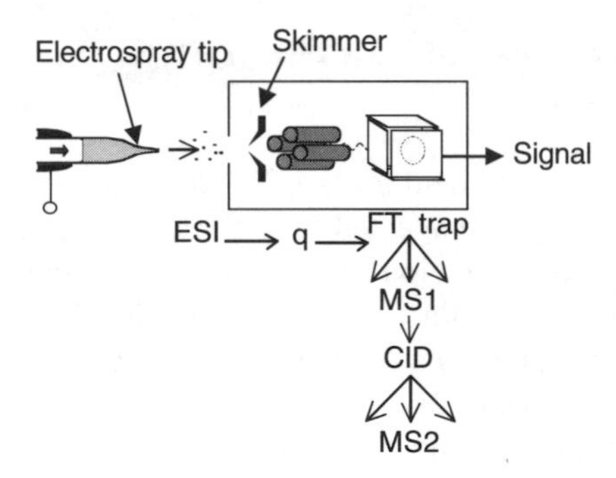

FIGURE 4.5 Continued. (d) ESI quadrupole time-of-flight mass spectrometer and (e) ESI Fourier transform ion cyclotron mass spectrometer. CID = collision-induced dissociation, Q = quadrupole mass analyzer, q = RF only ion guide, D = detector, P = pusher electrode, MS1 = first stage of mass spectrometry, MS2 = second stage of mass spectrometry, and PSD = post source decay.

Quadupole Time-of-Flight (Qq-TOF) Mass Spectrometer

The Qq-TOF mass spectrometer, a variant of the quadrupole analyzer and TOF analyzer listed previously, is shown schematically in Figure 4.5(d). The Qq-TOF mass spectrometer is capable of MS and MS–MS scanning and can record a high-resolution product spectrum. For MS–MS scanning, an ion can be selected in Q1, induced to fragment in q2, and the product ions can be analyzed in the orthogonal TOF analyzer. This instrument can scan very rapidly and produce medium- to high-resolution spectra throughout the entire mass range with excellent mass accuracy.

Fourier Transform Ion Cyclotron (FT) Mass Spectrometer

The Fourier transform ion cyclotron (FT) mass spectrometer shown in Figure 4.5(e) is a trapping mass spectrometer that can be coupled to an ESI source[30] with many stages of differential pumping. The FT mass spectrometer is capable of both MS and MS^n scanning. This mass analyzer is operated to measure the number of ion cycles that an ion makes in the trapping cell for a given period of time. An ion's frequency, if measured long enough, can be determined very accurately and, thus, this instrument can result in a mass spectrum with extremely high resolution. Excellent peptide sequence results have been obtained by using a nano-LC-ESI-FT mass spectrometer from 10 amoles of peptide loaded onto a capillary column.[31] The FT mass spectrometer is capable of the highest mass resolution of all the mass spectrometers listed here, but analysis times will increase as well.

SAMPLE PREPARATION FOR MALDI AND ESI

One of the most important, but often ignored, aspects of obtaining a quality mass spectrum of a peptide or protein is the preparation of the sample.

WORKING AT THE FEMTOMOLE LEVEL

It should be noted that during the preparation of any protein or peptide sample, some of the analyte will be permanently adsorbed to the walls of the vial, to packing material, or to any transfer device used such as a pipet tip. The loss of sample is of critical concern when working below approximately 100 fmole/μl. Protein adsorption by some vials is so great that the sample will not be observed in the mass spectrum. In addition, contamination from airborne protein particles, such as human or sheep keratin, can cause misleading results when working at these low levels. Covering your pipet tips and vials is recommended for protein identification work.

SAMPLE VOLUMES REQUIRED FOR ESI AND MALDI-MS ANALYSIS

The liquid volume of a sample required for analysis depends on the ionization technique, MALDI or ESI, and the introduction technique (see Table 4.1). The following statements assume that we are analyzing a sample near the detection limit of the analyte in a specific mass spectrometer. For MALDI-MS, the researcher typically spots 0.1 to 1 μL onto the MALDI sample plate. Thus, a minimum starting volume of 1 of 5 μL of sample is recommended. For ESI, the required sample volume is primarily dependent on the sample introduction technique. If the researcher uses a nanoflow electrospray technique, capillary LC, or capillary electrophoresis, then typically a 1-μL volume is required. However, larger sample volumes are recommended for ease of handling. If the volume is small, then the analysis may be limited to one experiment when additional MS or MS–MS experiments are desired. For higher flow rate ESI sources, the researcher should supply 50 μL or more for direct infusion experiments or for loading 5 to 20 μL onto an analytical LC column.

ANALYTE CONCENTRATIONS REQUIRED FOR MALDI AND ESI-MS

Typically, for MALDI-MS, 0.1 to 2 pmole/μL for a 1000-amu peptide and 1 to 10 pmole/μL for a 50,000-amu protein will produce a *S/N* level of greater than 20. Of course, lower concentrations can also produce excellent results. In MALDI, it is necessary to lift the analyte into the gas phase in order to accelerate it down the flight tube. To achieve this, the correct ratio of analyte to matrix is essential; too little matrix will not provide the necessary conditions for lift. Reduced sensitivity is often due to an apparent decreased ionization efficiency of samples at high molecular weight and reduced detector sensitivity at high m/z. As a result, a 10,000- to 100,000-amu protein can be readily analyzed by MALDI-TOF-MS, but the analysis of a 500,000-amu protein is not routine. Resorting to higher concentrations for the analysis of large or small molecules is not usually the correct procedure to achieve improved signal if the analyte is already in the concentration range mentioned previously. Instead, the proper action would be to increase the purity of the analyte.

Sample Quality for ESI or MALDI

To obtain optimal results, the key to analyzing peptide mixtures at the femtomole or attomole levels may not be to buy the latest high-priced mass spectrometer but to improve sample preparation! Several techniques are available to purify and concentrate your peptide or protein sample for an improved analysis. Sample preparation can make the difference between a failed analysis and acquiring a great mass spectrum. Both MALDI-MS and ESI-MS require the analyte to be pure to obtain the highest quality results. Ideally, the protein of interest should be the only analyte present in the sample being ionized. The remaining sample volume should be made up of water, a volatile solvent such as methanol, ethanol, or acetonitrile, and a volatile acid such as acetic acid, formic acid, or trifluoroacetic acid typically at 0.1% (v/v). Salts such as NaCl, buffers, detergents, gel particles, and other molecules in the sample other than the previously mentioned components are avoided or kept at a minimum concentration. Nonvolatile acids, buffers, detergents, and solvents should be avoided. Of course, when analyzing some proteins you may find that they will not stay in solution without a certain concentration of a stabilizing compound or buffer. In this case, you may need to experiment with the sample preparation to ensure that the protein remains in solution with a minimum concentration of additional solvating molecules.

MALDI is much more tolerant of a poorly prepared sample than ESI. In MALDI, however, the signal height produced will be greatly reduced in high salt concentrations. The reduced ion signal is owing partially to the distribution of the charge among additional sodiated species that broaden the peak and reduce its height. Buffers, such as phosphate buffer, also tend to reduce signal strength, but they can be tolerated at 25 m*M* in many examples.

Because ESI is less tolerant of sample conditions, sample preparation is even more critical than in MALDI. In addition, the solvent used for the ESI process is extremely important, unlike in MALDI, where the solvent evaporates away before the analysis. Nonvolatile solvents like water or a 1% concentration of sodium-dodecyl-sulfate (SDS) tend to produce poor sprays and a poor or noisy ion current. A sample in 99% water will not spray as well as a sample prepared in 50% methanol. A peptide sample prepared using acetic acid will also produce better results than one using trifluoroacetic acid (TFA) probably due to a counter ion effect. Nonvolatile salts or small molecules will eventually plug the entrance hole to the mass spectrometer and so they should be avoided.

Pipet Tip Packing Bed Purification Method

One method for desalting and concentrating a sample involves the use of a micropipette tip that has been manufactured to hold a small amount of packing material (C_8, C_{18},..., etc.) in the first few millimeters of the tip.[32] In this purification procedure, the packing material is first conditioned with an organic solvent and then aspirated with 99.9% water, 0.1% acid. The tip is now ready to receive a peptide sample. A 10-μL peptide sample is aspirated into the packing material and then the packing bed is washed with water containing 0.1% acid. This latter step reduces the level of salts and buffers. Finally, a highly organic solvent like 50 : 50, acetonitrile : water containing 0.1% TFA is applied to the packing material to release the concentrated peptides. The 1- to 2-μL sample is collected into a vial for ESI-MS for nanoflow electrospray or spotted directly onto a sample plate with matrix for MALDI-MS.

Capillary Liquid Chromatography MS

Capillary chromatography-MS, a more direct and sophisticated chromatographic method used to concentrate and purify peptides from a mixture, is shown in Figure 4.6. Capillary chromatography allows for low detection limits because there is less packing material in which to lose the peptide sample. In the capillary-LC-MS method, a 10-cm section of a 100-μm i.d. fused silica column is packed with

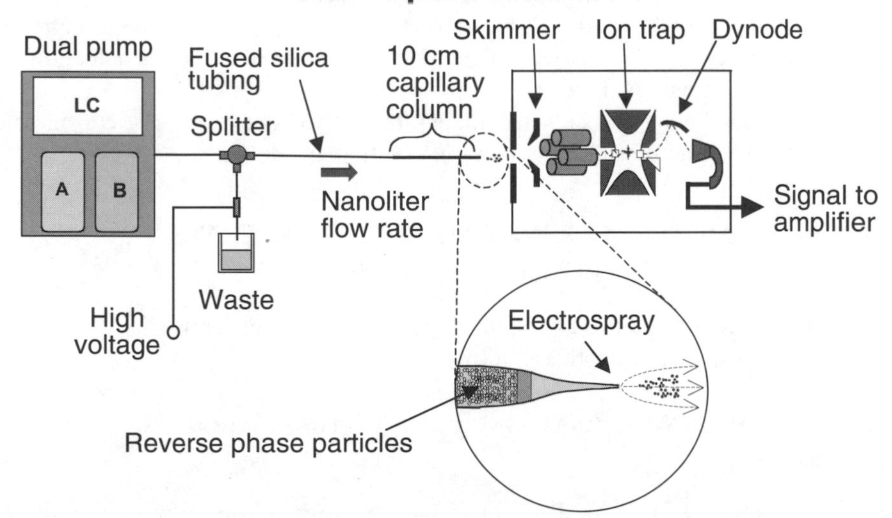

FIGURE 4.6 Capillary liquid chromatography electrospray ionization mass spectrometry (cap-LC-ESI-MS) used to concentrate, separate, and analyze low amounts (fmoles) of peptides or proteins. The packed capillary column length is typically 10 cm long and the column flow rate is maintained at 100 to 1000 nL/min. The inset shows the electrospray tip formed at the end of a 100-μm i.d. fused silica capillary column.

10 μm, C_{18} particles conditioned and washed with 98% water, 1.9% acetonitrile, and 0.1% acetic acid (solvent A). The sample is either pressure loaded (1 μl) or injection loaded onto the head of the capillary column. Many peptides will concentrate on the head of the column, preferring the C_{18} packing material to the 98% aqueous liquid phase. Next, a dual pump gradient LC system is used to change the composition of the mobile phase from 98% aqueous (solvent A) to 98% organic (solvent B). As the concentration of the organic phase increases, peptides adsorbed to the packing material at the head of the column will begin to move down the column and spend more time in the liquid phase. Eventually, the column separates the peptides and the effluent is electrosprayed directly from the capillary column needle tip into the mass spectrometer as shown in the inset of Figure 4.6. This latter technique is capable of detecting a mixture of peptides at the 10-femtomole level ($S/N > 10$). Additional preparation steps can further lower the detection limits. Packed capillary columns of internal diameter of 100 μm, however, have a small capacity (<1 pmole) and so more concentrated samples should be diluted for good chromatography.

Capillary Electrophoresis ESI-MS

Capillary electrophoresis (CE) MS is another technique used to separate and measure the *m/z* ratios of a mixture of peptides and proteins. In this method, the peptide mixture is separated by different migration rates through the electrophoresis media and the effluent is again directly sprayed into the mass spectrometer using a micro ESI device. This method is capable of femtomole or lower detection limits and is further discussed in a review.[33] In 1996, a group at the University of Washington described a solid phase extraction (SPE) capillary electrophoresis MS–MS approach for the analysis of peptides and showed limits of detection at the 100 s of attomole level.[34]

The two capillary separation techniques mentioned above are not considered routine, however, partially because they require homemade parts.

MOLECULAR WEIGHT DETERMINATIONS OF PEPTIDES AND PROTEINS

This section defines MS terms and explains good MS practices for use in determining the molecular weights of peptides and proteins.

Multiply Charged Molecular Ions and the Determination of Charge State

When using either ESI or MALDI sources in the positive ion mode, a peptide or protein will be protonated in the form $[M + H]^{+}$, $[M + 2H]^{+2}$, $[M + 3H]^{+3}$ … $[M + nH]^{+n}$ as shown in Figures 4.2 and 4.4 and/or deprotonated in the negative ion mode to form $[M - H]^{-}$, $[M - 2H]^{-2}$, $[M - 3H]^{-3}$ … $[M - nH]^{-n}$. The researcher must determine the charge state of the ions to determine the molecular weight. Once the charge state is determined, one can correct for the proton addition or subtraction in the molecular weight calculations of the neutral protein or peptide. It should be noted that in many samples, the mass spectrum will show some evidence of sodium or potassium even when the researcher who prepared the sample claims, "there is no salt in my sample because I desalted," or "no salt was ever used in the preparation of the sample." Sodium is present everywhere (see the small peaks in Figures 4.2 and 4.4). Even when sodium is not added to the sample, it can be observed and glass and plastic ware are the probable sources. As a result, sodium or potassium or other cation addition will distribute the charge among the many ion species such as $[M + Na]^+$, $[M + Na + H]^{+2}$, $[M + 2H + Na]^{+3}$ or $[M + K]^+$, $[M + H + K]^{+2}$, $[M + 2H + K]^{+3}$, and so on. At high salt concentrations, this effect reduces the overall signal-to-noise ratio in the mass spectrum. Proton bound dimers $[M + H + M]^+$ are also found to be a common ion type found in both a MALDI and an ESI mass spectrum.

As mentioned previously, to determine the molecular weight of a protein from a mass spectrum, the researcher must first determine what peaks are due to the protein in question and then determine the charge state of the various *m/z* peaks. The charge state is determined in two ways: by isotope separation or by multiple-charge ion calculations. If your mass spectrometer can resolve the isotope peaks, you may simply measure the *m/z* difference between the adjacent isotope peaks and calculate the charge state. For example, if the *m/z* separation between two isotope peaks is 1 amu, the ion is singly charged. Likewise, if the separation is 0.5, 0.33, or 0.25 amu, the charge state is +2, +3, or +4, respectively, for positive ions. Figure 4.7(a) shows an ESI-quadrupole ion trap (QIT) mass spectrum of interleukin-8 (rat) (MW_{ave} = 7845.3 amu) infused at 10 pmole/uL and the multiply charged ion distribution $[M + H]^+$, $[M + 2H]^{+2}$, $[M + 3H]^{+3}$ …, and so on. Figure 4.7(b) shows the high-resolution mass spectrum of the +4 charge state that is used to calculate the $MW_{monoiso}$ = 7840.38 amu with a mass accuracy error of 0.004% (theoretical $MW_{monoiso}$ = 7840.09 amu).

The second method used to determine the charge state of an ion makes use of all of the multiply charged ions observed in a spectrum such as in Figure 4.7(a). Both MALDI and ESI produce multiply charged ions, but MALDI typically shows only the singly charged ion as the base peak for peptides or proteins. The MALDI process appears to be too energetic to produce an abundance of highly charged proteins. The ESI source is a much softer ionization technique and generates a multiply charged ion distribution consisting of an average of one charge per 600 to 1200 amu. For example, if your protein weighs 45,000 amu, the ESI spectrum might show an average charge state of +45. This average charge state in ESI will, of course, be protein and solution dependent. Algorithms have been developed to calculate the molecular weight using each peak in the distribution of multiply charged peaks. For example, by using any two adjacent multiply charged peaks, m_n/z_n and m_{n+1}/z_n+1, from the mass spectrum, the formula:

$$MW = \left(\frac{m_n}{z_n}\right)z_n - z_n = \left(\frac{m_{n+1}}{(z_n+1)}\right)(z_n+1) - (z_n+1)$$

can be used to calculate the charge state of all the ions by solving for z_n (use Figures 4.2, 4.7(a), 4.9(a), or 4.9(b) to demonstrate this). Molecular weight calculations

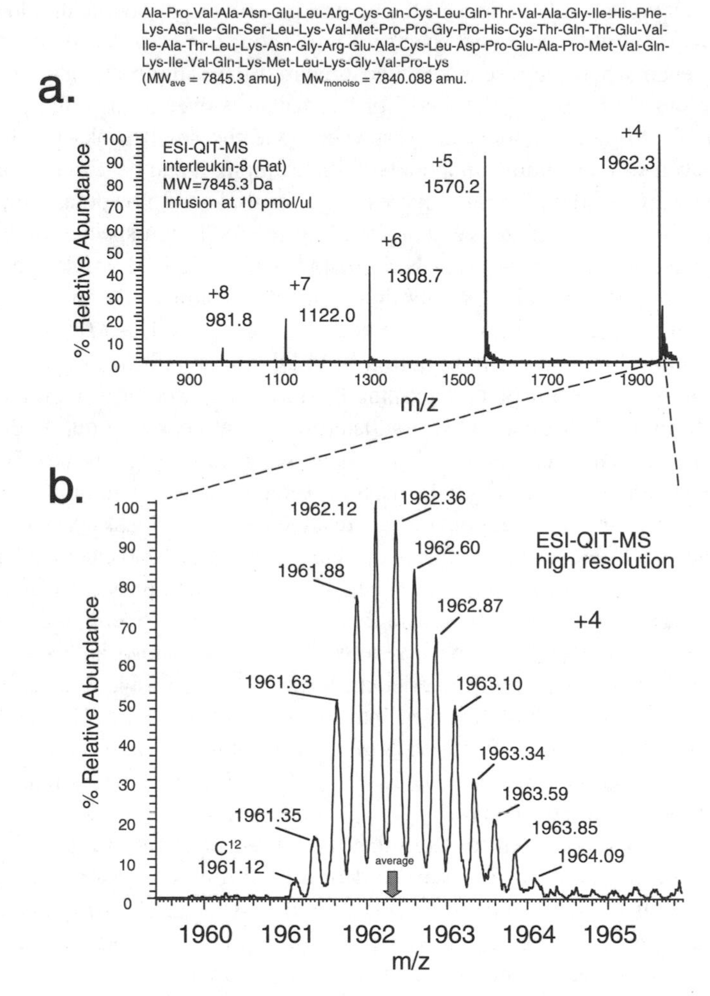

FIGURE 4.7 ESI-quadrupole ion trap (QIT) high-resolution mass spectrum of interleukin-8 (rat) (MW_{ave} = 7845.3 amu) infused at 10 pmole/µl. (a) the amino acid sequence and the multiply-charged ion distribution $[M + H]^{+}$, $[M + 2H]^{+2}$, $[M + 3H]^{+3}$, . . . , and so on; (b) a high-resolution mass spectrum of the +4 charge state used to calculate the $MW_{monoiso}$ = 7840.38 amu (theoretical MW = 7840.09 amu). (Data from Sanders, Finnigan Corporation, 1995. With permission.)

using all of the peaks in an ESI mass spectrum with a high signal-to-noise ratio can give a mass accuracy value of 0.005%.

Average vs. Exact Monoisotopic Molecular Weights

Mass spectrometry can determine the molecular weights of peptides and proteins with mass accuracies orders of magnitude better than the molecular weights determined by gel electrophoresis. It is important to note that in determining molecular weights, we need to distinguish between the monoisotopic weight and the average molecular weight. Most mass spectrometers are only capable of returning an average molecular weight for proteins above 10 to 20 kDa. Most chemical manufacturers only list the average molecular weights for proteins. As shown in Figure 4.7(b), the theoretical MW_{ave} for interleukin-8 from rat is 7845.3 amu and the monoisotopic molecular weight, $MW_{monoiso}$ = 7840.09. It should be noted that at peptide molecular weights greater than 2000 amu, the ^{12}C monoisotopic ion is no longer the highest peak in the isotope distribution as shown by the small ^{12}C peak in Figure 4.7(b).

Resolution

A common question asked of the mass spectrometrist is how close in m/z can two peaks at m/z = 1 000 or m/z = 20,000 be and still be distinguished with confidence? The detailed answer is not as simple as one would like because it is dependent on the sample and instrument resolution. First, there are different definitions that exist for the resolving power of a mass spectrometer. A common definition, $R = m/\Delta m$ at mass m, is used in this chapter. The value of $\Delta m/z$ is the measurement of the mass peak at full width at half maximum (FWHM). If the FWHM for the C^{12} monoisotopic peak for a 1000-amu peptide is 0.5 amu, then R = 2000, and for a 20,000-amu protein with the same FWHM, R = 40,000. For the preceding example, the isotope peaks would be distinguishable. In other words, for m/z up to 10,000 amu, where z = 1, it is possible to obtain monoisotopic information with a mass spectrometer capable of resolution between 20,000 to 30,000. At high mass (>10,000 amu), however, most mass spectrometers can only acquire average molecular weight data, not monoisotopic data. The Fourier transform mass spectrometer is capable of R = 1,000,000 at m/z = 1000.

It is important to remember that a mass spectrometer's realized measured resolution is dependent on many variables that may not allow the operator to observe the true intrinsic resolution of the instruments. For example, the digitization rate of the ion signal must be high enough to measure the intrinsic resolving power of the mass spectrometer. A poorly tuned instrument, an improperly prepared sample with high salts and buffers, and overlapping isotope distributions from two or more ions are all factors that may impair obtaining the intrinsic or highest resolution of the mass spectrometer.

Mass Accuracy

Today, most mass spectrometers used for protein analysis can achieve mass accuracy errors of ±0.01% = ±100 ppm. A mass accuracy of 10 ppm means that one could measure a 1000-amu peptide to ±0.01 amu. For a 10,000-amu protein, one could measure the m/z to ±0.1 amu. This mass accuracy error is considered excellent.

The PerSeptive Biosystems Voyager STR MALDI-TOF-MS, which is used at Carnegie Mellon University, has delayed extraction and a high mass detector. This mass spectrometer can achieve a mass accuracy of ±5 ppm for peptides near 1000 amu. One must be careful when talking about mass accuracy, however. People often refer to the best achievable by the instrument and not what the average user routinely obtains. In addition to the mass spectrometer, mass accuracy depends on several factors: how one calibrates, the software algorithm for calibration, the analyte, the matrix, scan rate, how the sample is run, the software that is used to centroid the mass peaks, and the ion statistics. Most good mass spectrometrists know the importance of these variables, but perhaps many good scientists using MS do not. For example, one might achieve better mass accuracy errors by manually centroiding the mass peaks rather than trusting the computer algorithm. Another common mistake when calibrating the mass spectrometer is to believe that a mass peak made up of few ions will have the same mass error as a strong mass peak made up of many ions.

Figures 4.8 and 4.9 show the mass spectra of three proteins acquired on a MALDI-TOF and ESI-QIT mass spectrometer, respectively. Both MALDI and ESI figures are of (a) cytochrome-*c* (equine), MW_{ave} = 12,360.1, (b) apomyoglobin (equine), MW_{ave} = 16951.6, and (c) serum albumin (bovine), MW_{ave} = 66,430.0. Both sets of data give molecular weights with mass accuracies of 0.005 to 0.01%. The base peaks in the MALDI data set in Figure 4.8 are due to the $[M + H]^+$ ion. Proton bound dimers and trimers are also observed at 2× and 3× the mass of the $[M + H]^+$ ion and adducts (possibly owing to the matrix) are observed in the insets of spectrum 4.8(a) and 4.8(b). Spectra 4.9(b) and 4.9(c) are from capillary-LC-MS of these proteins with the final LC gradient concentration of 98% acetonitrile, 2% water, and 0.1% acetic acid. Electrosprayed proteins give a multiply charged ion distribution that can be deconvoluted to calculate the molecular weights as shown in the insets. Note that the inset for the ESI mass spectrum of serum albumin (bovine) shows that the sample was not pure, as seen by the broad deconvoluted mass peak.

PROTEIN SEQUENCING

Direct protein sequencing by mass spectrometry is currently not possible. Although multiply charged peaks from ESI can be fragmented, this method does not produce the complete information required for easy interpretation of the sequence. Another sequencing approach would be to digest the protein sequentially from the C- or N-terminal end using enzymes, but this technique is not capable of complete sequencing of high molecular weight proteins. This technique is discussed later under peptide ladder sequencing. Protein sequencing is possible, however, by first breaking the protein down into smaller peptides using enzymes and then sequencing these individual peptides.

Peptide Sequencing by Mass Spectrometry–Mass Spectrometry

If a protein is digested into peptide fragments with an enzyme, the amino acid sequence of the individual fragments can be determined by MS–MS. The MS–MS analysis step can be done directly from the mixture, or this sample can be separated by capillary liquid chromatography. In the MS–MS technique, two mass analysis steps are performed (see triple quadrupole section). First, an ion is selected or isolated for MS–MS analysis. In the ESI process at 0.1 *M* acid, the most abundant peptide molecular ion for a molecular weight between 700 to 2400 amu is usually the $[M + 2H]^{+2}$ or the $[M + H]^{+3}$ ion, so one of these two ions is typically chosen. Because the $[M + H]^+$ ion is the base peak in MALDI-MS, it

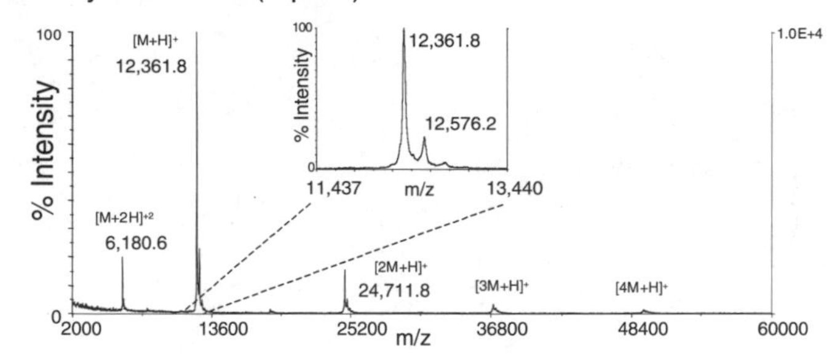

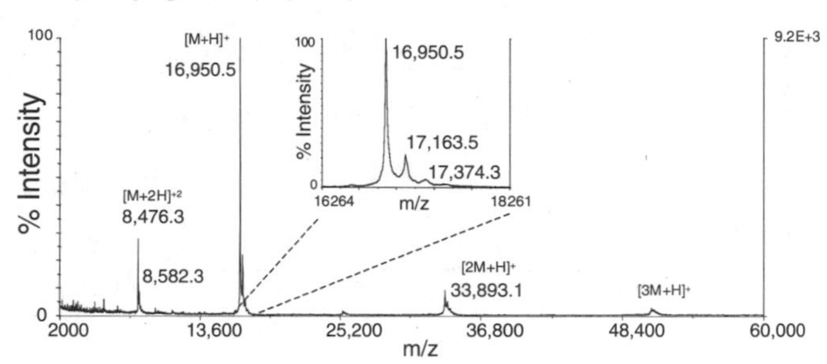

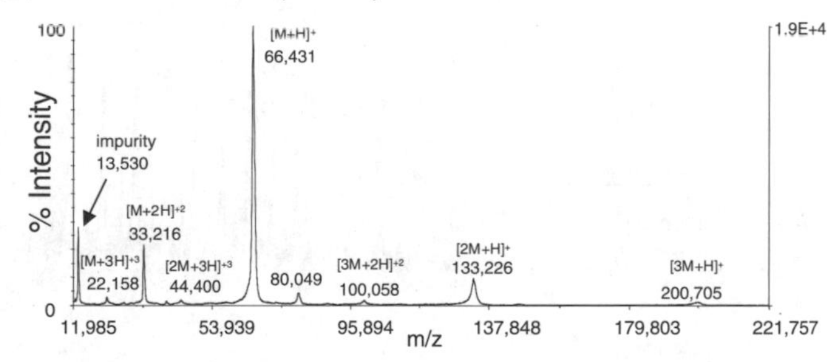

FIGURE 4.8 MALDI-TOF mass spectra of three proteins in a 10 mg/ml sinapinic acid and 0.1 to 0.3% acetic acid matrix: (a) cytochrome-*c* (equine), MW_{ave} = 12,360.1; (b) apomyoglobin (equine), MW_{ave} = 16951.6; and (c) serum albumin (bovine), MW_{ave} = 66,430.0. The base peak is typically the $[M + H]^+$ ion. Proton bound dimers and trimers are also observed at 2× and 3× the $[M + H]^+$ mass and adducts are observed in the insets of spectrum (a) and (b). Mass accuracies of 0.005 to 0.01% are possible for data with good resolution and strong signal to noise.

would be chosen for the precursor mass. Next, the selected ion is isolated in the first mass analysis step (MS1) and, then, fragmented typically by collision-induced dissociation (CID). The peptide fragments from the CID step are then analyzed by the second mass analysis step (MS2). Interpreting the MS–MS product spectrum of fragment ions and elucidating the amino acid sequence has

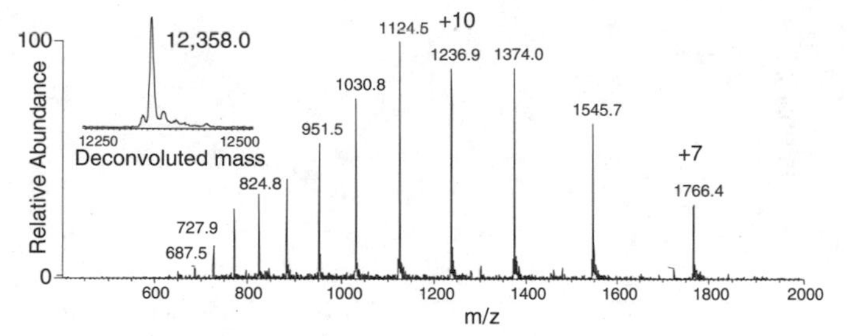

b. Apomyoglobin (equine)

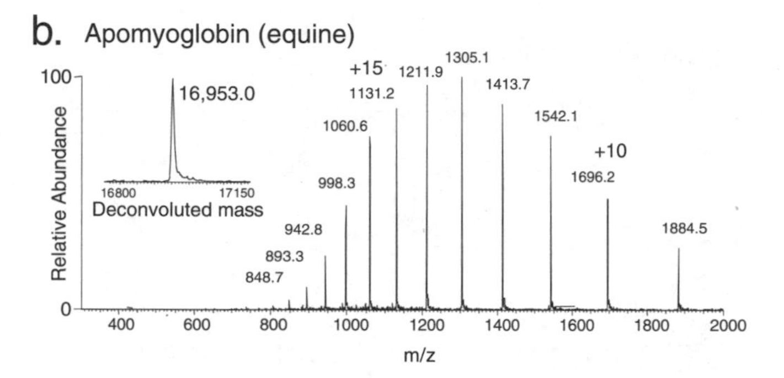

c. Serum albumin (bovine)

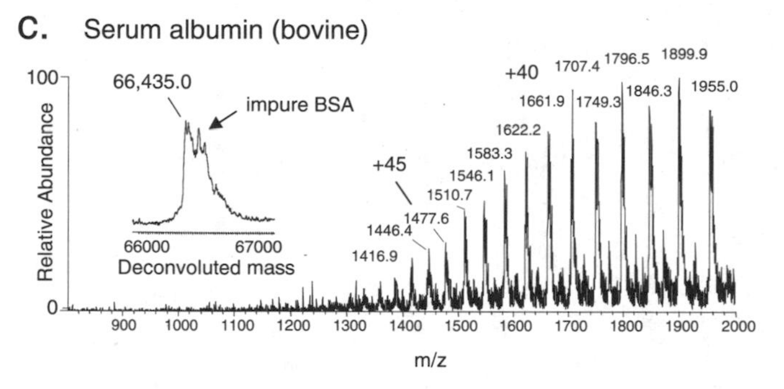

FIGURE 4.9 ESI ion trap mass spectra of three proteins: (a) cytochrome-*c* (equine), MW_{ave} = 12,360.1; (b) apomyoglobin (equine), MW_{ave} = 16951.6; and (c) 8 pmoles serum albumin (bovine), MW_{ave} = 66,430.0. Spectra (b) and (c) are from capillary-LC-MS of protein with the final gradient concentration of 98% acetonitrile, 2% water, and 0.1% acetic acid. Electrosprayed proteins give a multiply-charged ion distribution that can be deconvoluted as shown in the insets to calculate the molecular weights with mass accuracies of 0.005 to 0.01%. Note that the serum albumin (bovine) sample is not pure as seen in the mass spectrum and the deconvoluted mass inset.

recently been termed *de novo* sequencing. Efficient determination of the amino acid sequence in this fashion takes considerable practice. Several chemical modification steps are often used to help determine the amino acid sequence. Interpretation of the MS–MS data allows for the partial or complete elucidation of the peptide sequence that can be used to piece together a more complete sequence

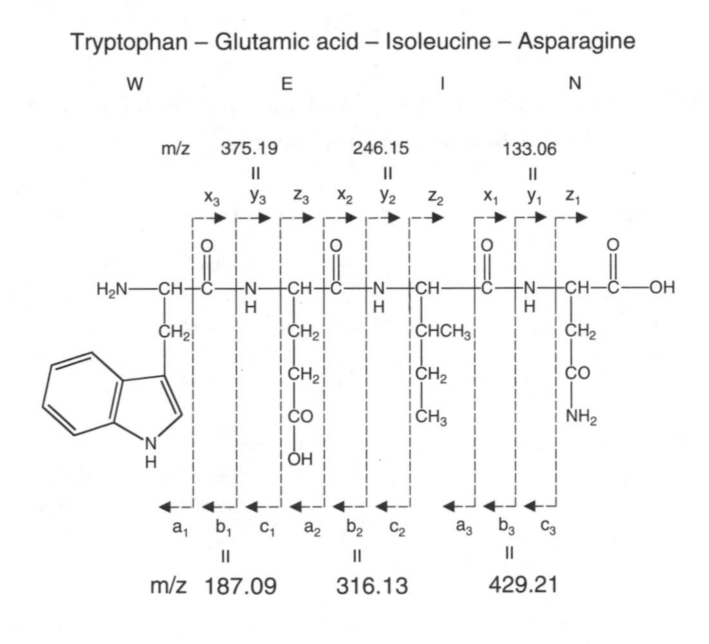

b_1 ion = $H_2NCH(R_1)CO^+$ $\qquad$ y_1 ion = $H_3N^+CH(R_4)COOH$

FIGURE 4.10 Peptide fragmentation nomenclature[35] for ions produced from collision-induced dissociation of the peptide WEIN. The arrowheads show the fragments that retain the charge. The chemical formulas for the b_1 and y_1 ions are noted and the respective masses for the *y*- and *b*-ion series are listed for the tetrapeptide WEIN.

of the protein. For example, Figure 4.10 shows a tetrapeptide, WEIN, and the fragmentation nomenclature for the N-terminal *a*, *b*, *c* ions and the C-terminal *x*, *y*, *z* ions as defined by Roepstorff and Fohlman.[35] Other ions will form resulting from side chain fragmentation. The dashed lines show the point of bond cleavage and the arrowheads show which fragments retain the charge for the various ions. The chemical formulas for the b_1 and y_1 ions are noted at the bottom of Figure 4.10. The respective masses for the *b*- and *y*-ion series are listed in the figure for the dissociation of the tetrapeptide WEIN.

Sequencing a protein is a considerable task. Proteins with 200 or more amino acids are common, and natural modifications need to be identified. Taking 10 femtomoles of a protein to a mass spectrometry lab and asking for a sequence determination will be met with resistance. Digesting the protein, creating a peptide map, and acquiring MS–MS data of the individual peptides for partial sequencing is a much more reasonable approach. Martin and co-workers[36] showed an excellent example of subfemtomole peptide sequencing, but most general service MS laboratories would not have the required experience, instrumentation, or interpretation skills. Peptide-sequencing knowledge is found in specialized laboratories that work routinely in this area, but even most of these laboratories will not consider the process trivial. These types of specialized MS laboratories are found in large pharmaceutical laboratories, research centers, and

TABLE 4.1
Comparison of Approximate Detection Limits for Peptides and Proteins Using Various MS and Biochemical Techniques

Technique	Detection Limit[a]	Minimum Volume Required
SDS-PAGE 2-D-Gel electrophoresis	10–100 fmoles	limit of transfer
Edman peptide sequencing	500 fmoles	limit of transfer
Capillary-LC-ESI-MS of peptides	100 amoles–10 fmoles	0.1–1 μL
Capillary-LC-ESI-MS of proteins	100 fmole	0.1–1 μL
Capillary-LC-ESI-MS-MS of peptides	100 amoles–1 fmole	0.1–1 μL
Cap-electrophoresis-ESI-MS of peptides	100–1000 amoles	0.1–1 μL
MALDI-MS of peptides (linear)	10 fmoles	limit of transfer
MALDI-MS of peptides (reflectron)	30 fmoles	limit of transfer
MALDI-MS of proteins (linear)	100–1000 fmoles	limit of transfer
Direct infusion ESI-MS of peptides	1 pmole	10–100 μL
Direct infusion ESI-MS of a protein	5 pmoles	10–100 μL

[a] These are very approximate detection limits with signal-to-noise equal to at least 10 and based on full-scan mode MS or MS–MS. Detection limits can vary widely with solution, operator techniques, type of mass spectrometer, column i.d., scan mode, and specific peptide or protein analyzed.

individual research groups at major universities. Table 4.1 shows a comparison of detection limits for various analytical techniques, some of which are used to sequence peptides.

Peptide Ladder Sequencing

The *de novo* method of sequencing that uses enzymes to digest the terminal amino acids from a peptide is called peptide ladder sequencing.[37–40] In this technique, an enzyme such as carboxypetidase-Y is used to digest one amino acid at a time from the C-terminus. No additional fragment peaks are formed. This method, unlike the MS–MS *de novo* technique, requires that the peptide be pure. In addition to the need for a pure peptide, another drawback of this technique is the somewhat lengthy sample preparation. The major advantage of this technique is the ease with which the mass spectrum can be interpreted and the partial sequence determined.

PROTEIN IDENTIFICATION

One of the most powerful techniques for protein identification uses mass spectrometry.[41–43] It is faster relative to older methods given that the protein sequence is known and logged into a database. The technique allows for protein identification at the 1-pmole level loaded onto a gel in skilled labs and <1 pmole in laboratories that have more extensive experience. If the analyst is careful and the mass spectra are of high quality, the mass spectrometry method is extremely powerful. Several

TABLE 4.2
Mass Spectrometry Protein Identification Internet Sites

Software Programs	Protein Database Search Engines
MS-Fit, MS-Digest, MS-Product[a]	*http://prospector.ucsf.edu/*
Peptide Search[b]	*http://www.embl-heidelberg.de/*
Mascot	*http://www.matrixscience.com*
SEQUEST™[c]	*http://fields.scripps.edu/sequest/index.html*
PROWL- ProFound, PepFrag, ...	*http://prowl.rockefeller.edu/PROWL/prowl.html*
Software Programs	**Educational Internet Software Used to Teach Protein Identification and MS**
Virtual Mass Spectrometry Laboratory*[d]	*http://mass-spec.chem.cmu.edu/VMSL/*

[a] See Reference 44.
[b] See Reference 45.
[c] See Reference 46.
[d] An interactive, Internet educational tool currently under development to teach both MS and experimental procedures such as protein identification. See Reference 47.

individuals have written software programs that use the collected MS and/or the MS–MS data to identify the top "hits" in a protein database. It should be noted that this is not a protein-sequencing technique, but instead, it is a protein identification method. Table 4.2 lists some helpful MS protein identification Internet sites and search engines.

Preparation of the Protein Sample for Identification

Although two-dimensional (2-D) gel electrophoresis is extremely powerful, it should be recognized that this technique will discriminate against hydrophobic proteins such as those found in membranes,[48] and in all cases some amount of protein will be unrecoverable from the gel. In this procedure, the protein mixture is purified by 2-D gel electrophoresis. After electrophoretic separation, the gel containing the desired protein is stained, excised from the larger gel slab, diced into small 1-mm cubes, and placed in a small vial (see Figure 4.11). Once the stain is washed from the sample, the protein disulfide bonds are reduced and the –SH groups are chemically blocked. Next, trypsin (other enzymes may be used) is added to cleave the protein at specific amino acid sequences while in the gel. Trypsin cleaves the amino acid bonds R–X and K–X, where R = arginine, K = lysine and X = any amino acid except for proline (P). Because different proteins have different amino acid sequences, a tryptic digest will result in a unique peptide fingerprint for each protein. The unique pattern occurs because the resulting peptide fragments from the digest will have a high probability of being different from a peptide mass fingerprint from a different protein. The mixture of peptides is extracted from the gel digest and often further purified for mass spectrometry analysis.

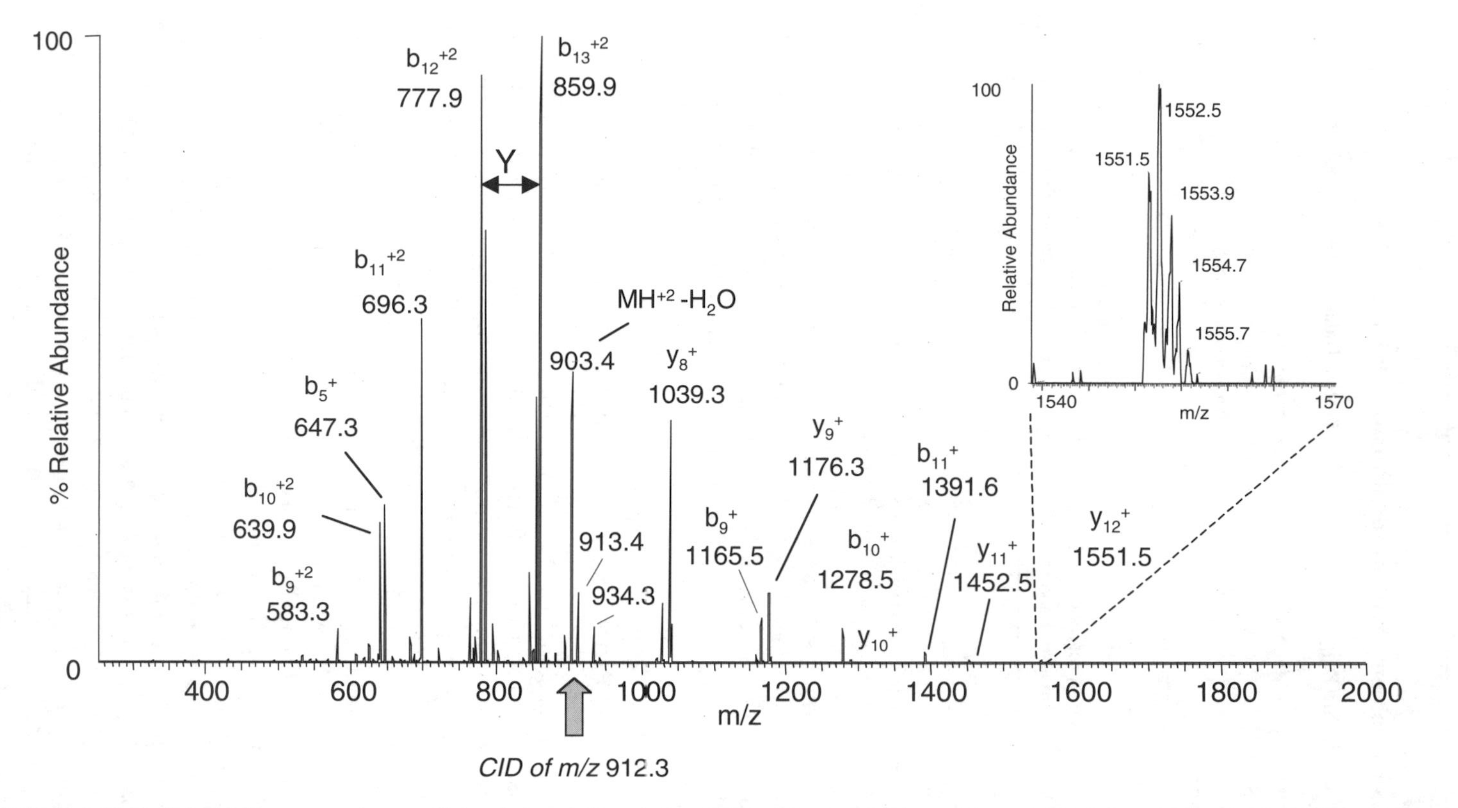

FIGURE 4.11 ESI-MS-MS spectrum from collision-induced dissociation (CID) of the doubly charged ion *m/z* 912.3 from 20 pmole/ul of rennin substrate (rat) in 0.1% acetic acid showing the *b*- and *y*-ion fragments from the peptide. The amino acid tyrosine (Y) is determined in the sequence using the doubly charged b_{12}^{+2} and b_{13}^{+2} ions. See Figure 4.2 for the full scan spectrum.

One should be aware of peptide contaminates observed in the spectrum. One's sample may be a mixture of proteins that will result in a mixture of unrelated tryptic peptides. It is also essential to prepare the sample in a clean environment. Otherwise, keratin contamination may be observed in the spectrum. Unused pipet tips should be covered and wool clothing should not be worn in the laboratory. In addition, trypsin will undergo autolysis and these tryptic peptides will often be observed in the mass spectrum.

Protein Identification by Peptide Mapping

The left side of Figure 4.12, shows the peptide mapping procedure for protein identification using MS scan to determine the molecular weights of all of the peptides. First, the peptides extracted from the in-gel tryptic digest are either spotted directly onto a MALDI-MS sample plate with matrix, injected into a nanoflow capillary for direct infusion electrospray-MS, or loaded onto a capillary column for nanoflow capillary-LC-ESI-MS. For samples at the femtomole/μL concentration, a cleanup (desalting and debuffering) and concentrating step is usually necessary to obtain useful results. The purification and cleanup can be done in one step using commercially available pipet tips[32] (described earlier under the section on sample preparation: for MALDI and ESI) or by performing capillary-LC-MS, capillary-CE-MS or a more advanced hybrid separation technique. The desired mass spectrum should have 20 or more peptide masses* where the isotopes are resolved, the signal is strong (i.e., good ion statistics), and the monoisotopic mass peaks have a mass accuracy error of <10 ppm. The set of peptide ions are entered into a search engine that will compare the data to the theoretically calculated tryptic peptide masses expected from all known proteins in the database.

Small mass accuracy errors from a TOF MS or FTMS are good for high-quality protein-mapping experiments for protein identification, and new developments with triple-quadupole technology are expected to achieve 10 to 20 ppm errors.[27] The reader may find it educational to enter the *m/z* data from the BSA tryptic peptide reflectron spectrum provided in Figure 4.13 into one of the search engines listed in Table 4.2. To provide the best match, the BSA digest spectrum was acquired in reflectron mode and only the monoisotopic masses are assigned.

Sequence Tag Approach to Protein Identification Using MS–MS Analysis

The mass spectrometry group at the European Molecular Biomolecule Laboratory[49] has developed a technique for protein identification that is based on digesting the protein into peptides, determining the molecular weights of these fragments, and determining a partial sequence of the peptide(s).[50] By knowing a peptide MW, the start and end mass for the sequenced portion of the peptide, and the known digestion procedure (e.g., with trypsin), one can search a protein database for a match. This procedure should be repeated using a second or third peptide from the protein for confirmation.

* A protein with a MW of 50,000 amu should have approximately 40 tryptic peptides >500 amu after complete digestion. For example, a bovine serum albumin (BSA) homolog, accession no. 2190337, MW = 69,323.9 amu, has 59 tryptic peptides >500 amu.

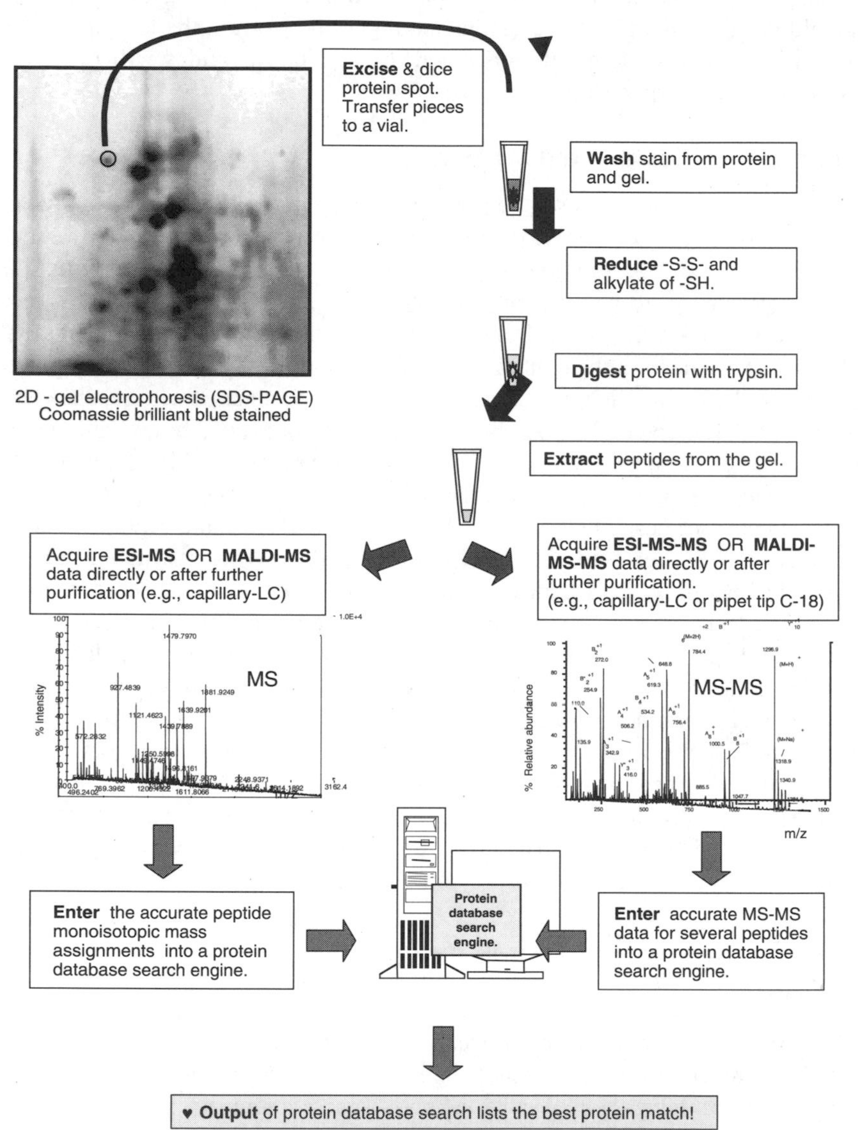

FIGURE 4.12 A flow chart showing the basic steps used to identify a protein. The protein is excised from a 2-D electrophoresis gel, the gel diced, and the stain is washed. The protein is reduced and digested in the gel. The newly formed peptides are extracted and analyzed using a MS and/or MS–MS scan mode. The resulting MS and MS–MS data are entered into a protein database search engine that performs the match.

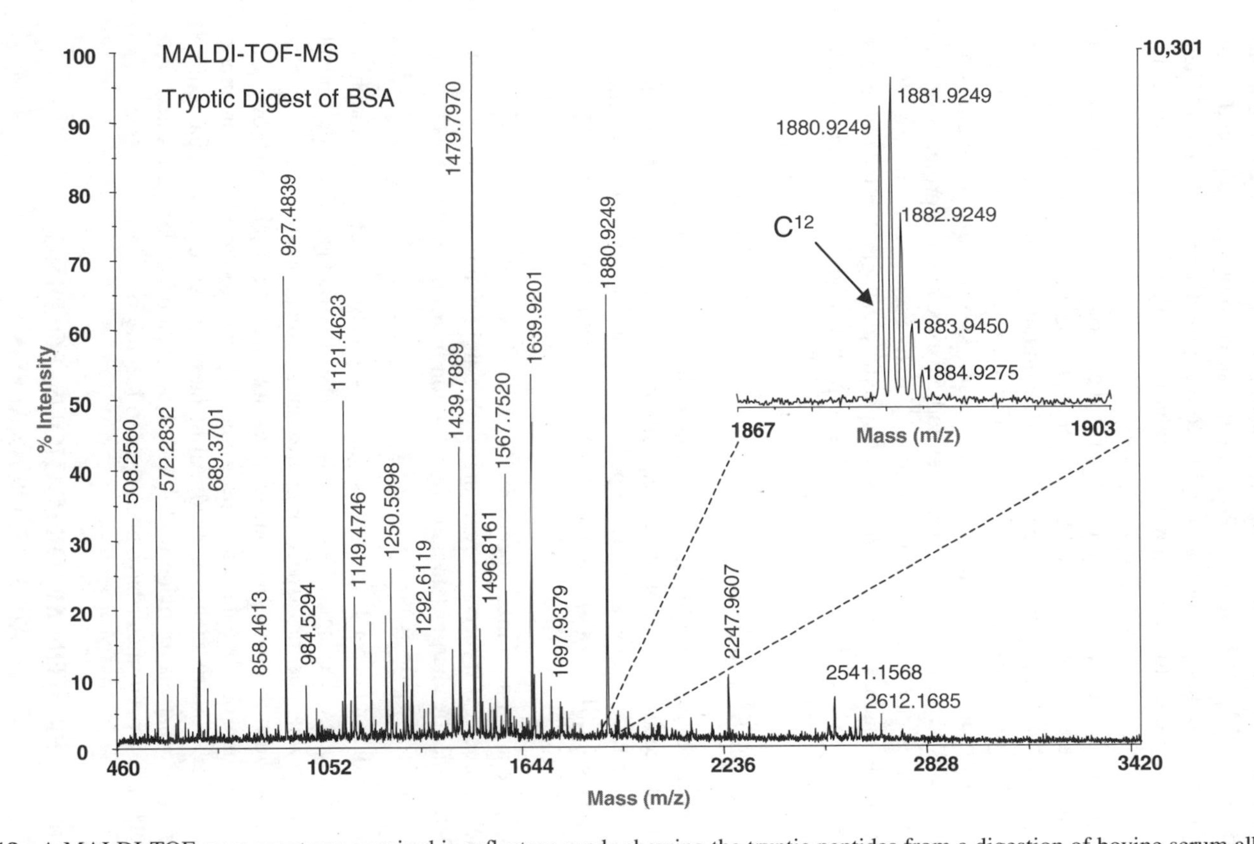

FIGURE 4.13 A MALDI-TOF mass spectrum acquired in reflectron mode showing the tryptic peptides from a digestion of bovine serum albumin. The *m/z* of the ^{12}C monoisotopic ions are assigned and can be entered into a protein database search engine to match to the theoretical trypsin digest of BSA.

SEQUEST™ Approach to Protein Identification Using MS–MS Analysis

A method developed by Eng and Yates uses a MS/MS cross-correlation algorithm called SEQUEST™.[51–53] In this method, an unknown protein is digested using trypsin and the resulting peptides are used to collect product MS–MS scans. These MS–MS scans are used to search a protein database for a match within the expected MS–MS data from the known tryptic peptides.

The protein identification techniques described previously show excellent matching capability to known proteins. Neither technique, however, will work if the protein in question is not in the protein database. With the human genome sequence essentially complete, we can expect efficiencies similar to those researchers now experience with yeast protein identification. In many cases, knowing that a protein is not in the database is sufficient for the researcher and, in some cases, a protein matched to another species can give useful information. In any regard, care must be taken when a match is made; keep in mind that nonspecific cleavages can lead to errors, proteins are usually modified, weak signals give poor confidence levels of a good match, and there are errors in the database.

Data-Dependent Scanning

Data dependent scanning is a MS technique that utilizes algorithms to automatically make real-time decisions to control the acquisition mode of the mass spectrometer. The data system executes the next scan type based on previously collected data.[54, 55] For example, an ESI mass spectrometer can be programmed to implement the following experiment.

1. Acquire a full-scan mass spectrum.
2. Select a peak if it meets a preset signal threshold requirement.
3. Determine the charge state and, thus, the ion *m/z* ratio, if possible.
4. Acquire a MS–MS scan.

The advantage of data-dependent scans is that they greatly increase the amount of information gained from a single LC-MS analysis. The disadvantage of data-dependent scanning occurs when the mass spectrometer executes a scan on an ion that provides no useful data; when the mass spectrometer does not execute a desired scan owing to the program settings or when the mass spectrometer spends too much time on an ion for which no more information is needed. All of these scans can take valuable time away from scans that may provide useful MS data. Data-dependent scans acquired using a QIT mass spectrometer have been used to identify proteins.[56, 57]

PROTEIN MODIFICATIONS, FOLDING, AND TISSUE ANALYSIS

Identifying post-translational modifications is another successful application of mass spectrometry. Mass spectrometry can determine covalent modification sites due to phosphorylation,[58–60] glycosylation,[61, 62] acylation, and sulfation. Mass spectrometry is a rapid and effective method used to detect these chemical changes. Because these post-translational modifications probably have biological significance, the elucidation of the modified position is extremely important. Both the number of additions and the location of the modification can often be identified using MS and MS–MS scans. Settineri and Burlingame have written a mass spectrometry review covering carbohydrate and glycoconjugate structure analysis.[63]

Mass spectrometry is not able to provide detailed information about the tertiary structure of proteins, but information is gained through hydrogen/deuterium (H/D) exchange reactions.[64, 65] Rate and equilibrium constants can also be determined for the folding and refolding process using H/D exchange reactions and MS.[66] Feng and Konish,[67] Douglas et. al.,[68] and Konermann et. al.[69] have studied the unfolding of proteins by observing the change in the relative abundance of the multiply charged peaks using ESI-MS. The ESI mass spectra of native, denatured, and partially renatured homomyoglobin show a distinctly different charge distribution. As a protein unfolds under acid conditions, the average distance between charges on the protein increases as the macromolecule becomes string like. In addition, as the unfolding occurs, additional sites for protonation are more accessible rather than blocked by the sterically hindered folded protein. Additional protons increase the average charge state and cause the multiply charged ion distribution to shift to lower *m/z*. Upon renaturing of, in this case, myoglobin, the ion distribution shifts back to higher *m/z* as the charge state decreases during the refolding process. Konermann has explored the kinetics of the folding process more closely for homomyglobin using ESI-MS.[68]

Tissue analysis[70–73] and imaging[74, 75] using MALDI-MS is a fascinating area that we may someday use as a diagnosis tool in medical laboratories. For nonimaging work, the tissue may be homogenized and analyzed by MALDI-MS. For MS imaging, the tissue is thinly sectioned for direct introduction into the mass spectrometer or the tissue can be blotted onto a membrane that is analyzed by MALDI-MS. For the imaging example, the laser beam is rastered across the membrane or tissue slice to create a map of the areas where a specific peptide is found. One might image a cancerous tumor by analyzing nonself peptides. Alternatively, an image of a newly found peptide or protein signal could be mapped for research purposes. This may become a useful method to locate peptides or proteins from many different tissues, but one limitation of this technique is sensitivity.

A FINAL THOUGHT

The data that one receives are dependent on the operator, the sample, and the mass spectrometer. Poor performance is usually the result of the operator or the sample, not the instrument. Likewise, great data are usually reflections of the operator and the sample preparation.

ACKNOWLEDGMENTS

I would like to thank Pamela Nakajima and Lisa Falenski for their editing and support during the writing of this manuscript. I thank Manimalha Balasubramani for her two-dimensional gel picture. I thank the NSF for funding Carnegie Mellon's

MALDI-TOF and ESI-ion trap mass spectrometers through grant numbers CHE-9808188 and DBI-9729351, respectively. These two mass spectrometers were used to collect much of the data found in this chapter.

REFERENCES

1. Barber, M., Bordoli, R. S., Garner, G. V., Gordon, D. B., Sedgwick, R. D., Tetler, L. W., and Tyler, A. N., *Biochem. J.*, 1981, 197, 401–404.
2. Barber, M., Bordoli, R. S., Sedgwick, R. D., and Tyler, A. N., *Nature,* 1981, 293, 270–275.
3. Schulten, H. R., *Int. J. Mass Spectrom. Ion Phys.*, 1979, 32, 97–283.
4. MacFarlane, R. D. and Torgerson, D. F., *Science,* 1976, 191, 920–923.
5. Aleksandrov, M. L., Gall, L. N., Krasnow, V. N., Nikolaev, V. I., Palenko, V. A., Shkurov, V. A., Baram, G. I., Gracher, M. A., Knorre, V. D., and Kusner, Y. S., *Bioorg. Khim.,* 1984, 10, 710.
6. Meng, C. K., Mann, M., and Fenn, J. B., *Z. Phys. D.,* 1988, 2, 95.
7. Fenn, J. B., Mann, M., Meng, C. K., and Wong, S. F., *Mass Spectrom. Rev.*, 1990, 9, 37.
8. Covey, T. R., Bonner, R. F., and Shushan, B. I., *Rapid Comm. Mass Spec.,* 1988, 2, 249–256.
9. Fenn, J. B., Mann, M., Meng, C. K., Wong, S. F., and Whitehouse, C. M., *Science,* 1989, 246, 64–71.
10. Karas, M. and Hillenkamp, F., *Anal. Chem.,* 1988, 60, 299–2301.
11. Karas, M., Bachmann, D., Bahr, U., and Hillenkamp, F., *Int. J. Mass Spectrom. Ion Process*., 1987, 78, 53–68.
12. Karas, M., Bahr, U., and Hillenkamp, F., *Int. J. Mass Spectrom. Ion Proces.*, 1989, 92, 231–242.
13. Dole, M., Mach, L. L., Hines, R. L., Mobley, R. C., Ferguson, L. D., and Alice, M. B. J., *Chem. Phys.,* 1968, 49, 2240.
14. Clegg, G. A. and Dole, M., *Biopolymers,* 1971, 10, 821–826.
15. Dole, M., Cox, H. L., and Gieniec, J., *Advances in Chemistry Series,* 1973, No. 125, *Polymer Molecular Weight Methods*, Mason, E. A., Ed., Electrospray Mass Spectroscopy, 1973, chap. 7, 73.
16. Gieniec, J., Mack, L. L., Nakamae, K., Gupta, C., Kumar, V., and Dole, M., *Biomed. Mass Spectrom.,* 1984, 11, 259–268.
17. Fenn, J. B., Mann, M., Meng, C. K., and Wong, S. F., *Mass Spectrom. Rev.*, 1990, 9, 37.
18. Aleksandrov, M. L., Gall, L. N., Krasnow, V. N., Nikolaev, V. I., Palenko, V. A., Shkurov, V. A., Baram, G. I., Gracher, M. A., Knorre, V. D., and Kusner, Y. S., *Bioorg. Khim.,* 1984, 10, 710.
19. Meng, C. K., Mann, M., and Fenn, J. B., *Z. Phys. D.*, 1988, 2, 95.
20. Dohmeier, D. M. and Jorgenson, J. W., *Microelectrospray Method and Apparatus,* U.S. Patent 5, 115, 131, 1992.
21. Lewis, K. C., Dohmeier, D. M., Jorgenson, J. W., Kaufmann, S. L., Zarrin, F., and Dorman, F. D., *Anal. Chem.*, 1994, 66, 2285–2292.
22. Wilm, M. S. and Mann, M., *Int. J. of Mass Spectrom. Ion Proces.*, 1994, 136, 167–180.
23. Wilm, M. and Mann, M., *Anal. Chem.*, 1996, 66, 1–8.
24. Hillenkamp, F., Karas M., and Giessmann, U., *Process and Device for the Laser Desorption of Analyte Molecular Ions, Especially of Biomolecules*, U.S. Patent 5, 118, 937, 1992.

25. Yates, J., *Rapid Commun. Mass Spectrom.*, 1993, 7(1), 20–26.
26. Schwartz, J. C. and Bier, M.E., *Rapid Commun. Mass Spectrom.*, 1993, *7*(1), 27–32.
27. Personal communication, Dr. Alan Schoen and Dr. Hans Schweingruber, ThermoFinnigan Corporation, San Jose, CA, 2001.
28. March, R. E. and Hughes, R. J., Eds., *Chemical Analysis*, Vol. 102, *Quadrupole Storage Mass Spectrometry*, Wiley Interscience, New York, 1989, 471.
29. Bier, M. E. and Schwartz, J. C., Electrospray ionization quadrupole ion trap mass spectrometry, in *Electrospray Ionization Mass Spectrometry: Fundamentals, Instrumentation, and Applications*, Cole, R. Ed., John Wiley & Sons, New York, 1997, chap 7.
30. Henry, K. D., Williams, E. R., Wang, B. H., McLafferty, F. W. Shabanowitz, J., and Hunt, D. F., *Proc. Natl. Acad. Sci. U.S.A.*, 1989, 86, 9075–9078.
31. Martin, S. E., Shabanowitz, J., Hunt, D. F., and Marto, J. A., *Anal. Chem.*, 2000, 72(18), 4266–4274.
32. ZIPTIPS™, Millipore Corp., Suprotip™, The Nest Group, Inc., Southborough, MA, and AMICON, Inc., Beverly, MA.
33. Figeys, D. and Aebersold, R., *Electrophoresis*, 1998, 19, 885–892.
34. Figeys, D., Ducret, A., Yates, J. R., and Aebersold, R., *Natl. Biotech.*, 1996, 14, 1579–1583.
35. Roepstorff, P. and Fohlman, J. *Biomed. Mass Spetrom.*, 1984, 11(11), 601.
36. Martin, S. E., Shabanowitz, J., Hunt, D. F., and Marto, J. A., *Anal. Chem.*, 2000, 72, 4266–4274.
37. Chait, B. T., Wang, R., Beavis, R. C., and Kent, S. B., *Science*, 1993, 262, 89–92.
38. Wang, R., Chait, B. T., and Kent S. B. H., *Techniques in Protein Chemistry*, Vol. V, Crabb, F. W., Ed., Academic, San Diego, CA, 1994, 19–26.
39. Patterson, D. H., Tarr, G. E., Regnier, F. E., and Martin, S. A., *Anal. Chem.*, 1995, 67, 3971–3978.
40. Bonetto, V., Bergman, A. C, Jornvall, H., and Sillard, R., *Anal. Chem.*, 1997, 69, 1315–1319.
41. Yates, J. R., *Electrophoresis*, 1998, 19, 893–900.
42. Haynes, P. A., Fripp, N., and Aebersold, R., *Electrophoresis*, 1998, 19, 939–945.
43. Stone, K. L., DeAngelis, R., LoPresti, M., Jones, J., and Papov, V. V., *Electrophoresis*, 1998, 19, 1046.
44. The University of California, San Francisco.
45. European Molecular Biology Laboratory, Meyerhofstr. 1, 69117 Heidelberg, Germany.
46. SEQUEST™, Eng, J. and Yates, J., Department of Cell Biology, SR11, Scripps Research Institute, La Jolla, CA.
47. Bier, M. E. Department of Chemistry, Carnegie Mellon University and Grabowski, J., Department of Chemistry, University of Pittsburgh, personal communication.
48. Chalmers, M. J. and Gaskel, S. J., *Curr. Opin. Biotechnol*,, 2000, 11(4), 384–390.
49. European Molecular Biology Laboratory, Meyerhofstr. 1, 69117 Heidelberg, Germany.
50. Mann, M. and Wilm, M., *Anal. Chem.*, 1994, 66, 4390–4399.
51. Eng, J. K., McCormack, A. L., and Yates, J. R., *J. Am. Soc. Mass Spectrom.*, 1994, 5, 976–989.
52. Yates, J. R., Eng, J. K., McCormack, A. L., and Schieltz, D., *Anal. Chem.*, 1995, 67, 1426–1436.
53. Yates, J. R., *J. Mass Spectrom.*, 1998, 33, 1–19.
54. Stahl, D. C., Martino, P. A., Swiderek, K. M., Davis, M. T., and Lee T. D., *Proc. 40th ASMS Conf. Mass Spectrom. Allied Topics*, Washington, D.C., 1992, 1801–1802.

55. Mylchreest, I., Campbell, C., Wheeler, K., and Wakefield, M., *Proc. 43rd Conf. Mass Spectrom. Allied Topics*, Atlanta, GA, 1995, 436.
56. Arnott, D., King, K., Bier, M., Land A., and Stults, J., *Proc. 43rd ASMS Conf. Mass Spectrom. Allied Topics*, Atlanta, GA, 1995, 31.
57. Yates, J. R., Eng, J., Schieltz, D., and Link, A., *Proc. 43rd Conf. Mass Spectrom. Allied Topics*, Atlanta, GA, 1995, 325.
58. Gillece-Castro, B. L., Arnott, D. P., Bier, M. E., Land, A. P., and Stults, J. T., *Proc. 43rd Conf. Mass Spectrom. Allied Topics*, Atlanta, GA, 1995, 302.
59. Cao, P. and Stults, J. T., *J. Chromatogr. A*, 1999, 853(1–2), 225–235.
60. Cao, P. and Stults, J. T., *Rapid Commun. Mass Spectrom.*, 2000, 14(17), 1600–1606.
61. Sogaard, M., Anderson, J. S., Roepstorff, P., and Svensson, P., *Biotechnology*, 1993, 11, 1162–1165.
62. Lapolla, A., Fedele, D., Plebani, M., Garbeglio, M., Seraglia, R., D'Alpaos, M., Arico, C. N., and Traldi, P., *Rapid Comm. Mass Spectrom.*, 1999, 13, 8–14.
63. Settineri, C. A. and Burlingame, A. L., *J. Chromatogr. Libr.*, 1995, 58, 447–514.
64. Zhang, Z. and Smith, D. L., *Protein Sci.*, 1993, 2, 522–531.
65. Deng, Y. and Smith, D. L., *J. Mol. Biol.*, 1999, 294(1), 247–258.
66. Deng, Y. and Smith, D. L., *Anal. Biochem.*, 1999, 276(2), 150–160.
67. Feng, R. and Konishi, Y., *J. Am. Soc. Mass Spectrom.*, 1993, 4, 638–645.
68. Konermann, L., Rosell, F. I., Mauk, A. G., and Douglas, D. J., *Biochemistry*, 1997, 36, (21), 6448–6454.
69. Lee, V. W. S., Chen, Y.-L., and Konermann, L., *Anal. Chem.*, 1999, 71(19), 4154–4159.
70. Jimenez, C. R., van Veelen, P. A., Li, K. W., Wildering, W. C., Geraerts, W. P. M., Tjaden, U. R., and van der Greef, J., *J. Neurochem.*, 1994, 62(1), 404–407.
71. Huff, T., Muller, C. S. G., and Hannappel, E., *FEBS Lett.*, 1997, 414, 39–44.
72. Chaurand, P., Stoeckli, M., and Caprioli, R. M., *Anal. Chem.*, 1999, 71(23), 5263–5270.
73. Jimenez, C. R., Li, K. W., Dreisewerd, K., Spijker, S., Kingston, R., Bateman, R. H., Burlingame, A. L., Smit, A. B., van Minnen, J., and Geraerts, W. P. M., *Biochemistry*, 1998, 37(7), 2070–2076.
74. Koomen, J. M., Stoeckli, M., and Caprioli, R. M., *J. Mass Spectrom.*, 2000, 35(2), 258–264.
75. Caprioli, R., Chaurand, P., Stoeckli, M., Norris, J. L., and Baxter, S. A., *48th Conf. Mass Spectrom. Allied Topics*, Long Beach, CA, 2000.

5 An Orientation to Edman Chemistry

John Hempel

CONTENTS

Introduction ... 104
A Brief History ... 104
Amino-Acid Identification Improvements ... 104
Improvements in Sample Processing ... 105
Overview ... 106
Sample Preparation ... 107
Sample Purity ... 108
Data Interpretation ... 108
Salt ... 109
Problem Amino Acids ... 110
Other Chemical Modifications ... 111
Site-Directed Labeling ... 111
How Far Can I Sequence? ... 112
Complications at the N-Terminus ... 112
Acetylation ... 112
Pyroglutamate ... 113
Overcoming Blockage ... 113
CNBr Cleavages at Methionine ... 113
Acid Lability of Asp-Pro Bonds ... 114
Asn-Gly ... 114
Cleavage at Trp ... 115
In-Gel Digestion ... 115
Other N-Terminal Anomalies ... 116
Ragged Ends ... 116
α-Amino Alkylation ... 116
Post-Translational Modifications at Internal Residues ... 117
Phosphorylation ... 117
Glycosylation ... 118
Summary ... 118
Acknowledgment ... 119
Abbreviations Used ... 119
References ... 119

0-8493-9453-8/02/$0.00+$1.50

INTRODUCTION

As of this writing, 50 years have passed since Pehr Edman published his first note on chemical degradation of peptide chains, "A Method for the Determination of the Amino Acid Sequence in Peptides."[1] As testament to the continuing relevance of this technique, the following job announcement was recently posted.

> The successful candidate will perform structural analysis on a wide variety of proteins using state of the art Edman sequencing and mass spectrometry instruments. The candidate will also be involved in development of new techniques to aid in protein characterization.
>
> The ideal candidate ... will have a bachelor/masters degree in (bio)chemistry with 3 to 5 years experience in protein sequencing, and protein characterization using HPLC, mass spectrometry and amino acid analysis.
>
> **Announcement for position 1021 (Genentech), March 8, 1999.**

A BRIEF HISTORY

Amino-Acid Identification Improvements

Edman's initial description[1] outlined the chemistry for a "*micromethod, the requirements of substance being approximately 10 mg amino acid residue per peptide bond*." The greatest drawback to the technique at the time was the inability to identify the phenylthiohydantoin (PTH) product directly and in a straightforward manner. The transparency of the PTH derivative in the visible spectrum was in marked contrast to that formed by Sanger's reagent, fluorodinitrobenzene. Still, because the near quantitative ability of the Edman reagent to remove one amino acid residue at a time had no equal or close second, strategies emerged for identifications, including hydrolysis of the derivative to the parent amino acid and identification by paper chromatography.

The ensuing two decades brought continual refinements in identification involving both indirect methods such as the Dansyl-Edman technique and direct identification of the PTH itself.[2] By the late 1970s, automated sequencers had become widespread and microsequence analysis was practiced on amounts <10 nmol, with direct identification of PTH derivatives by a variety of methodologies involving gas, paper, and thin-layer chromatography, and mass spectrometry.[3, 4] Still, subtractive manual techniques, including the Dansyl-Edman approach, were widely practiced.[5–7] These required removal of an aliquot of the peptide at each cycle, representing mandatory loss of material at each cycle.

With nondestructive methods in which the only material deliberately removed is the PTH derivative released at each cycle of the degradation, losses occur through extraction of varying amounts of sample at each cycle, during the solvent wash steps for removal of unspent reagents and by-products. Difficulties, particularly in retaining short peptides in automated sequencers, were first tackled through solid-phase approaches, in which the peptide was chemically coupled to an inert support.[8] This approach lost favor in liquid phase spinning-cup instruments, when it was found that peptide loss could be greatly prevented by pretreating the cup with Polybrene,[9] a polyquarternary amine. The use of polybrene on the glass fiber sample supports continues in contemporary sequencing.

HPLC finally provided a practical means to separate and directly identify all 20 PTH derivatives.[10] Generally, all involve gradients of aqueous sodium acetate at pH ~4 against acetonitrile, on C-18 silica columns. HPLC also greatly reduced the amount of sample required, shifting efforts back to the degradation end of the equation for increasing sensitivity.

Numerous modifications of Edman's reagent, phenylisothiocyanate (PITC), to create other isothiocyanate reagents have been investigated, but a review of all is beyond the scope of this chapter. In the end, none replaced PITC itself, with one possible exception. The methyl yellow derivative dimethylaminoazobenzene isothiocyanate (DABITC)[11] provided the ability to form a PTH derivative that absorbed in the visible spectrum, which lent itself to use in manual sequencing strategies. However, with only about 50% coupling efficiency, remaining free N-termini needed to be scavenged at each cycle with the traditional Edman reagent, thus complicating the methodology. Still, this approach afforded the ability to sequence multiple samples by hand, generally one cycle per day. Residue identifications were made by two-dimensional chromatography on small (3 × 3-cm) polyamide thin-layer plates, thus circumventing a large investment in an automated sequencer and HPLC equipment (the co-generated, transparent conventional PTH derivatives did not interfere with the visible DABTH derivatives). However, as the definition of "microsequencing" came to imply the low picomole range, DABITC largely faded from use. An alternative strategy, capitalizing on aminolysis of the thiazolinone[12] with a fluorophoric amine to further enhance sensitivity,[13] failed to progress.

Also, in the late 1970s, parallel efforts focused on increasing sequencer sensitivity through use of biosynthetically generated proteins in which one or several radio-labeled amino acids had been incorporated.[14] The high specific activity of ^{35}S made ^{35}S-methionine and cysteine attractive residues to incorporate, but their relative rarity and the short half-life of ^{35}S acted against this becoming a commonplace approach, although incorporation of ^{35}S-methionine was useful in locating peptides providing overlaps to CNBr fragments. Further, for a variety of reasons,[14] it was not possible to incorporate labeled versions of all residues in one biosynthetic effort, nor could a uniform level of specific activity be achieved for all amino acids. These were not the only factors complicating analysis of labeled amino acids.[14] The continued increase in sensitivity of sequencers, as well as alternate strategies toward the goal of obtaining sequence through use of recombinant DNA techniques has made this approach little more than a historical footnote now. An alternate radiosequencing strategy, involving use of ^{35}S-labeled Edman reagent, was abandoned due to the large molar excess required in coupling, difficulties in purification of the reagent, and danger of laboratory contamination.

Improvements in Sample Processing

Automation of the process of converting ATZ to PTH derivatives and miniaturization of the reaction cell,[15] with demonstration of the ability to bind sample onto a (polybrene-

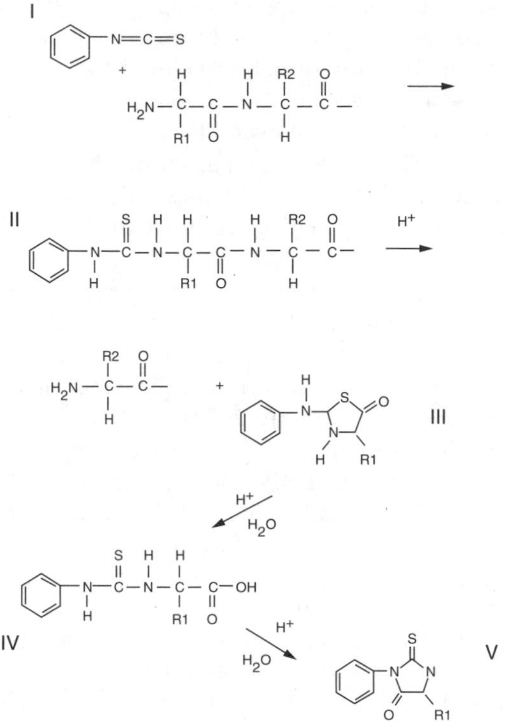

SCHEME 5.1 Sequential steps of Edman degradation.

treated) glass fiber filter and retain it during the degradation, culminated in the development of the gas-phase sequencer.[16] This advancement capitalized on the volatility of most reagents and solvents for their complete removal at each step of the degradation, but required introduction of a volatile base (initially triethyl-amine) for the coupling step. The addition of online injection of derivatives to an HPLC completed the needed hardware. All of these features combined to bring sequencer sensitivities into the 10-pmol level.[17] Ensuing developments have been even more technical—minimization of line volumes and dead volumes in delivery valves, and a shift to microbore HPLC together with an increase in detector cell sensitivity, bringing routine sensitivities to below 1 pmol in many instruments. Still, after 50 years, the chemical requirements have remained the same: Coupling of the Edman reagent, phenylthiocyanate (I) to the unionized α-amine of the protein or peptide, acidolysis of the phenylthiocarbamylated N-terminus (II), to yield the anilinothiazolinone (ATZ) amino acid derivative (III), conversion of the ATZ in aqueous acid to the more stable phenlthio-hydantoin (V), via the phenylthiocarbamyl (PTH) amino acid intermediate (IV).

OVERVIEW

This chapter is intended as a broad overview to give researchers with a prospective need for protein sequence analysis some feeling of the strengths/weaknesses of

Edman sequencing. Greater detail including specific protocols may be found in other volumes, including Allen,[18] Bhoun's earlier volume in this series,[19] *Current Protocols in Protein Science*,[20] and *Protein Sequencing Protocols*.[21]

Sample Preparation

Sequencing of intact proteins is now most often performed on material transblotted to PVDF membranes and stained.[22] PVDF is essential because nitrocellulose membranes dissolve in the organic solvents used for extractions during the degradation. Submitting a sample blotted to nitrocellulose is a good way to become blacklisted by the manager of your sequence laboratory. That having been noted, transblotting has become so commonplace that it is sometimes assumed that this is an essential prerequisite. Of course, if the sample is already pure, or nearly so, transblotting is not necessary. Proteins and peptides can be applied to pretreated glass fiber disks and loaded directly to the sequencer without being bound to PVDF, as long as they do not contain buffer salts (see below).

If the sample is impure, as is commonly the case, gel electrophoresis usually offers the quickest means of separating components of a mixture. Traditional Laemmli gels are fine, and it is likely that virtually any other variation on the electrophoretic step will work equally well. However, residual free radicals and unpolymerized acrylamide in freshly cast gels can react with free amino groups. Traditionally, this has been a cause for concern, because any α-amino group reacting with another compound prior to exposure to PITC will be unable to form a PTC derivative (II). The simplest precaution is simply to cast the gel the day before running it.[23] Do not forget to flush the wells of the gel with running buffer before loading the sample. Some prefer prerunning with thioglycolate-containing buffer as well. Also, reserving one lane for a control protein of approxi-mately the same size as your band of interest (and known to have a free N-terminus) is a good precaution. Such control bands can be sequenced to confirm instrument performance if results are not obtained from the unknown.

For transblotting, tris-glycine buffers can be used, but this buffer has two disadvantages. First, residual glycine can compromise interpretation of the early cycles of the degradation. In addition, proteins with unusually high isoelectric points either may not transfer well or, because there is no SDS in the transfer buffer, they migrate in the opposite direction. To overcome both of these potential problems, transblotting with CAPS buffer at pH 11 is a favored method.[22] Protocols summarized by ProMega[24] also perform well.

For staining the blotted membrane, a variety of stains including Amido Black and Ponceau S are compatible with Edman chemistry. However, without a special reason for using another stain, Coomassie R-250 is recommended, because advice on whether sufficient material is likely to be present is best based on a common stain. Because most blots submitted to sequence laboratories are stained with Coomassie, sequencing personnel are quite accustomed to looking at Coomassie-stained membranes and they have a feel for the amount of material present based on stain intensity. It does not help to use a dye with a sensitivity higher than that of the laboratory sequencing facility to which you submit your sample.

Other preparatory details include the use of high-quality methanol for destaining, as histological grades may contain traces of formaldehyde, which can also react with amino groups causing N-terminal blocking and premodification of cysteine (for both, see below). Also, use gloves when handling membranes, and leave it to your sequencing facility to cut the bands from your blot.

Various purification protocols may call for a protein to be exposed to urea, which leads to a common problem. Urea decomposes to cyanate, and cyanate irreversibly reacts with amino groups to form carbamyl derivatives that can only be removed by destructive methods.[25] This is one form of N-terminal blockage rendering the protein refractive to Edman degradation (see below). Because the pKa value of an α-amino group is lower than the ε-amino group of lysine, α-amino groups will be preferentially derivatized. An easy way of avoiding this potential problem is to substitute guanidinium chloride for urea in the purification protocol, but this is often not possible, particularly if the urea is used at an electrophoretic or ion-exchange chromatographic step. If urea must be used, the problem may be overcome if it is understood beforehand. First, do not assume that you will avoid problems by purchasing the highest quality, most expensive urea. As noted above, cyanate is a product of urea decomposition. The presence of cyanate can be detected using a conductivity meter. A cyanate-free urea solution has a conductivity similar to that of Milli-Q water. By passing a urea solution (prior to blending with buffer) through a mixed-bed ion exchange resin (e.g., Dowex 501-X8 or Amberlite MB-1) several times, this level of conductance is easily achieved. At this stage, a scavenger such as methylamine[26] can be added to consume any cyanate forming during exposure of the protein to the solution. Stock solutions of urea without further buffer salts may be used at a later date by again passing the solution over the ion exchanger as described previously. Alternatively, some investigators freeze aliquots of deionized urea stock solutions together with a few beads of the ion exchanger, which seems to yield a good quality solution when thawed later. This author has found that a small bottle of 8 *M* urea containing mixed-bed beads and kept at 4°C also maintains a solution of appropriately low conductivity for months at a time.

Sample Purity

It should be obvious that the cleaner a preparation is, the better the resulting information will be. However, this does not mean that sequencing impure material is pointless. Even with the advent of micro techniques that enable sequencing from gel-purified material, co-migrating proteins are still encountered. In the worst scenario, one protein is N-terminally blocked (see the following). If the blocked protein is actually the one of interest, long-lasting confusion may follow easily.

Data Interpretation

If fairly unambiguous sequence data can be obtained, database searching usually follows. BLAST searches (*http://www.ncbi.nlm.nih.gov/BLAST/*) can be rapidly performed online and probably represent the most expedient means of identifying an unknown protein, as discussed further in this volume. (See Chapter 6 in this volume.) When performing searches, Doolittle's primer, *Of Urfs and Orfs*,[27] contains many valuable tips, particularly Chapter 2, entitled "So You Found Something!"

If your sample contains a mixture of two proteins, both with free N-termini, and the mixture is roughly equimolar, it is sometimes possible to identify both proteins. A variety of search programs, such as MS-Edman (*http://falcon.ludwig.ucl.ac.ukucsfhtml3.2/msedman.htm*), are capable of using equivocal assignments at each cycle to find potential matches. Alternatively, if the relative amounts of two proteins are skewed to around 70/30, it is usually possible to call the two sequences based on the relative abundance of the two PTH signals. Because most sequencers are equipped with software to automatically calculate molar amounts of each amino acid recovered at each cycle, it might seem that this should be a simple matter of noting the two residues of highest and next highest amount at each cycle. In practice, it is frequently not so straightforward.

Several factors can complicate interpretation of simultaneously sequenced peptides. First, not all residues report as well as others, as detailed below. Thus, a cycle with a weakly reporting residue from the major peptide and strongly reporting residue in the minor peptide may present difficulties in assigning the residues to the correct sequences. These problems are further complicated when the same residue occurs in each peptide at the same cycle, or consecutive identical residues (e.g., Ala–Ala) occur in one peptide, or identical residues occur in consecutive cycles but from different peptides. These situations can sometimes be interpreted by comparison of yields of residues with similar stabilities in other cycles, particularly if those cycles are close to the one in question.

Salt

Salts interfere with sequencing, particularly if they alter the pH during coupling or cleavage, because efficient coupling depends on an un-ionized α-amino group and efficient cleavage (acidolysis) requires a strong acid pH. For this reason, samples for sequencing should be salt-free. To help illustrate why this is important, consider that the application of 5 pmol of a protein in solution at a concentration of, say, 1 pmol/μl in 20 m*M* buffer would mean the application of 100 nmoles of buffer salt, a 20,000-fold molar excess of buffer salt over protein. Even though the quantities of coupling base and cleavage acid may to some extent overcome this, it is best to simply remove all buffers and salts before sequencing.

Whereas blotted samples are salt-free after destaining, samples in solution must be in a volatile solvent that can be easily removed by drying. Dilute (1 m*M*) HCl is a good solvent because it is volatile and proteins tend to remain soluble in it. Although it frequently runs counter to the instincts of those trained to avoid conditions that may denature proteins, sequence can be perfectly well obtained from a suspension of precipitated protein. As a series of chemical reactions, Edman degradation is independent of native folding. Acetic or formic acids (~1%) are also frequently used to apply samples for sequencing, as is ammonium bicarbonate. However, the latter is less volatile and repeated dryings may be needed for complete removal.

As a widely employed alternative to solvent exchanges, many sequencing laboratories simply apply the sample, as supplied by the investigator, directly to a ProSorb® filter (Applied Biosystems) regardless of the buffer solution it is dissolved in. These filters consist of a PVDF membrane attached to a holder with an absorbent material on one side that draws the protein solution across the PVDF membrane, binding proteins/peptides in the process.

Problem Amino Acids

Cysteine presents special problems for sequencing. Generally, cysteine decomposes during Edman degradation, leaving insufficient evidence to enable positive identification. The disulfide crosslink cystine, on the other hand, may sometimes be identified directly[28] by its release at the cycle containing the second half-cystine moiety, but this is not routinely dealt with by most sequencing laboratories. Ordinarily, to positively identify cysteine residues, reduction and alkylation of the sample prior to sequencing are necessary, and a variety of methodologies exist to achieve this end. PTH-derivatives of S-carboxymethyl or S-carboxamidomethyl cysteine (formed after reduction and alkylation with iodoacetate or iodoacetamide, respectively) are sometimes distinguished from other PTH derivatives only with difficulty, and use of the ^{14}C-labeled reagents was once widespread to ensure positive identification. Because of these difficulties, alternative alkylating agents were surveyed, resulting in the discovery that alkylation with N-isopropyliodoacetamide yielded a PTH derivative with an unambiguous HPLC elution position,[26] and this became the favored reagent in some laboratories. Many others favor S-pyridylethylation with 4-vinyl pyridine, because the derivative can be formed directly on blotted material.[29]

The most practical approach is to first find out which method your sequencing laboratory favors. If the objective is to find a sequence for probe design, the di-degeneracy of Cys codons makes positive identification of cysteine residues valuable. If the objective is just to identify the protein via a database search from directly sequencing the intact protein, positive identification of cysteine may be less essential as long as 8 to 10 cycles of unambiguous residue identifications can be made.

Cysteine is not the only residue that presents analytical difficulties, although the other problematic residues do not require modification of the sample prior to analysis. Serine, threonine, histidine, and arginine also present difficulties. With serine and threonine, some dehydration of the sidechains will occur. Deacylation of their trifluoroacetate esters formed during repeated cleavage steps[7] is actually responsible for the problem. Serine yields dehydroalanine that is then susceptible to a variety of addition products, and this tendency is greater than with threonine. The degree of decomposition varies widely, depending on the proximity of the residue to the N-terminus and on technical aspects of the degradation, such as the length of the cleavage step and whether the anilinothiazolinone dries completely before conversion.

With histidine and arginine, the problem is the intrinsic polarity of the derivative. This affects the solubility of the derivative in the solvent (ethyl acetate or chlorobutane) used to transfer the derivative from the reaction cell to the conversion flask. With some systems, trifluoroacetic acid (TFA) is added to this solvent to help counteract the problem, although the tenacious attraction of TFA to proteins/peptides and their derivatives[7] indicates that His and Arg derivatives have already formed the TFA salt during the cleavage reaction and continue to exist largely in this state. The elution of PTH-histidine and arginine can be also be compromised by any free (uncapped) anionic $-SiO_3$ groups on the reversed-phase column, resulting in peak broadening.

Lysine may also present problems because its ε-amine couples with phenylisothiocyanate to form the ε-phenylthiocarbamyl derivative, which is sensitive to any trace of peroxides in the sequencer solvents that may decrease its yield. Its chromatographic elution close to leucine frequently requires special attention to gradient conditions as well. Tryptophan, tyrosine, and methionine are sensitive to severe oxidation, although this is usually an infrequent problem, especially with the latter two. Those residues with sidechains containing only carbon and hydrogen (e.g., Ala, Val, Ile, Leu, and Phe) are the most stable and are not known to present any complications from the Edman degradation procedure.

In addition to stability issues, cyclization to the ATZ and cleavage occurs considerably more slowly with proline than with other residues, and conversion of the ATZ of glycine to the PTH is particularly slow.[7] If a sequencer is inadvertently allowed to continue running without the conversion reagent (25% TFA), the problem is often detected by seeing an additional, earlier elution peak at each glycine cycle, while other residues are unaffected. The earlier peak is the more polar (open chain) PTC derivative. All other residues seem to convert quite well without 25% TFA, perhaps the result of tenaciously bound residual TFA.[7]

Other Chemical Modifications

Apart from cysteine, as noted above, no other residues require chemical modification prior to sequencing to enable positive identification, although lysine residues are occasionally acylated prior to cleavage of the protein chain. Citraconylation (with citraconic anhydride)[30] affords a facile and reversible means of blocking lysines that enables restriction of tryptic cleavage to arginine residues only. Succinylation[30] irreversibly provides the same restriction of tryptic specificity. Two additional advantages of succinylation are that succinylated proteins and peptides are typically more soluble and, relative to the issues discussed above with lysine, the PTH derivative of succinyl-lysine is more stable and unambiguously identified. In working with a green fluorescent protein, this author found succinylation the only practical means of keeping the protein unfolded for effective proteolysis.[31]

Site-Directed Labeling

As already noted, while wholesale sequencing based on radioactive labels is a nonproductive approach, identification of sites of selective modification using radiolabeled reagents is still valid; however, many sequencers in current use are no longer equipped with a fraction collector, which presents an extra hurdle to clear. If the primary structure of the target protein is already known, as is generally the case at this stage of an investigation, it is sometimes possible to combine knowledge of the specificity of the cleavage agent(s) with the reactivity of the reagent, to identify the site of modification from directly sequencing whole digests without separation of peptides. As an elegant example of this approach, the sites of reaction of two photoaffinity labels were mapped with confidence on the glucocorticoid receptor after sequencing whole tryptic, chymotryptic, and CNBr digests.[32] The assignments were deduced entirely from the cycles containing the released label together with the cleavage specificities.

How Far Can I Sequence?

This is a common question. The answer is, "It all depends." The Edman reaction, like virtually all other chemical reactions, does not go to completion. A plot of pmol PTH-amino acid recovered at each cycle ideally follows a linear log decay with the slope equal to the repetitive yield, where values in the 92 to 95% range are common. It should be apparent that the more quantitative the performance of the reaction, or the higher the repetitive yield, the greater the possible length of sequence data uncompromised by rising backgrounds of trailing amino acids. Thus, the simple answer to the question is to say that it depends on the repetitive yield.

Aside from aspects of reagent quality and other sequencer-related issues, with all else optimized the number of residues possible to obtain from intact proteins also seems to depend on the individual protein. The reasons are usually unknown. The length of the chain being degraded is an important factor because, from the standpoint of the reaction, it is reasonable to assume that the remaining residues act as a contaminant to the Edman reactions, probably by retaining excess counterions from the coupling base and cleavage acid as well as water. After sequencing a large number of peptides, all isolated by nominally identical reversed-phase HPLC procedures, the longest runs in the author's experience came from CNBr fragments of ~65 residues, with calls beyond 50 residues.[33]

With shorter peptides, the polarity of the C-terminal residues is of great significance. The more hydrophobic the C-terminus of the peptide, the greater the amount of the peptide lost from the sequencing support with each wash cycle of the degradation.

To try to eliminate this problem, solid-phase sequencing techniques,[8] in which the peptide is first coupled to an inert support prior to sequencing, were devised to circumvent these problems. The first generation of solid-phase techniques involved chemically coupling proteins and peptides to derivatized polystyrene beads. Repetitive yields were high, but tailoring the coupling of each sample to the support was a drawback, and this approach fell from use. This approach underwent a renaissance of sorts about 10 years ago with the advent of the Milli-Gen sequencer, which was designed for use with chemically derivatized membranes as inert supports. With the discontinuance of that instrument and the membranes it required, the technique has again faded from the scene, but it could always be revived.

Complications at the N-Terminus

Acetylation

Many proteins, maybe even the majority in eukaryotes by some estimates, have "blocked" N-termini, which render them refractive to coupling with PITC. By far the most frequent blocking modification is acetylation. Contrary to frequent assumptions, this can occur in prokaryotes as well.[34] The function of protein acetylation, assuming there is one, is obscure, but is presumed to relate to prevention of degradation by aminopeptidases. Whether a protein is acetylated or not seems dependent on the nature of residues in the first few positions.[35,36]

If your protein sample has an acetylated N-terminus, it will not yield any information from Edman degradation of the intact protein. Enzymes have been described that remove N-acetyl groups from peptide chains, but, in general, these have been found to function well only on peptides and not on intact proteins. Procedures have been described for cleavage of blocked proteins into peptides followed by modification of all newly revealed peptide N-termini (e.g., by succinylation) and then incubation of the mixture with a deacetylase.[37] The entire mixture can then be subjected to Edman degradation. Only the deacetylated peptide representing the original N-terminus will provide sequence.

As an alternative to this complicated approach, some success has been achieved with direct chemical deactylation in the case of proteins with N-acetyl serine and threonine residues (serine is actually the most commonly N-acetylated residue[35]) by treatment with trifluoroacetic acid for an extended period.[37,38] This approach capitalizes on acid-catalyzed N $\rightarrow$ O acyl shifting.

Pyroglutamate

An additional N-terminal modification also rendering the protein refractive to Edman degradation can occur when the N-terminal residue is a glutamine. The residue can cyclize to pyroglutamate by attack of the α-amine on the side-chain amide carbonyl carbon, with loss of NH_3. Pyroglutamyl peptidases have been described that cleave off the pyroglutamate residue, but, as with deacetylases, these seem to function well only on peptides and not on intact proteins. Further, the pyroglutamyl peptidases are notorious for poor stability. This is likely attributable to their being thiol proteases and, therefore, sensitive to the usual culprits of air oxidation and heavy metals. Despite the recommendation of a digestion buffer containing 5 m*M* DTT and 10 m*M* EDTA,[20] it is suspected that the additional recommendation to flush the digestion mixture with nitrogen before incubation is frequently ignored. This author's experience with other sensitive thiol enzymes suggests that this is likely a serious omission. Failure to modify the thiol groups of the substrate protein prior to incubation with the peptidase may also compromise activity.

Overcoming Blockage

CNBr Cleavages at Methionine

When a degradation produces no result, the first question is usually whether the material was blocked or whether insufficient material was applied. Further, while knowing the N-terminal sequence of an intact protein may be useful, more frequently the objective is simple identification of the protein, and for that, a sequence from anywhere in the chain will suffice. This can be addressed by *in situ* CNBr cleavage of the remaining material on the sequencer membrane. Because CNBr cleaves at Met residues, which are infrequent—just over 2.3% of all residues[39] — a limited number of fragments can be expected. Because about six fragments may be expected from a 25-kDa protein with an average amino-acid composition, re-electrophoresis and blotting the separated mixture of fragments will be necessary to obtain a clean sequence. Otherwise, by directly sequencing the mixture of fragments on the membrane, the newly revealed N-termini will sequence simultaneously. Little useful sequence information can be expected from this approach, but the level of response will provide an indication of the amount of material on the sample support that would have sequenced had the N-terminus of the mature protein not been blocked at the outset.

Acid Lability of Asp-Pro Bonds

Another chemical cleavage method expected to produce more limited numbers of fragments and which can be applied *in situ* on membranes takes advantage of the greater acid lability of Asp-Pro peptide bonds.[20, 40] Asp and Pro are neither abnormally over- nor underrepresented in proteins, thus the frequency of Asp-Pro linkages is about 1 in 400. With proteins of unknown sequence, this approach is probably best used only when there is good probability of encountering at least one such linkage (i.e., with proteins of at least 300 residues or ~33 kDa).

Asn-Gly

Asn-Gly sequences are susceptible to cleavage by alkaline hydroxylamine, owing to the tendency of this residue combination to form a cyclic di-imide between the side-chain carbonyl carbon of the asparagine and the amide nitrogen of the glycine.[41] (The added flexibility of the main chain at glycine, removing steric constraints against formation of the ring structure, accounts for the specificity of the second rcsiduc, although minor cleavages at Asn-*X*, where *X* is usually a small residue, are sometimes seen.) Formation of the cyclic imide may also occur in peptides during sequencing. In this situation, any subsequent ring opening before coupling with the original Asn residue will yield aspartate in place of asparagine. But an Asp derivative will only be released if the imide is hydrolyzed between the original side-chain carbonyl carbon and the amide nitrogen of the glycine. If, instead, the original peptide bond hydrolyzes, a β-peptide bond results. From the standpoint of Edman chemistry, if either a cyclic imide or a β-peptide bond is encountered, no cleavage takes place, and degradation of that particular chain stops. For these reasons, if Asn-Gly is encountered during Edman degradation, a drop in yield *may* be seen. However, given the frequency of Asn-Gly linkages, about 1.2 in 400,[39] this is, at worst, an infrequent concern.

The low expected frequency of Asn-Gly does, however, offer the possibility of obtaining a single cleavage in a blocked protein chain of 400 to 500 residues. Because the hydroxylamine cleavage reaction may be performed on PVDF-adsorbed material, and the procedure is not complicated, any effort with hydroxylamine on a blocked sample of sufficient size is probably time well spent. In terms of the methodologies, attempting cleavage at either Asp-Pro or Asn-Gly is not complicated, particularly if the sample is blotted, and the frequencies of these residue pairs is about equal. However, the hydroxylamine cleavage requires preparation of 4.5 *M* LiOH, and LiOH rapidly forms the poorly soluble carbonate salt on exposure to air, whereas cleavage at Asp-Pro simply entails incubation in 88% formic acid overnight at 45°C; thus, any success won through cleavage at Asp-Pro will be gained with less effort.

It should be noted that similar possibilities for cleavage do not exist with Gln-Gly sequences. (This is in keeping with Jane and David Richardson's description of asparagines as "quirky and opinionated" in their excellent review of the roles of amino acids in protein tertiary structure,[42] while glutamine was considered a "relatively indifferent plain vanilla residue that goes reasonably well with almost anything.") Multiple glutamine residues, however, can complicate the efficiency of the degradation, presumably through some level of cyclization to pyroglutamate as each new N-terminal Gln is revealed during the degradation.

Cleavage at Trp

A variety of reagents have been described for chemical cleavage at tryptophan residues; all can be classified as oxidative halogenations, and some have been applied to transblotted material. Given the frequency of tryptophan at ~1.2%, a more limited number of fragments should be expected vs. cleavage at Met. However, none of the reactions appears to be robust, and differing percentages of cleavage at different tryptophans in the same protein should be expected. A further disadvantage is that the stench of the more popular reagent, BPNS-skatole, will win you the instant enmity of your co-workers. For these reasons, attempting to cleave at Trp should probably be a last resort.

In-Gel Digestion

Rather than deal with even the possibility of a blocked N-terminus, many laboratories prefer instead to perform in-gel digestion on polyacrylamide gel-separated bands from the start. Even with a free N-terminus, an in-gel digest offers the ability to obtain sequence from numerous internal sites vs. just one from the intact protein without further work. Given the odds of N-terminal blockage from samples from eukaryotic sources, in-gel digestion offers a means of preventing loss of material to an unsuccessful degradation. Further, by skirting the blotting process, transfer losses at that step are avoided.

A variety of methods have been published over the past decade, but the most popular versions involve washing any residual SDS from the stained gel pieces with 50% acetonitrile followed by desiccation.[43] The dried gel pieces are rehydrated with an initial aliquot of enzyme followed by further additions of buffer as the gel pieces re-swell, drawing the protease further into the polyacrylamide. Then, after a suitable incubation interval, the resultant peptides are leached from the gel pieces with 60% acetonitrile, concentrated, and applied to HPLC separation.

The protease most commonly used for in-gel digests is endoprotease Lys-C, which cleaves after the carbonyl group of lysine residues with high fidelity. The enzyme from *Achromobacter lyticus*, available from Wako (Richmond, VA) is preferred by most laboratories for this purpose. [44] There are two reasons. First, lysine residues occur with a frequency such that an abundance of peptides ~8 to 20 residues long can be expected. A sequence of 8 residues is usually sufficient to provide good identification of a protein if its sequence exists in a database, and an exact match of upwards of 20 residues is generally considered unequivocal identification. The corollary to this is that peptides of this length generally present no problem to separate on a C-18 reversed-phase column. Second, the protease is both robust and rugged, and of a size (~25 kDa) that enables it to penetrate the pores of typical polyacrylamide gels. An additional advantage of Lys-C is that its specificity is reliable enough that a lysine residue preceding the first residue of each peptide can be inferred with good confidence. This can be a particular advantage if the objective is to obtain enough sequence to clone the protein because only two codons exist for lysine.

Other N-Terminal Anomalies

Ragged Ends

A sample that passes the hurdle of N-terminal blockage may still present a problem. The sample may consist of a mixture of products from the same gene that starts at different points. For instance, some portion of proteins in your preparation might start at position 1 and the other at position 2, and this may reflect partial N-terminal proteolysis. Depending on the ratios of the two chains, the degradation may seem either to preview the residue in the subsequent cycle or to suffer from abnormally large lag or carryover of the derivative from the preceding cycle. The chief clue that this is the case and not a problem with the degradation is the appearance of two residues in the first cycle. In these cases, the eventual rise in overall background coupled with lagging residues can diminish the number of cycles that can be called confidently. To overcome such problems, or in any other case of Edman sequencing multiple chains simultaneously, if a proline residue is encountered in the run, a second sample can be run and stopped at the end of the cycle that releases the residue preceding proline. At this point, fluorescamine or o-phthalaldehyde (OPA) may be applied to the sample.[45,46] These reagents irreversibly react with primary amines (all amino acids except proline) but not secondary amines (proline), thus blocking all other protein species in the reaction cell, while leaving all chains with proline currently at the N-terminus for further cycles of Edman degradation.

α-Amino Alkylation

An uncommon modification that complicates Edman sequencing is N-alkylation.[47,48] In contrast to N-acylation, which generates an amide bond, N-alkylation converts the α-amino group to a secondary amine, and internal secondary amines couple with PITC just as primary amines do.[49] This is fortunate. Otherwise, Edman degradation would stop at every proline residue because proline is a cyclic secondary amine. However, the empirical consequence of coupling to a noncyclic secondary amine is that a significant degree of spontaneous cyclization and cleavage occurs during the coupling phase of the reaction, thus liberating a new N-terminus which will also couple with PITC still present. The amount of spontaneous cleavage will be reflected in the amount of the PTH derivative representing the second residue in the chain, while the derivative of the first residue will have an anomalous HPLC mobility owing to the alkyl substituent remaining with its PTH derivative. Furthermore, if the alkyl substituent is more than a methyl group, the PTH may not be seen at all if hydrophobicity is added sufficient to cause elution at a percent organic solvent for elution higher than the end of the standard chromatographic gradient. The good news is that N-α-alkylation is rare and you are unlikely ever to encounter it.

POST-TRANSLATIONAL MODIFICATIONS AT INTERNAL RESIDUES

The two most common post-translational modifications of widespread interest are likely to be phosphorylation and glycosylation. With identification of a phosphorylation site (see below), it is generally expected that it should be possible to sequence a ^{32}P-labeled sample and follow the amount of label released at each cycle of the degradation. In principle, there is no problem with this approach, but several issues make this approach less straightforward than would be expected.

PHOSPHORYLATION

Identification of protein phosphorylation sites is relevant in many systems. The usual approach is to first determine whether Ser, Thr, or Tyr phosphorylation applies to your system of interest. This is typically approached by hydrolysis of the ^{32}P-labeled protein in 6 N HCl followed by two-dimensional chromatography of the hydrolysate and autoradiography of the chromatography plate with comparison to stained, non-labeled standards.[50, 51] Preliminary qualitative information may also be of value to rule in or out phosphoserine on the basis of its alkali lability, in contrast to the other two phosphorylated residues.[23, 50] Serine is widely acknowledged as the most frequently phosphorylated residue, although estimates of its frequency are not easily found. This is likely owing in part to the fact that phosphorylation is typically a dynamic process owing to the continuous action of kinases and phosphatases resulting in nonstoichiometric modification of any particular residue of a given protein at any particular time in a cell. Still, from the analytical standpoint, the frequency of serine phosphorylation is fortunate because it is also the most easily modified for subsequent assay. Direct identification of a ^{32}P-labeled PTH derivative by simply measuring the radioactivity of the PTH derivatives from each cycle may seem to be the logical approach, but this is compromised by the instability of phosphoserine which undergoes β-elimination to dehydroalanine.[52] Throughout Edman degradation of a phosphorylated peptide, phosphate is released. It is not uncommon to recover $<5\%$ of the label originally applied at the cycle actually corresponding to the phosphorylated residue. Added to this is the short half-life of ^{32}P. For these reasons, it is analytically advantageous to first convert all phosphorylated serines to S-ethyl cysteine using ethanethiol.[53] This approach also circumvents the use of ^{32}P, enabling the derivative to be detected as an ordinary amino acid with unique chromatographic mobility.

Unfortunately, phosphothreonine, which also undergoes β-elimination and modification by ethanethiol, is converted to a derivative that presents analytical complications.[53] Often, if it can be established that a phosphorylated protein contains only one type of phosphorylated residue, identification can be made on the basis that the sequence of the labeled peptide corresponds to a segment with only a single Ser, Thr, or Tyr. When this deductive approach is not possible, direct detection of the labeled residue by measuring radioactivity of each cycle is additionally complicated by the extreme polarity of the phosphorylated PTH derivative, hindering its extraction with the organic solvents used in the sequencer. To offset this problem, TFA or phosphate salts are sometimes added to the extraction solvent when running phosphopeptides. This strategy is particularly applicable to tyrosine, which does not undergo β-elimination and is, therefore, not susceptible to modification with ethanethiol.

Glycosylation

A frequent question is whether Edman degradation stops when a glycosylated residue is encountered. Because glycosylation involves modification of side chains and not peptide bonds, the answer is no. When a glycosylated residue is encountered in a degradation, the residue cyclizes and cleaves normally (although, perhaps, less efficiently). However, the polarity of the glycosylated PTH derivative is so extreme that it usually will not extract from the reaction cell, and results in a blank cycle where no residue assignment can be made. If the sequence can be matched in a database or otherwise and the blank cycle corresponds to a Ser or Asn residue, glycosylation of that residue can be inferred, particularly if, in the case of N-linked glycosylations (linked to an asparagine residue), that Asn occurs in an Asn-X-Thr/Ser sequence. As a means of more directly identifying N-linked sites, pretreatment of the sample with the endoglycosidase Endo-F or Endo-H removes all but the terminal N-acetylglucosamine.[18] In these cases, the glycosylated PTH derivative may show up as an unknown peak on the HPLC chromatograms.

The above discussion of two frequent post-translational modifications illustrates the issues that need to be considered in identifying other post-translationally modified amino acid residues.

- Is the modified residue stable in conditions it will be subjected to during Edman degradation?
- Is its polarity such that it can be efficiently transferred from the reaction cell to the conversion flask and then to the HPLC?

SUMMARY

The intent of this chapter has been to explain the capabilities of Edman chemistry as well as to outline some of the precautions necessary when contemplating use of the methodology. Because, in this limited space, exhaustive treatment of any topic is impossible, references have been selected largely to expand treatment of the various topic.

It should be apparent that the textbook example of Edman analysis of a pure protein sample with an unmodified N-terminus is often not attainable without additional effort. Still, combining an investigator's expertise in sample purification with the expertise of personnel in sequencing laboratories should result in positive outcomes from this technology at very low levels of sample.

ACKNOWLEDGMENT

The author is particularly grateful to Dr. Henry Krutzsch (National Institutes of Health) for his helpful comments.

ABBREVIATIONS USED

ATZ, anilinothiazolinone
CAPS, 3-[cyclohexylamino]-1-propanesulfonic acid
DABITC, dimethylaminoazobenzene isothiocyanate
DABTH, dimethylaminoazobenzene thiohydantoin
OPA, o-phthalaldehyde
PTC, phenylthiocarbamyl
PTH, phenylthiohydantoin
PVDF, polyvinylidine difluoride
TFA, trifluoroacetic acid

REFERENCES

1. Edman, P., A method for the determination of the amino acid sequence in peptides, *Arch. Biochem.*, 22, 475–476, 1949.
2. Schroeder, W., Degradation of peptides by the Edman method with direct identification of the PTH-amino acid, *Meth. Enzymol.*, 11, 445–475, 1967.
3. Edman, P. and Henschen, A., Sequence determination, in *Protein Sequence Determination*, Needleman, S. B., Ed., Springer-Verlag, New York; also listed as *Mol. Biol. Biochem. Biophys.,* 8, 232–279, 1975.
4. Pisano, J. J., Analysis of amino acid phenylthiohydantoins by gas chromatography and high-performance liquid chromatography, in *Protein Sequence Determination*, Needleman, S. B., Ed. Springer-Verlag, New York, also listed as *Mol. Biol. Biochem. Biophys.*, 8, 280–296, 1975.
5. Konigsberg, W., Subtractive Edman degradation, *Meth. Enzymol.*, 25, 326–332, 1972.
6. Gray, W. R., Sequence analysis with dansyl chloride, *Meth. Enzymol.*, 25, 332–344, 1972.
7. Tarr, G. E., Improved manual sequencing methods, *Meth. Enzymol.*, 47, 335–357, 1977.
8. Laursen, R. A., Automatic solid-phase Edman degradation, *Meth. Enzymol.*, 25, 344–359, 1972.
9. Tarr, G., Beecher, J. F., Bell, H., and McKean, D. J., Polyquaternary amines prevent peptide loss from sequenators, *Anal. Biochem.*, 84, 622–627, 1978.
10. Zimmerman, C. L., Appella, E., and Pisano, J. J., Rapid analysis of amino acid phenylthiohydantoins by high performance liquid chromatography, *Anal. Biochem.*, 77, 569–573, 1977.
11. Chang, J. Y., Manual micro-sequence analysis of polypeptides using dimethylaminoazobenzene isothiocyanate, *Meth. Enzymol.*, 91, 455–466, 1983.
12. Inman, J. K. and Appella, E., Identification of anilinothiazolinones after rapid conversion to N-α-phenylthiocarbamylamino acid methylamides, *Meth. Enzymol.*, 47, 374–384, 1977.
13. Farnsworth, V., Porton Instruments presentation, Smith-Klein Beecham, Philadelphia, October, 1991.

14. Coligan, J. E., Gates, F. T., III, Kimball, E. S., and Maloy, W. L., Radiochemical sequence analysis of biosynthetically labeled proteins, *Meth. Enzymol.*, 91, 413–434, 1983.
15. Wittman-Liebold, B., Graffunder, H., and Kohls, H., A device coupled to a modified sequenator for the automated conversion of anilinothiazolinones into PTH amino acids, *Anal. Biochem.*, 75, 621–633, 1976.
16. Hunkapiller, M. W., Hewick, R. M., Dreyer, W. J., and Hood, L. E., High-sensitivity sequencing with a gas-phase sequenator, *Meth. Enzymol.*, 91, 399–413, 1983.
17. Walsh, K. A, Ericsson, L. H., Parmelee, D. C., and Titani, K., Advances in protein sequencing, *Ann, Rev. Biochem.*, 50, 261–284, 1981.
18. Allen, G., *Sequencing of Protein and Peptides in Laboratory Techniques in Biochemistry and Molecular Biology*, Vol. 9, Elsevier, Amsterdam, 1989, 414 pp.
19. Bhoun, A. S., Ed., *Protein/Peptide Sequence Analysis*, CRC Press, Boca Raton, FL, 1997, 256 pp.
20. Coligan, J. E., Dunn, B. M., Ploegh, H. L., Speicher, S. W., and Wingfield, P. T., Eds., Chemical analysis, in *Current Protocols in Protein Science*, John Wiley and Sons, New York, 1997.
21. Smith, B. J. Ed., *Protein Sequencing Protocols*, Humana Press, Totowa, NJ, 1997.
22. LeGendre, N., Mansfield, M., Weiss, A., and Matsudaira, P., Purification of proteins and peptides by SDS-PAGE, in *A Practical Guide to Protein and Peptide Purification for Microsequencing*, 2nd ed., Matsudaira, P., Ed., Academic Press, San Diego, 1998, 71–101.
23. Smith, B. J., SDS-polyacrylamide gel electrophoresis for N-terminal protein sequencing, in *Protein Sequencing Protocols*, Smith, B. J., Ed., Humana Press, Totowa, NJ, chap. 2, 1997, 17–24.
24. ProMega Corp., *Probe-Design Separation System*, Technical Manual, 1991.
25. Stark, G. R., Use of cyanate for determining amino-terminal residues in proteins, *Meth. Enzymol.*, 11, 125–138, 1967.
26. Krutzsch, H. C. and Inman, J. K., N-isopropyliodoacetamide in the reduction and alkylation of proteins: use in microsequence analysis, *Anal. Biochem.*, 209, 109–116, 1993.
27. Doolittle, R., *Of Urfs and Orfs*, University Science Books, Mill Valley, CA, 1987.
28. Marti, T., Rösselet, S. J., Titani, K., and Walsh, K., Identification of disulfide-bridged substructures within human von Willebrand factor, *Biochemistry*, 26, 8099–8109, 1987.
29. Carne, A. F. and Carne, U. S., Chemical modification of proteins for sequence analysis, *Protein Sequencing Protocols*, Smith, B. J., Ed., Humana Press, Totowa, NJ, 1997, chap. 26, pp. 271–284.
30. Means, G. E. and Feeney, R. E., *Chemical Modification of Proteins*, Holden-Day, San Francisco, 1971.
31. Hempel, J., Unpublished.
32. Carlstedt-Duke, J, Strömstedt, P.-E., Persson, B., Cederlund, E., Gustafsson, J.-Å., and Jörnvall, H., Identification of hormone-interacting amino acid residues within the steroid-binding domain of the glucocorticoid receptor in relation to other steroid hormone receptors, *J. Biol. Chem.*, 263, 6842–6848, 1988.
33. Hempel, J., von Bahr-Lindström, H., and Jörnvall, H., Aldehyde dehydrogenase from human liver: primary structure of the cytoplasmic isoenzyme, *Eur. J. Biochem.*, 141, 21–35, 1984.

34. Klein, M. L., Bartley, T. D., Davis, J. M., Whiteley, D. W., and Lu, H. S., Isolation and structural characterization of three forms of recombinant consensus α-interferon, *Arch. Biochem. Biophys.*, 276, 531–537, 1990.
35. Persson, B., Flinta, C., von Heijne, G., and Jörnvall, H., Structures of N-terminally acetylated proteins, *Eur. J. Biochem.*, 152, 523–527, 1985.
36. Flinta, C., Persson, B., Jörnvall, H., and von Heijne, G., Sequence determinants of cytosolic N-terminal protein processing, *Eur. J. Biochem.*, 154, 193–196, 1986.
37. Hirano, H., Komatsu, S., and Tsunasawa, S., On-membrane deblocking of proteins, in *Protein Sequencing Protocols*, Smith, B. J., Ed., Humana Press, Totowa, NJ, 1997, chap. 27, 285–292.
38. Gheorge, M. T. and Bergman, T., Deacetylation and internal cleavage of polypeptides for N-terminal sequence analysis, in *Methods in Protein Structure Analysis*, Atassi, M. Z. and Appella, E., Eds., Plenum Press, New York, 1994.
39. Based on the cumulative compositions of 195, 891 proteins in the PIR databank (67.9 million residues) at this writing.
40. Landon, M., Cleavage at aspartyl-prolyl bonds, *Meth. Enzymol.*, 47, 145–149, 1977.
41. Bornstein, P. and Balian, G., Cleavage at Asn-Gly bonds with hydroxylamine, *Meth. Enzymol.*, 47, 132–145, 1977.
42. Richardson, J. S. and Richardson, D. C., Principles and patterns of protein conformation, in *Prediction of Protein Structure and Principles of Protein Conformation*, Fasman, G. D., Ed., Plenum Press, New York, 1989, pp. 1–98.
43. Hellman, U., Wernstedt, C., Góñez, J., and Heldin, C.-I., Improvement of an "In-gel" digestion procedure for the micropreparation of internal protein fragments for amino acid sequencing, *Anal. Biochem.*, 224, 451–455, 1994.
44. Commentary posted on the Association of Biological Resource Facilities (ABRF) electronic bulletin board, archives available at *http://www.abrf.org.*
45. Bhown, A. S., Bennett, J. C., Morgan, P. H., and Mole, J. E., Use of fluorescamine as an effective blocking reagent to reduce the background in protein sequence analyses by the Beckman automated sequenator, *Anal. Biochem.*, 112, 158–162, 1981.
46. Spiess, J., Rivier, J., and Vale, W., Sequence analysis of rat hypothalamic corticotropin-releasing factor with the o-phthalaldehyde strategy, *Biochemistry*, 22, 4341–4346, 1988.
47. Hermodson, M. A., Chen, K. C. S., and Buchanan, T. M., *Neisseria* pili proteins: amino-terminal amino acid sequences and identification of an unusual amino acid, *Biochemistry*, 17, 442–445, 1978.
48. Chang, J.-Y., A novel Edman-type degradation: direct formation of the thiohydantoin ring in alkaline solution by reaction of Edman-type reagents with N-monomethyl amino acids, *FEBS Lett.*, 91, 63–68, 1978.
49. Hempel, J., Nilsson, K., Larsson, K., and Jörnvall, H., Internal chain cleavage and product heterogeneity during Edman degradation of isosteric peptide analogs lacking the α-carbonyl function, *FEBS Lett.*, 194, 333–337, 1986.
50. Duclos, B., Marcandier, S., and Cozzone, A. J., Chemical properties and separation of phosphoamino acids by thin-layer chromatography and/or electrophoresis, *Meth. Enzymol.*, 201, 10–21, 1991.
51. Aitken, A. and Learmonth, M., Analysis of sites of protein phosphorylation, in *Protein Sequencing Protocols*, Smith, B. J., Ed., Humana Press, Totowa, NJ, 1997, chap. 28, pp. 293–306.
52. Meyer, H. E., Hoffmann-Posorske, E., and Heilmeyer, L. M. G., Determination and location of phosphoserine in proteins and peptides by conversion to S-ethylcysteine, *Meth. Enzymol.*, 201, 169–185, 1991.

53. Meyer, H. E., Eisermann, B., Heber, M., Hoffmann-Posorske, E., Korte, H., Weigt, C., Wegner, A., Hutton, T., Donella-Deana, A., and Perich, J. W., Strategies for nonradioactive methods in the localization of phosphorylated amino acids in proteins, *FASEB J.*, 7, 776–782, 1993.

6 Computer Modeling of Protein Structure

Jonathan D. Hirst

CONTENTS

Introduction 123
Sequence Analysis 125
Motifs 125
Sequence Alignment 127
Substitution Matrices 127
Pairwise Alignment 127
Multiple Sequence Alignment 129
Secondary Structure Prediction 129
Early Methods 130
Recent Methods 130
Transmembrane Proteins 131
Tertiary Structure Prediction 132
Comparative Modeling 133
Fold Recognition 134
Fold Libraries 135
Potentials for Fold Recognition 135
Fold Recognition Algorithms 135
De Novo Folding 136
Results 136
Comparative Modeling 137
Fold Recognition 137
De Novo Prediction 137
Conclusions 138
Recommended Reading 138
References 138

INTRODUCTION

Protein structure is most accurately revealed by the experimental techniques discussed elsewhere in this book. That this is true is, in part, a reflection on our limited understanding of protein structure. Nevertheless, computer modeling has much to offer the protein chemist. Theoretical methods continue to advance, urged on by

0-8493-9453-8/02/$0.00+$1.50

genome sequencing efforts, which fuel the disparity between the amount of sequence data and the amount of structural data. Computer hardware is becoming cheaper and faster, and the World Wide Web has put most scientists just a mouse click away from access to some of the latest methods in protein structure prediction.

The hierarchical nature of protein structure suggests an obvious form for this chapter and for the computer modeling of protein structure (Figure 6.1). We begin with sequence analysis, progress to secondary structure analysis, and conclude with tertiary structure prediction. We will forgo the customary introduction to protein sequence and structure, which can be found in many excellent textbooks.[1,2] The confines of this chapter also preclude the level of detailed discussion available in longer review articles[3,4] and books.[5] We do not attempt an exhaustive listing of available resources and algorithms. Instead, we provide a more practical sample of some of the more established methods. The appearance of a particular method in this chapter is not necessarily an endorsement. In protein structure prediction, there are few guarantees. We can, however, safely recommend that the readers who wish to model protein structure would be well advised to develop a consensus from several approaches and consult the literature cited in this chapter for further details.

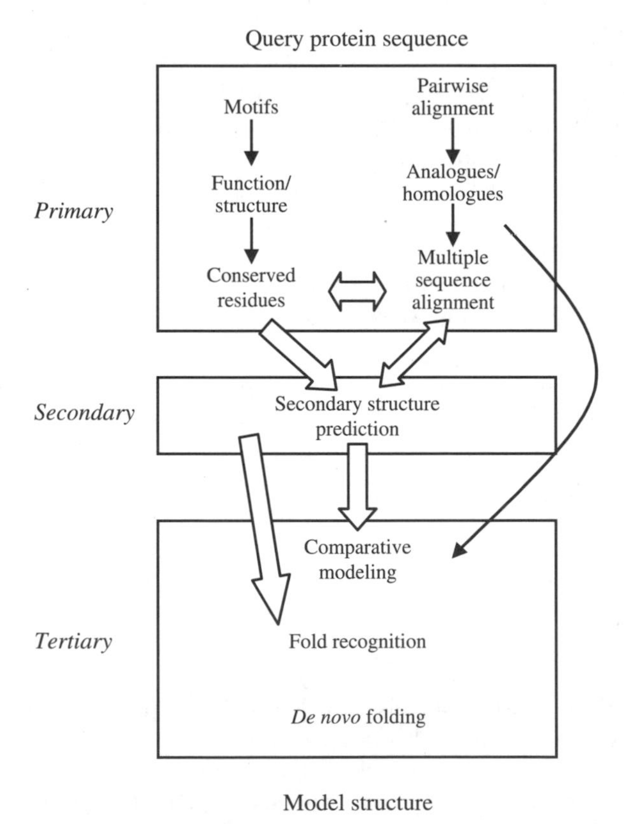

FIGURE 6.1 Modeling approaches: from sequence to structure, block arrows indicate where information from one approach may be utilized in another.

SEQUENCE ANALYSIS

The first resources for computer modeling of protein structure are the nucleic acid and protein sequence databases (see Table 6.1), curated by the European Molecular Biology Laboratory (EMBL) in Europe, the National Center for Biotechnology Information (GenBank at the NCBI) in the United States, and the DNA Database of Japan (DDBJ) in Japan. These databases are accessible via the Internet, and most likely one's own scientific institution maintains a local version, which is updated through CD-ROMs released quarterly. Perhaps the predominant protein sequence database is SWISS-PROT.[6, 7] Others include the nonredundant protein sequence database (OWL)[8] and the protein identification resource database (PIR).[9, 10]

Searches of these databases are the first steps toward learning if a novel protein sequence has any relationship to known proteins that may be exploited. If there is a relationship, and there may well not be, it may be a common sequence signature from which one may infer common structure or function. Less likely, but much more revealing, the novel protein sequence may have a significant level of similarity with a known protein, possibly indicating that the proteins are homologous (i.e., related by divergent evolution). Throughout this chapter, we will see that one adopts different modeling strategies depending on the strength of the relationship between the protein to be modeled and known proteins. Generally, the stronger this relationship, the more reliable the model.

MOTIFS

A sequence motif is a well-conserved group of amino acids in a specific region of the protein sequence. Sequences with a common motif may be otherwise quite dissimilar. As many of these motifs are involved in catalysis or ligand binding,[11–13] their presence in a protein can indicate putative function. An example of a GTP/ATP-binding motif[14] is shown in Figure 6.2. Scanning a new sequence against motif databases can thus uncover useful information relating to function or structure.

The PROSITE database[13, 15] of protein sequence motifs is a standard. The motifs are represented by patterns and profiles. Patterns are denoted using the single letter amino acid alphabet, with each position separated by a hyphen, e.g., A-D-E. X is used if any residue is allowed. If several residues are allowed at a given position, they are grouped in square brackets (e.g., [AC]-D-E). Disallowed residues are grouped in curly brackets: A-D-{FGH}. A repeat is indicated by its length after the

TABLE 6.1
Some Sequence Databases

Database	Web Sites
GenBank	*http://www.ncbi.nlm.nih.gov/GenBank/Overview.html*
EMBL	*http://www.ebi.ac.uk/embl.html*
DDBJ	*http://www.ddbj.nig.ac.jp*
SWISS-PROT	*http://www.expasy.ch*
OWL	*http://bmbsgi11.leeds.ac.uk/bmb5dp/owl.html*
PIR	*http://www-nbrf.georgetown.edu*

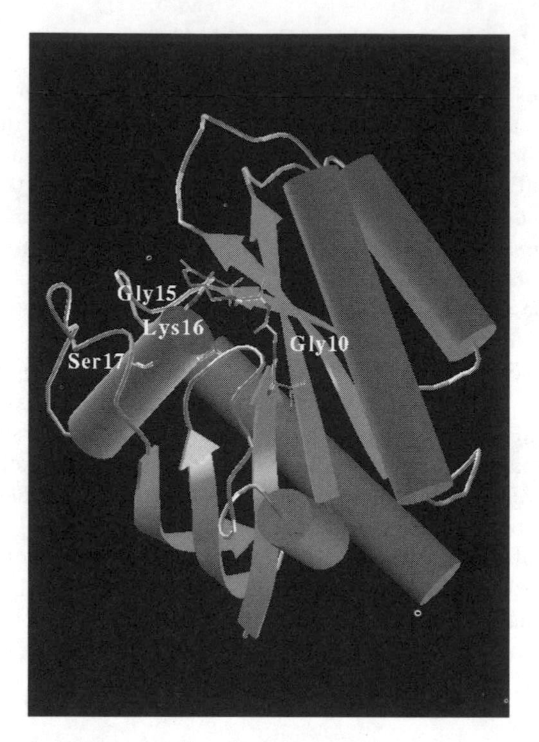

FIGURE 6.2 A GTP-/ATP-binding motif, the P-loop, G-X(4)-G-K-[ST], in human ras 21 protein.[99] This figure and Figure 6.4 were produced using the graphics program PREPI, available upon request from its developer, Dr. Suhail Islam (Imperial Cancer Research Fund).

residue: A-P(4)-E, or by a range: A-P(4, 8)-E. Profiles are, in a sense, a generalization of patterns and are defined by a weighting of each residue type at each position in the motif. The sites and patterns in PROSITE are well documented and have a relatively high specificity. Release 15, July 1998, contains 1025 documented entries that describe 1361 different patterns, rules, and profiles. PROSITE may be searched online, by downloading the database by ftp, or within sequence analysis packages.

Another database is PRINTS,[16, 17] a compendium of motifs identified from conserved regions of multiple familial sequence alignments. Such a definition can provide greater sensitivity than a single PROSITE pattern. A conceptually related database, BLOCKS, comprises multiply aligned ungapped segments from highly conserved regions of proteins.[18, 19] Relevant Web sites are given in Table 6.2.

As well as these general motif databases, there are some more specialized sequence analysis tools for identifying particular structural or functional features. One well-studied example is that of coiled coils. The program COILS[20] allows one to compare a sequence against profiles derived from known parallel two-stranded coiled coils, to assess the likelihood that the query sequence is indeed a coiled coil. More involved analyses based on pairwise interactions have been developed, such as PAIRCOIL.[21] Several families of coiled coils are actually well recognized by their PROSITE motifs. However, the leucine zipper PROSITE motif, L-X(6)-L-X(6)-L-X(6)-L, is poorly discriminating, and for this specific family of coiled coils, the program TRESPASSER provides a more refined identification

TABLE 6.2
Sequence Motif Databases

Database	Web Sites
PROSITE	*http://www.expasy.org/prosite*
	http://www.2.ebi.ac.uk/ppsearch
PRINTS	*http://www.biochem.ucl.ac.uk/bsm/ddbrowser*
BLOCKS	*http://www.bioinf.man.ac.uk/dbbrowser*

tool.[22] Thus, additional information can sometimes allow one to use more specialized analysis tools.

Sequence Alignment

Sequence alignment is an invaluable tool and often provides a starting point for more involved modeling. An alignment of two sequences is a measure of their similarity. A high similarity may lead to the inference that there are structural characteristics in common between the two sequences.

Substitution Matrices

Usually, alignment methods go beyond a simple counting of the identical residues to capture in a quantitative fashion the idea that some mutations are more disruptive (and, thus, less likely) than others. For example, glycine to tryptophan is an unlikely mutation, whereas leucine to isoleucine is much more tolerable. An all vs. all comparison of amino acids gives a 20×20 symmetric matrix, called a substitution matrix. Based on the substitution frequencies observed in protein sequences related by evolution, Dayhoff[23] developed a series of substitution matrices, known as PAM matrices. The evolutionary distance is reflected in the percentage of acceptable point mutations per 10^8 years (PAM). For a given evolutionary distance (say 250 PAM), normalizing the mutation probability matrix and taking the logarithm of each element give a log odds matrix often referred to as the PAM 250 Dayhoff matrix. The distance of 250 PAM is widely used because it corresponds to the limit of detection of distant relationships, where about 80% of amino acid positions have mutated. The PAM 250 Dayhoff matrix has been popular for many years, although there are now some, perhaps, better and more modern alternatives, such as the PET91 matrix[24] and the BLOSUM matrices.[25] As usual, for an important modeling project, it is worth exploring several approaches.

Pairwise Alignment

The simplest comparison of two sequences does not consider gaps in the alignments, corresponding to insertions or deletions (collectively known as indels) in one sequence relative to the other. This type of alignment is rarely useful, and we turn directly to consider alignments with gaps. A standard approach is the dynamic programming algorithm, developed for protein sequence comparison by Needleman and Wunsch.[26] Consider two protein sequences, A and B of lengths m and n, respectively. In dynamic programming, a matrix, **M**, is constructed with *m* rows and *n* columns. The elements of the matrix, M_{ij}, give the similarity between residue *i* in sequence *A* and residue *j* from sequence *B*, for example, based on the PAM 250 Dayhoff matrix. Different pathways through the matrix correspond to different alignments, and the score of an alignment is the sum of the matrix elements along the pathway. Dynamic programming finds the optimal alignment by working along each sequence successively, finding the highest score for aligning the subsequences $A_{1,\ldots,i}$ with $B_{1,\ldots,j}$.

A gap corresponds to matching a residue from one sequence with nothing from the other sequence. Gaps normally receive a negative score, called the gap penalty. Without gap penalties, the optimal Needleman-Wunsch alignment can contain overly long gaps. The Smith-Waterman algorithm provides for gaps[27] and is relatively sensitive, although also quite slow. FASTA[28] is a faster, but more approximate algorithm that has gained popularity. Short runs of identity between two sequences are used to identify a subset of promising alignments, which are then analyzed more rigorously. Another popular tool is BLAST,[29] which is very fast. It is based on the statistics of ungapped sequence alignments,[30] and, thus, has the drawback of not allowing gaps. Figure 6.3 shows an alignment of myoglobin with hemoglobin.

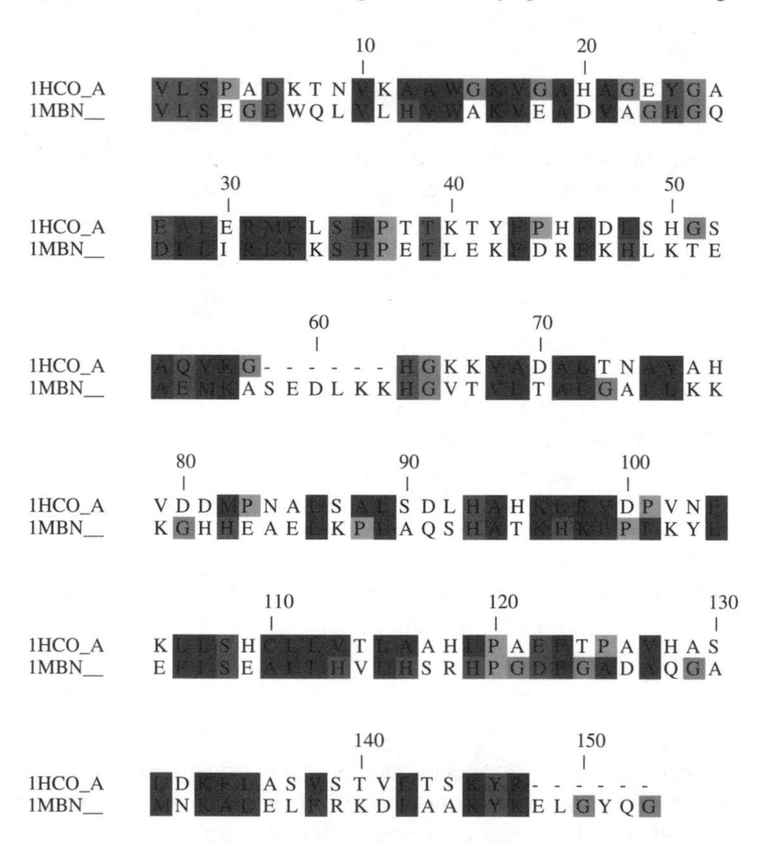

FIGURE 6.3 An alignment of myoglobin and hemoglobin sequences generated by the ClustalW sequence alignment interface at the European Bioinformatics Institute (EBI) server (see Table 6.3). Residues are color coded by type.

The significance of an alignment score may be assessed by comparison with a distribution of alignment scores with random sequences of the same length and amino acid composition. For a protein of about 100 to 200 residues, a score lying 15 standard deviations above the mean of the distribution is a close to perfect alignment. Scores above 5 standard deviations are often good enough to be useful. Percentage identity is another measure of quality, but its significance varies with the sequence length. A given percentage identity is more significant for a longer sequence.

Different alignment searches can give different results. So it is worth searching several databases, to develop the fullest list of significant matches. A search against the Protein Data Bank (PDB)[31] is also worthwhile, so that any similarity to a protein of known structure is readily apparent. The results from the alignment searches should be manually screened to eliminate obvious errors. If there are enough genuine matches, a multiple sequence alignment is the next step.

Multiple Sequence Alignment

Trends are often more apparent in a multiple sequence alignment than in a pairwise alignment, and they build confidence in inferences about the presence of domains; the location of conserved, functionally important residues, and the prediction of secondary structure elements. The most practical multiple alignment algorithms are extensions of pairwise methods.[32] All sequences are compared pairwise and the pairwise data are clustered. The multiple alignment is then built starting with the most similar pair and extended through decreasing similarity. Once a pairwise alignment is made, then it is fixed and its average is used in subsequent steps in the multiple alignment. Methods for multiple sequence alignment include Clustal W[33] and AMAS.[34] Table 6.3 lists some useful Web sites.

SECONDARY STRUCTURE PREDICTION

The most obvious structural building blocks in proteins are the elements of secondary structure, the α-helix and the β-strand. These elements arise, in part, through the formation of intramolecular hydrogen bonds. Secondary structure prediction usually attempts to assign the conformational state of a residue as helix (H), extended (E), or coil (C), based on the local sequence (i.e., the surrounding residues). However, other ordered structures, such as β-turns[35] and Ω-loops, do exist.[36] Secondary

TABLE 6.3
Alignment Methods

Database	Web Sites
FASTA	*http://www2.ebi.ac.uk/fasta3*
BLAST	*http://www.ncbi.nlm.nih.gov/BLAST/*
	http://www2.ebi.ac.uk/blast2/
Clustal W	*http://www2.ebi.ac.uk/clustalw*
AMAS	*http://barton.ebi.ac.uk/servers/amas_server.html*

structure elements can often be well conserved in sequence alignments, and an accurate prediction of secondary structure can be a critical step in the development of a reasonable model of the tertiary structure. Secondary structure prediction is a long-standing problem that continues to attract interest, although recent improvements have been relatively modest.[37] We focus first on prediction methods for globular proteins and then consider transmembrane proteins.

Early Methods

Three popular early methods are due to Chou and Fasman,[38] Garnier et al.,[39] and Lim.[40] The Chou-Fasman method[38] is based on statistical propensities of residues to form an α-helix or a β-strand, combined with a series of rules. Many methods, like Chou-Fasman, are based on statistical analyses of structural databases. The GOR method[39] adopts an information theory approach. Although the theory is a little daunting, this is transparent to the user and the method has a firm statistical foundation that permitted extension to more subtle analyses of a larger structural database.[41] Methods may also be based on the physico-chemical principles underlying protein structure. For example, α-helices often have a distinct pattern of hydrophobic residues, leading to a hydrophobic face that can facilitate the packing of the helix in the protein. A series of such rules forms the basis of the Lim method.[40]

These early methods were compared to one another in a study[42] that alerted the scientific community to the importance of careful validation of protein structure prediction tools. Predictive performance should be assessed on proteins not used in the development of a method. In other words, the data should be divided into a training and a testing set. This division may be done multiple times in several independent trials, such as in crossvalidation, to obtain a more robust estimate of the predictive accuracy. It may, in fact, be useful also to exclude from the evaluation homologues of proteins in the training set. It has been further suggested (somewhat controversially) that methods can only be properly assessed on genuinely unknown structures, with predictions only confirmed subsequently by experiment.[4] We will return to this point in the Results section of this chapter. While the issues in this paragraph are directly pertinent to developers of prediction tools, users should also be aware of them, so that they can judge new methods with the appropriate circumspection.

Recent Methods

The literature on methods for secondary structure prediction is too extensive for us to review here. A host of methods have been applied to the problem and much of the relevant work is cited in review articles.[3, 37, 43, 44] Here we focus on a few methods that can be accessed through Web servers (see Table 6.4). The methods are all capable of utilizing multiple sequence alignments, which can enhance the accuracy of the prediction by about 10%.[45]

One widely used modern method is PHD.[46] The program incorporates evolutionary information, in the form of multiple sequence alignments, to develop a neural network method that predicts secondary structure with an accuracy >70%. Knowledge of neural networks is not necessary for the practical user of this method, and the methodology is not discussed in detail here. Instead, the interested reader should consult reviews of the use of neural networks in protein structure prediction.[47–49]

TABLE 6.4
Secondary Structure Prediction Methods

Method	Web Sites
PHD	*http://www.embl-heidelberg.de/predictprotein/ppDoPredDef.html*
PROF (DSC sequel)	*http://www.aber.ac.uk/~phiwww/prof*
PREDATOR	*http://www.embl-heidelberg.de/cgi/predator_serv.pl*

The general scheme of PHD illustrates how different sources of information can be incorporated into secondary structure prediction. The program PHD uses three levels of neural networks. The input to the first level is the frequency with which each amino acid occurs at each position in a 13-residue window in the alignment. The target output is the secondary structure class of the central residue. The second neural network takes account of the correlation of secondary structure between consecutive residues. It is unlikely that a single β-strand residue occurs within a stretch of helical residues. The second neural network, thus, predicts the secondary structure of a residue, given the secondary structure predictions from the first neural network for that residue and its neighbors. A number of predictions are made based on different parameter settings, and the last neural network combines these predictions in a jury decision, to give a final prediction.

Neural networks, in principle, provide a powerful means for detecting nonlinear relationships in data. It is not clear, however, that the protein structure database is large enough for nonlinear relationships to be manifested in a general fashion. The basis of predictions made by linear methods is much more readily understood, and it appears that such methods can be used without a loss of accuracy. The algorithm DSC is one such method.[50] It is based on a linear discriminant analysis of a variety of properties of the sequence. For each residue the following are calculated: the GOR information parameters, the distance to the end of the chain, moments of hydrophobicity, the presence of insertions or deletions in the multiple alignment, and the degree of conservation at that position. The results of the analysis are smoothed and filtered using some simple rules to remove unrealistic predictions. Another useful prediction method, PREDATOR, is based on long-range contact potentials.[51] For prediction from a set of sequences, multiple sequence information is incorporated through the best nonoverlapping pairwise alignments.

The prediction methods can be sensitive to the quality of the multiple sequence alignment. It is worthwhile paying particular attention to the multiple sequence alignment to avoid propagating errors in the development of a structural model. The general strategy of using several methods to construct a consensus prediction applies both in multiple sequence alignment and in secondary structure prediction.

Transmembrane Proteins

Most of this chapter is devoted to the modeling of globular proteins because very few structures of membrane proteins have been experimentally determined. To date, the structures that have been determined belong to only two different classes, helical

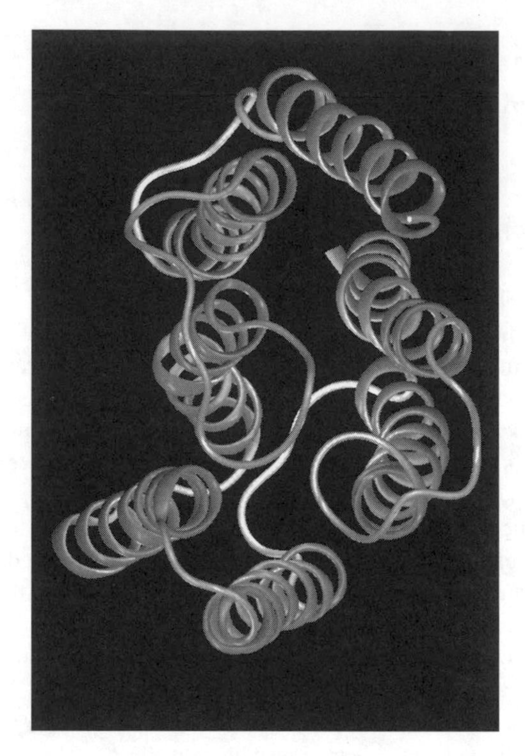

FIGURE 6.4 PREPI ribbon diagram of bacteriorhodopsin—a transmembrane seven helical bundle.

bundles (e.g., bacteriorhodopsin,[52]) (Figure 6.4) or β-barrels (e.g., porin). Modeling of membrane proteins often focuses on the prediction of transmembrane helices. These helices are identifiable from plots of the hydrophobicity along the amino-acid sequence. A classic scale of hydrophobicity is that of Kyte and Doolittle,[53] where isoleucine is the most hydrophobic residue and arginine is the least. The positive inside rule,[54, 55] which requires that the majority of positively charged residues (arginine and lysine) are on the cytoplasmic side (interior) of the transmembrane segment, may then be used to enhance the prediction. Such an approach, as adopted in the TOP-PRED algorithm,[56] and other more involved methods[57, 58] can achieve predictions of transmembrane helices that reach 95% accuracy.

TERTIARY STRUCTURE PREDICTION

In tertiary structure prediction, there are three distinct scenarios. The sequence to be modeled may be closely related to a protein whose structure has been determined experimentally. Here, there is the prospect of developing a relatively accurate structural model through comparative modeling. This prospect drops markedly with the strength of the relationship between the sequence to be modeled and known structures. As this relationship weakens, one moves from the realm of comparative modeling to an approach known as fold recognition. Here, one attempts to utilize weak or subtle similarities between proteins with the same tertiary structure or fold. Analyses of known protein structures suggest that there are a limited number of different possible folds.[59] As experimental structure determinations reduce the number of undiscovered folds, new sequences are more likely to be related, albeit perhaps distantly, to a protein with known fold. Improving the sensitivity and accuracy of fold recognition is currently an active area of research. Finally, it may well be that the sequence and structure databases offer no help. For small proteins, one might consider *de novo* protein folding to build a model. However, this approach is very much an area of research, not application. We briefly discuss recent efforts in this exciting field, but *de novo* folding is not yet a practical method for modeling protein structure. Table 6.5 lists some resources for teritary structure prediction.

Comparative Modeling

Proteins with a degree of similarity may be homologous, related by divergent evolution, or analogous, related by convergent evolution. Comparative modeling is a broader term for homology modeling, which avoids the connotation that the proteins must have a common ancestor. One of the earliest examples of this approach was the model structure of α-lactalbumin based on the x-ray structure of hen egg-white lysozyme.[60] An archetypal study was the work of Greer on serine proteases.[61] Recently, comparative modeling has been applied to entire genomes. All atom structural models have been built for over 1000 proteins (17%) in the *Saccharomyces cerevisiae* yeast genome.[62] The fraction of sequences out of the genome gives an idea of the likely applicability of comparative modeling, and the sheer number of models reflects the speed and automation of current methods.

The comparative modeling approach has been reviewed in detail.[63] The process begins with the identification of known structures related to the sequence to be modeled. A sequence alignment is then constructed. A model of the backbone

TABLE 6.5
Tertiary Structure Prediction Methods and Resources

Method/Resource	Web Sites
PDB	*http://www.rcsb.org*
PROCHECK	*http://www.biochem.ucl.ac.uk/~roman/procheck/procheck.html*
COMPOSER	*http://www-cryst.bioc.cam.ac.uk*
MODELLER	*http://guitar.rockefeller.edu/modeller/modeller.html*
SWISS-MODEL	*http://www.expasy.ch/swissmod/SWISS-MODEL.html*
SCOP	*http://scop.mrc-lmb.cam.ac.uk/scop*
CATH	*http://www.biochem.ucl.ac.uk/bsm/cath*
FSSP	*http://www2.ebi.ac.uk/dali/fssp/fssp.html*
MMDB	*http://www.ncbi.nlm.nih.gov/Structure/MMDB/mmdb.html*
THREADER	*http://insulin.brunel.ac.uk/threader/threader.html*
TOPITS	*http://www.embl-heidelberg.de/predictprotein/ppDoPredDef.html*
CASP	*http://predictioncenter.llnl.gov/casp2/Casp2.html*
	http://predictioncenter.llnl.gov/casp3

is created, based on the alignment. Side chains are added and their conformations modeled. The modeled structure is refined using a molecular mechanics force field. Finally, the reliability of the model is estimated, based on the properties of known structures.

The quality of the alignment is again a central issue and is worth particular attention. Modeling of structure in loop regions is difficult, because of greater flexibility and the lack of regularity, particularly for loops longer than eight residues. Short loops may be modeled through direct evaluation of possible geometries[64, 65] or by reference to known loop structures from the PDB.[66, 67] Another tricky element is modeling the rotameric states of side chains. This is also often based on conformational preferences observed in the PDB.[68, 69] Good combinations of side-chain rotamers can be identified using a heuristic search called dead-end elimination.[70]

Once a model has been created, a variety of stereochemical properties should be confirmed to lie within acceptable bounds, including the distribution main chain dihedral angles, the planarity of peptide bonds, bond lengths, and angles. These and other properties can be checked with the program PROCHECK.[71] Energy minimization is then applied to relieve unfavorable contacts or so-called van der Waals clashes. To retain the basic structure of the model, minimization is often performed in a gradual fashion, whereby the positions of the main chain atoms are initially constrained and only allowed to move in a subsequent minimization. As a final check, inverse folding methods discussed in the next section can be applied to identify regions containing possible errors, as exemplified by the approach of Eisenberg and co-workers using their three-dimensional profiles.[72] Here, the compatibility of a sequence with its structure is assessed based on the structural environment presented to each residue. Compatibility may be calculated as a function of residue number, permitting the identification of incompatible regions.

Homology modeling can be performed within several different packages, including COMPOSER (Tripos), MODELLER (Molecular Simulations Inc.), and Insight-Homology. It is also possible now to submit a sequence or an alignment directly to a fully automatic procedure SWISS-MODEL on the World Wide Web. It is simply a matter of pasting the desired sequence into a Web page, clicking the "submit" button and waiting for the email from the server, which is usually quite prompt. Where the sequence homology is greater than 50%, this procedure yields excellent results. Below 50%, one should try to send an alignment of the best possible quality.

Fold Recognition

Fold recognition, inverse protein folding, and threading are all terms to describe the modeling of protein structure through the evaluation of a sequence in a set of particular structures.[37,73] The inverse folding problem, "What sequences adopt a given structure?" is, in a sense, a reverse formulation of the folding problem, "What structure does this given sequence adopt?" In the context of fold recognition, one already has a particular sequence of interest and if it is related to a known fold, then a search through the vast space of protein conformation may be reduced to a search through a set or library of experimentally determined structures. This search may be thought of as the evaluation of the sequence threaded through each structure in turn.[74] With the conformational search essentially eliminated in the fold recognition approach, half the problem is solved. The other half of the problem, an accurate potential for discriminating correct folds from incorrect folds, is not yet solved completely. However, there has been sufficient progress in the field to make fold recognition a potentially useful modeling tool.

Fold Libraries

The starting point for fold recognition is the classification of protein structures into folds.[59] The fold or topology of a protein is a general description of the type of secondary structure elements in a protein and its architecture or how the secondary structure elements are arranged (e.g., β-barrels, helical bundles). The fold also describes the order of the secondary structure elements and their directions. Several databases have been created that classify protein structures. A manually constructed hierarchical classification forms the basis of the SCOP database.[75, 76] A more, although not completely, automated process is used to construct the CATH database.[77, 78] Fully automated procedures are used to generate FSSP[79] and the Molecular Modeling Database (MMDB).[80]

Potentials for Fold Recognition

The potential functions used to evaluate or score a sequence in a structure are usually empirical potentials of mean force estimated from statistical analysis of the distribution of residue–residue distances in the PDB.[81] The potentials are less detailed than full atomic potentials, often treating each residue as a single interaction site. Solvation is the key aspect that these potentials try to capture. Many different potentials have been developed,[3] and new variants continue to be tested.[73, 82] Most potentials appear to be comparable.

Fold Recognition Algorithms

A straightforward and popular method uses dynamic programming (see the section on pairwise alignment) and a simplification known as the frozen approximation.[83, 84] An alignment matrix for the dynamic programming treatment is calculated by replacing amino acids in the template structure one at a time with amino acids from the sequence to be modeled. The frozen approximation is the assumption that the field of the template structure will also favor the correct replacement from the threaded sequence, so that for each substitution, the rest of the template structure is unchanged. More sophisticated methods that avoid this approximation do not seem to perform better than the frozen approximation. This is probably more a reflection on the lack of sensitivity of current methods, rather than the adequacy of this seemingly severe approximation.

There are a variety of fold recognition programs that are available, and the field is expanding rapidly. A sample includes THREADER,[74] TOPITS[85] and Match-

Maker.[83] The field is too young for a definitive standard, and the methods tend to perform better in the hands of experts, although this, perhaps, is changing.[86] The recurring recommendation is that one develops a consensus prediction from several methods with several homologous sequences, if they are available.

De Novo Folding

De novo prediction of tertiary structure might be described as the least practical, but most direct approach, to the modeling of protein structure. *De novo* folding is also known as *ab initio* folding, which is somewhat of a misnomer because the potentials are not based on first principles but are empirically derived from known structures. However, the *ab initio* label is probably destined to last. *De novo* folding usually comprises a simplified potential, a means of reducing conformational space, and a method for searching this space. The potentials are often similar to those used for fold recognition, although greater care needs to be paid to their physical correctness. Conformational space may be reduced by restricting the protein chain to lie at points on a lattice and conformations that lie on this lattice may be searched using the Monte Carlo method.[87] This type of approach has shown some promise for simulating the folding of small α-helical and mixed α/β proteins.[88] An alternative to a discrete Cartesian space is to search dihedral angle space.[89] Very recently, a search method called the conformational space annealing method[90] was applied to the *de novo* folding of the 10–55 fragment of staphylococcal protein A and apo calbindin D9K using a united residue force field.[91]

RESULTS

The assessment and comparison of different methods for the computer modeling of protein structure has been to some degree bedeviled by the lack of standard measures and benchmarks (as alluded to earlier). Many scientists in the field are, of course, perfectly capable of making objective assessments of their own work. However, there are clear merits afforded by predictions for which the experimental result is confirmed only subsequently. Partly in recognition of the preceding points, the protein structure prediction community increasingly participates in a scientific meeting/competition on the Critical Assessment of Techniques for Protein Structure Prediction (CASP). This community-wide experiment provides a convenient summary of the results that one might expect from computer modeling of protein structure. The third of these meetings was held in 1998. The only published analysis of the CASP3 results to date is fairly preliminary,[92] so our discussion is largely based on the published CASP2 results.[93]

The organizers of CASP2 obtained 34 protein structures that were yet to be released into the public domain. Participants received the sequences of these 34 targets and were allowed to submit selected predictions. Based on the target, predictions were classified into comparative modeling, fold recognition, *de novo* prediction, and docking. We summarize the results in the first three categories.

Comparative Modeling

Comparative modeling at CASP2 was assessed based on nine target structures.[94] The sequence identity with the principal parent ranged from 20% (a difficult case) to 85% (an easy case). The difficulty of the modeling problem was reflected in the accuracy of the sequence alignments, which ranged from 100 to <10%. The quality of the structural model is highly dependent on the quality of the alignment. For the easiest target, the majority of approaches gave a root mean square deviation (RMSD) between the true structure and the model of <0.5 Å. For most other targets, models of 3 Å RMSD or better were often obtained. The quality of the side-chain placement was dependent on the quality of the backbone model. Below 20% sequence identity, the models become less accurate. In one case, a poor sequence alignment led to a model with an RMSD of 19 Å. Another cautionary note is that the highest sequence identity parent is not necessarily the best starting point for comparative modeling.[95] Similar trends were seen in CASP3, although for each target in CASP3, at least one prediction was made where the RMSD between the positions of the α-carbon atoms in the prediction and the experimentally determined structure was within 2 Å for 60% or more of the residues.

Fold Recognition

A total of seven CASP2 targets had previously known folds and were thus amenable to fold recognition approaches. As the RMSD between two structures increases to 5 Å and beyond, assessing their similarities in a definitive manner becomes problematic. This difficulty somewhat hampers the assessment of fold recognition methods. In his assessment of the results from CASP2, Levitt,[96] based the accuracy of individual models on the number of correctly aligned residues to a correctly recognized fold. The accuracy of his methods was based on weighted averages of fold recognition and alignment accuracy, with some accounting for the difficulty of the targets. The better predictions led to structural models with a RMSD for the α-carbon atoms in the range of 5 to 10 Å, with up to 80% of residues correctly aligned to the correct fold. Poorer predictions failed to identify the correct fold or had very large RMSD differences from the true structure. The most robust method was based on THREADER,[74] which gave reasonable models for five of the amenable targets. The most accurate models were predicted manually from expert knowledge,[97] suggesting that further information could be utilized in automatic approaches.

De Novo Prediction

The CASP2 assessment in this category[98] confirmed that the best secondary structure prediction methods are about 70% accurate, and that *de novo* prediction of tertiary structure is not yet a practical method for modeling protein structure. Out of 12 targets, only the structure of one, pig NK-lysin was predicted with any accuracy. In CASP3, the structures of 5 out of 15 targets were predicted with at least 60% of residues within 6 Å RMSD of the true structure. This suggests that methods are improving. However, the growth of sequence and structure databases and the consequent greater applicability and accuracy of comparative modeling and fold recognition methods may preclude *de novo* prediction from having a practical impact on structure prediction.

CONCLUSIONS

In this chapter, we have presented a broad overview of the current state of computer modeling of protein structure, and provided some pointers to the relevant literature. By browsing some of the Internet sites and related links, the curious reader can learn more about what is accessible from his or her personal computer. The Internet may be quite adequate for straightforward modeling problems. For nontrivial projects, modeling is probably best performed on local machines. A standard workstation or a high-end personal computer with graphics capabilities would be needed. A lot of useful software may be obtained on request or downloaded by ftp. There are also various commercial packages that can be very helpful. Expert advice is also bound to be helpful. Finally, one should consult the reviews cited in this chapter and the references therein.

RECOMMENDED READING

Sternberg, M. J. E., Ed., *Protein Structure Prediction*: *A Practical Approach*, IRL Press, Oxford, 1996.

Vásquez, M., Némethy, G., and Scheraga, H.A., Conformational energy calculations on polypeptides and proteins, *Chem. Rev.*, 94, 2183, 1994.

REFERENCES

1. Creighton, T. E., *Proteins—Structures and Molecular Principles*, W. H. Freeman, New York, 1983.
2. Cantor, C. R. and Schimmel, P. R., *Biophysical Chemistry.* Vol. II, W. H. Freeman, New York, 1980.
3. Eisenhaber, F., Persson, B., and Argos, P., Protein structure prediction: recognition of primary, secondary, and tertiary structural features from amino acid sequence, *Crit. Rev. Biochem. Mol. Biol.*, 30, 1, 1995.
4. Benner, S. A., Cannarozzi, G., Gerloff, D., Turcotte, M., and Chelvanayagam, G., *Bona fide* predictions of protein secondary structure using transparent analyses of multiple sequence alignments, *Chem. Rev.*, 97, 2725, 1997.
5. Sternberg, M. J. E., Ed., *Protein Structure Prediction. A Practical Approach*, IRL Press, Oxford, 1996.
6. Bairoch, A. and Boeckmann, B., The SWISS-PROT protein sequence data bank, *Nucleic Acids Res.*, 19, 2247, 1991.
7. Bairoch, A. and Apweiler, R., The SWISS-PROT protein sequence data bank and its supplement TrEMBL in 1999, *Nucleic Acids Res.*, 27, 49, 1999.
8. Sidman, K. E., George, D. G., Barker, W. C., and Hunt, L. T., The protein identification resource (PIR), *Nucleic Acids Res.*, 16, 1869, 1988.
9. Bleasby, A. J. and Wootton, J. C., Construction of validated, non-redundant composite protein sequence databases, *Protein Eng.*, 3, 153, 1990.

10. Barker, W. C., Garavelli, J. S., McGarvey, P. B., Marzec, C. R., Orcutt, B. C., Srinivasarao, G. Y., Yeh, L.-S. L., Ledley, R. S., Mewes, H.-W., Pfeiffer, F., Tsugita, A., and Wu, C., The PIR-international protein sequence database, *Nucleic Acids Res.*, 27, 39, 1999.
11. Sternberg, M. J. E., Library of common protein motifs, *Nature*, 349, 111, 1991.
12. Hodgman, T. C., The elucidation of protein function by sequence motif analysis, *Comput. Appl. Biosci.*, 5, 1, 1989.
13. Bairoch, A., *Prosite: a Dictionary of Protein Sites and Patterns*, 5th ed., Department de Biochimie Medicale, Universite de Geneve, Geneva, 1990.
14. Saraste, M., Sibbald, P. R., and Wittinghofer, A., The P-loop—a common motif in ATP- and GTP-binding proteins, *Trends Bio. Sci.*, 15, 430, 1990.
15. Hofmann, K., Bucher, P., Falquet, L., and Bairoch, A., The PROSITE database its status in 1999, *Nucleic Acids Res.*, 27, 215, 1999.
16. Attwood, T. K. and Beck, M. E., PRINTS—a protein motif fingerprint database, *Prot. Eng.*, 7, 841, 1994.
17. Attwood, T. K., Flower, D. R., Lewis, A. P., Mabey, J. E., Morgan, S. R., Scordis, P., Selley, J., and Wright, W., PRINTS prepares for the new millennium, *Nucleic Acids Res.*, 27, 220, 1999.
18. Henikoff, S. and Henikoff, J. G., Protein family classification based on searching a database of blocks, *Genomics*, 1, 97, 1994.
19. Henikoff, J. S., Henikoff, S., and Pietrokovsko, S., New features of the blocks database servers, *Nucleic Acids Res.*, 27, 226, 1999.
20. Lupas, A., van Dyke, M., and Stock, J., Predicting coiled coils from protein sequences, *Science*, 252, 1162, 1991.
21. Berger, B., Wilson, D. B., Wolf, E., Tonchev, T., Milla, M., and Kim, P. S., Predicting coiled coils by use of pairwise residues correlations, *Proc. Natl. Acad. Sci. U.S.A.*, 92, 8259, 1995.
22. Hirst, J. D., Vieth, M., Skolnick, J., and Brooks, C. L., III., Predicting leucine zipper structures from sequence, *Prot. Eng.*, 9, 657, 1996.
23. Dayhoff, M. O., Atlas of protein sequence and structure, in *Atlas of Protein Sequence and Structure*., National Biomedical Research Foundation, Washington, D.C., 1, 1978.
24. Jones, D. T., Taylor, W. R., and Thornton, J. M., The rapid generation of mutation data matrices from protein sequences, *Comput. Appl. Biosci.*, 8, 275, 1992.
25. Henikoff, S. and Henikoff, J. G., Amino acid substitution matrices from protein blocks, *Proc. Natl. Acad. Sci. U.S.A.*, 89, 10915, 1992.
26. Needleman, S. B. and Wunsch, C. D., A general method applicable to the search for similarities in the amino acid sequences of two proteins, *J. Mol. Biol.*, 48, 443, 1970.
27. Smith, T. F. and Waterman, M. S., Identification of common molecular substructures, *J. Mol. Biol.*, 147, 195, 1981.
28. Pearson, W. R. and Lipman, D. J., Improved tools for biological sequence comparison, *Proc. Natl. Acad. Sci. U.S.A.*, 85, 2444, 1998.
29. Altschul, S. F., Gish, W., Miller, W., Myers, E. W., and Lipman, D. J., Basic local alignment search tool, *J. Mol. Biol.*, 215, 403, 1990.
30. Karlin, S. and Altschul, S. F., Methods for assessing the statistical significance of molecular sequence features by using general scoring schemes, *Proc. Natl. Acad. Sci. U.S.A.*, 87, 2264, 1990.
31. Bernstein, F. C., Koetzle, T. F., Williams, G. J. B., Meer, E. F., Brice, M. D., Rodgers, J. R., Kennard, O., Shimanouchi, T., and Tasumi, M., The Protein Databank: a computer-based archival file for macromolecular structures, *J. Mol. Biol.*, 112, 535, 1977.

32. Barton, G. J. and Sternberg, M. J. E., A strategy for the rapid multiple alignment of protein sequences—confidence levels from tertiary structure comparisons, *J. Mol. Biol.*, 198, 327, 1987.
33. Thompson, J. D., Higgins, D. G., and Gibson, T. J., CLUSTAL W: improving the sensitivity of progressive multiple sequence alignment through sequence weighting, positions-specific gap penalties and weight matrix choice, *Nucleic Acids Res.*, 22, 4673, 1994.
34. Livingstone, C. D. and Barton, G. D., Protein sequence alignments: a strategy for the hierarchical analysis of residue conservation, *Comput. Appl. Biosci.*, 9, 745, 1993.
35. Wilmot, C. M. and Thornton, J. M., Analysis and prediction of the different types of β-turn in proteins, *J. Mol. Biol.*, 203, 221, 1988.
36. Leszczynski, J. F. and Rose, G. D., Loops in globular proteins: a novel category of secondary structure, *Science*, 234, 849, 1986.
37. Westhead, D. R. and Thornton, J. M., Protein structure prediction, *Curr. Opin. Biotech.*, 9, 383, 1998.
38. Chou, P. Y. and Fasman, G. D., Prediction of protein conformation, *Biochemistry*, 13, 222, 1974.
39. Garnier, J., Osguthorpe, D. J., and Robson, B., Analysis of the accuracy and implications of simple methods for predicting the secondary structure of globular proteins, *J. Mol. Biol.*, 120, 97, 1978.
40. Lim, V. I., Algorithms for prediction of alpha-helical and beta-structural regions in globular proteins, *J. Mol. Biol.*, 88, 873, 1974.
41. Gibrat, J. F., Garnier, J., and Robson, B., Further developments of protein secondary structure prediction using information theory. New parameters and consideration of residue pairs, *J. Mol. Biol.*, 198, 425, 1987.
42. Kabsch, W. and Sander, C., How good are predictions of protein secondary structure?, *FEBS Lett.*, 155, 179, 1983.
43. Sternberg, M. J. E., Secondary structure prediction, *Curr. Opin. Struct. Biol.*, 2, 237, 1992.
44. Barton, G. J., Protein secondary structure prediction, *Curr. Opin. Struct. Biol.*, 5, 372, 1995.
45. Zvelebil, M. J. J. M., Barton, G. J., Taylor, W. R., and Sternberg, M. J. E., Prediction of protein secondary structure and active sites using the alignment of homologous sequences, *J. Mol. Biol.*, 195, 957, 1987.
46. Rost, B. and Sander, C., Prediction of protein secondary structure at better than 70% accuracy, *J. Mol. Biol.*, 232, 584, 1993.
47. Hirst, J. D. and Sternberg, M. J. E., Prediction of structural and functional features of protein and nucleic acid sequences by artificial neural networks, *Biochemistry*, 31, 7211, 1992.
48. Presnell, S. R. and Cohen, F. E., Artificial neural networks for pattern recognition in biochemical sequences, *Ann. Rev. Biophys. Biomol. Struct.*, 22, 283, 1993.
49. Burns, J. A. and Whitesides, G. M., Feed-forward neural networks in chemistry: mathematical systems for classification and pattern recognition, *Chem. Rev.*, 93, 2583, 1993.
50. King, R. D. and Sternberg, M. J. E., Identification and application of the concepts important for accurate and reliable protein secondary structure prediction, *Prot. Sci.*, 5, 2298, 1996.
51. Frishman, D. and Argos, P., Incorporation of non-local interactions in protein secondary structure prediction from the amino acid sequence, *Prot. Eng.*, 9, 133, 1996.

52. Grigorieff, N., Ceska, T. A., Downing, K. H., Baldwin, J. M., and Henderson, R., Electron-crystallographic refinement of the structure of bacteriorhodopsin, *J. Mol. Biol.*, 259, 393, 1996.
53. Kyte, J. and Doolittle, R. F., A simple method for displaying the hydropathic character of a protein, *J. Mol. Biol.*, 157, 105, 1982.
54. von Heijne, G., Membrane proteins—the amino acid composition of membrane penetrating segments, *Eur. J. Biochem.*, 120, 275, 1981.
55. von Heijne, G. and Gavel, Y., Topogenic signals in integral membrane proteins, *Eur. J. Biochem.*, 174, 671, 1988.
56. von Heijne, G., Membrane protein structure prediction, *J. Mol. Biol.*, 225, 487, 1992.
57. Rost, B., Casadio, R., Fariselli, P., and Sander, C., Transmembrane helices predicted at 95% accuracy, *Prot. Sci.*, 4, 521, 1995.
58. Jones, D. T., Taylor, W. R., and Thornton, J. M., A model recognition approach to the prediction of all-helical membrane structure and topology, *Biochemistry*, 33, 3038, 1994.
59. Swindells, M. B., Orengo, C. A., Jones, D. T., Hutchinson, E. G., and Thornton, J. M., Contemporary approaches to protein structure classification, *BioEssays*, 20, 884, 1998.
60. Browne, W. J., North, A. C. T., Phillips, D. C., Brew, K., Vanaman, T. C., and Hill, R. L., A possible three-dimensional structure of bovine α-lactalbumin based on that of hen's egg-white lysozyme, *J. Mol. Biol.*, 42, 65, 1969.
61. Greer, J., Comparative model-building of the mammalian serine proteases, *J. Mol. Biol.*, 153, 1027, 1981.
62. Sánchez, R. and Sali, A., Large-scale protein structure modeling of the *Saccharomyces cerevisiae* genome, *Proc. Natl. Acad. Sci. U.S.A.*, 95, 13597, 1998.
63. Johnson, M. S., Srinivasan, N., Sowdhamini, R., and Blundell, T. L., Knowledge based protein modelling, *Crit. Rev. Biochem. Mol. Biol.*, 29, 1, 1994.
64. Go, N. and Scheraga, H. A., Ring closure and local conformational deformations of chain molecules, *Macromolecules*, 3, 178, 1970.
65. Bruccoleri, R. E. and Karplus, M., Prediction of the folding of short polypeptide segments by uniform conformational sampling, *Biopolymers*, 26, 137, 1987.
66. Jones, T. A. and Thirup, S., Using known substructures in protein model building and crystallography, *EMBO J.*, 5, 819, 1986.
67. Rufino, S. D., Donate, 1. E., Canard, L. H. C., and Blundell, T., Predicting the conformational class of short and medium sized loops connecting regular secondary structures: application to comparative modelling, *J. Mol. Biol.*, 267, 352, 1997.
68. Ponder, J. W. and Richards, F. M., Tertiary templates for proteins. Use of packing criteria in the enumeration of allowed sequences for different structural classes, *J. Mol. Biol.*, 193, 775, 1987.
69. Dunbrack, R. K. and Karplus, M., Backbone-dependent rotamer library for proteins. Application to side-chain prediction, *J. Mol. Biol.*, 230, 543, 1993.
70. De Maeyer, M., Desmet, J., and Lasters, I., All in one: a highly detailed rotamer library improves both accuracy and speed in the modelling of sidechains by dead-end-elimination, *Fold. Des.*, 2, 53, 1997.
71. Laskowski, P. A., MacArthur, M. W., Moss, D. S., and Thornton, J. M., PROCHECK: a program to check the stereochemical quality of protein structures, *J. Appl. Crystallogr.*, 2, 283, 1993.
72. Lüthy, R., Bowie, J. U., and Eisenberg, D., Assessment of protein models with three-dimensional profiles, *Nature*, 356, 83, 1992.

73. Jones, D. T., Progress in protein structure prediction, *Curr. Opin. Struct. Biol.*, 7, 377, 1997.
74. Jones, D. T., Taylor, W. R., and Thornton, J. M., A new approach to protein fold recognition, *Nature*, 358, 86, 1992.
75. Murzin, A. G., Brenner, S. E., Hubbard, T., and Chothia, C., SCOP: a structural classification of proteins database for the investigation of sequences and structures, *J. Mol. Biol.*, 247, 536, 1995.
76. Hubbard, T. J. P., Ailey, B., Brenner, S. E., Murzin, A. G., and Chothia, C., SCOP: a structural classification of proteins database, *Nucleic Acids Res.*, 27, 254, 1999.
77. Orengo, C. A., Michie, A. D., Jones, S., Jones, D. T., Swindells, M. B., and Thornton, J. M., CATH—a hierarchic classification of protein domain structures, *Structure*, 5, 1093, 1997.
78. Orengo, C. A., Pearl, F. M. G., Bray, J. E., Todd, A. E., Martin, A. C., Lo Conte, L., and Thornton, J. M., The CATH database provides insights into protein structure/function relationships, *Nucleic Acids Res.*, 27, 275, 1999.
79. Holm, L. and Sander, C., The FSSP database of structurally aligned protein fold families, *Nucleic Acids Res.*, 24, 206, 1996.
80. Gibrat, J. F., Madej, T., and Bryant, S. H., Surprising similarities in structure comparison, *Curr. Opin. Struct. Biol.*, 6, 377, 1995.
81. Sippl, M. J., Calculation of conformational ensembles from potentials of mean force. An approach to the knowledge-based prediction of local structures in globular proteins, *J. Mol. Biol.*, 213, 859, 1990.
82. Jones, D. T. and Thornton, J. M., Potential energy functions for threading, *Curr. Opin. Struct. Biol.*, 6, 210, 1996.
83. Godzik, A. and Skolnick, J., Sequence-structure matching in globular proteins: application to supersecondary and tertiary structure determination, *Proc. Natl. Acad. Sci. U.S.A.*, 89, 12098, 1992.
84. Ouzounis, C., Sander, C., Scharf, M., and Schneider, R., Prediction of protein structure by evaluation of sequence-structure fitness. Aligning sequences to contact profiles derived from three-dimensional structures, *J. Mol. Biol.*, 232, 805, 1993.
85. Rost, B. and Sander, C., Progress of 1D protein structure prediction at last, *Proteins*, 23, 295, 1995.
86. Edwards, Y. J. K. and Perkins, S. J., Assessment of protein fold predictions from sequence information: the predicted α/β doubly wound fold of the van Willebrand factor type A domain is similar to its crystal structure, *J. Mol. Biol.*, 260, 277, 1996.
87. Hinds, D. A. and Levitt, M., A lattice model for protein structure prediction at low resolution, *Proc. Natl. Acad. Sci. U.S.A.*, 89, 2536, 1992.
88. Kolinski, A. and Skolnick, J., Monte Carlo simulations of protein folding. II. Application to protein A, ROP, and crambin, *Proteins*, 18, 353, 1994.
89. Li, Z. Q. and Scheraga, H. A., Monte-Carlo-minimization approach to the multiple-minima problem in protein folding, *Proc. Natl. Acad. Sci. U.S.A.,* 84, 6611, 1987.
90. Lee, J., Scheraga, H. A., and Rackovsky, S., New optimization method for conformational energy calculations on polypeptides: conformational space annealing, *J. Comput. Chem.*, 18, 1222, 1997.
91. Lee, J., Liwo, A., and Scheraga, H. A., Energy-based *de novo* protein folding by conformational space annealing and an off-lattice united-residue force field: application to the 10–55 fragment of staphylococcal protein A and to apo calbindin D9K, *Proc. Natl. Acad. Sci. U.S.A.*, 96, 2025, 1999.
92. Koehl, P. and Levitt, M., A brighter future for protein structure prediction, *Nat. Struct. Biol.*, 6, 108, 1999.

93. Moult, J., Hubbard, T., Bryant, S. H., Fidelis, K., and Pedersen, J. T., Critical assessment of methods of protein structure prediction (CASP): round II, *Proteins,* Suppl. 1, 2, 1997.
94. Martin, A. C. R., MacArthur, M. W., and Thornton, J. M., Assessment of comparative modeling in CASP2, *Proteins*, Suppl. 1, 14, 1997.
95. Bates, P. A., Jackson, R. M., and Sternberg, M. J. E., Model building by comparison: a combination of expert knowledge and computer automation, *Proteins*, Suppl. 1, 59, 1997.
96. Levitt, M., Competitive assessment of protein fold recognition and alignment accuracy, *Proteins*, Suppl. 1, 92, 1997.
97. Murzin, A. and Bateman, A., Distant homology recognition using structural classification of proteins, *Proteins*, Suppl. 1, 105, 1997.
98. Lesk, A. M., CASP2: report on *ab initio* predictions, *Proteins*, Suppl. 1, 151, 1997.
99. Milburn, M. V., Tong, L., deVos, A. M., Brünger, A., Yamaizumi, Z., Nishimura, S., and Kim, S.-H., Molecular switch for signal transduction: structural differences between active and inactive forms of protooncogenic ras proteins, *Science*, 247, 939, 1990.

7 Genetic Analysis of Proteins

Jan Borén and Claes M. Gustafsson

CONTENTS

Introduction 146
Genetic Study of Multiprotein Complexes in *Saccharomyces cerevisiae* 146
Multiprotein Complexes 146
Gene Disruption in Yeast 149
Tagging Proteins in Yeast 150
Measurement of Global Transcription in Yeast 152
Studies of the Med2 Subunit of the Yeast Mediator Complex 153
Genetics to Study Single Proteins in Transgenic Mice 155
Designing Transgenes 155
Modification of Transgenes 155
Preparation of DNA for Micro-injection into the Pronucleus 157
Micro-injection of DNA into the Zygote 158
Analysis of Transgenic Animals 159
Analysis of the *In Vivo* Expression of the Transgene and the Phenotype of the Mouse 160
Structure–Function Studies of the Human ApoB 160
Studies of the Atherogenicity of ApoB 161
An Approach for Studying Gene Regulation by Distant DNA Elements 162
ApoB-Containing Lipoproteins Secreted by the Heart 163
Sample Protocols 163
Protocol 1 Large-Scale Protein Preparation from Yeast 163
Protocol 2 PCR-Mediated Gene Disruption in Yeast 164
Protocol 3 Transformation of Yeast with Lithium Acetate 164
Solutions 164
Method 165
Protocol 4 Rapid Preparation of DNA from Yeast 165
Protocol 5 Analysis of Total Cell Proteins from Yeast 166
Required Solutions 166
Method 166
Protocol 6 Isolation of Undegraded RNA from Yeast 166
Precautions to Minimize RNase Contamination 167

Required Solutions 167
Method 167
Protocol 7 RARE Cleavage of Large DNA 168
Protocol 8 P1 and BAC Miniprep Protocol with the Modified Alkaline Lysis Method 168
Protocol 9 Isolation of DNA Fragments for Pronuclear Injection 169
Protocol 10 Preparation of P1 Bacteriophage and BAC DNA for Pronuclear Injection 170
Preparation of P1 Bacteriophage and BAC DNA 170
Preparative Pulsed-Field Electrophoresis 171
Protocol 11 Mini DNA Preparation for PCR and/or Southern Analysis 172
Protocol 12 Quick Preparation of DNA for PCR Analysis 173
Buffers for Lysis before PCR 173
References 174

INTRODUCTION

Complete genome sequences of eukaryotic cells are now starting to become available. The genome of *Saccharomyces cerevisiae* was published in 1996, and recently the genome of the multicellular nematode *Caenorhabditis elegans* was presented.[1,2] At the present time, sequencing data for mouse and human genomes are accumulating at a rapid pace, and soon we will know the sequence of every gene in these genomes as well. Most of the genes identified in eukaryotic DNA sequencing efforts are uncharacterized open reading frames (ORFs). These are potential protein-coding regions for which no information exists regarding function. Sometimes the deduced amino acid sequences of the ORFs give clues about function, but often they do not. With "reverse genetics," it is possible to start with a cloned gene and, through various genetic methods, deduce the function of its product. Such techniques have an enormous impact on the biological and medical sciences today, and new methods are constantly being developed.

Here we will review some of the genetic techniques used to analyze protein function in two systems. In the first part, we will discuss some basic methods used in *S. cerevisiae* (e.g., protein tagging, gene disruption) and describe how these techniques can be used to study the function of individual subunits in multiprotein complexes. In the second part, we will review the use of transgenic mice to study protein function. In particular, techniques used for *in vitro* manipulation of large genes and the introduction of these into the mouse genome will be discussed.

GENETIC STUDY OF MULTIPROTEIN COMPLEXES IN *SACCHAROMYCES CEREVISIAE*

MULTIPROTEIN COMPLEXES

Many of the basic processes in the eukaryotic cell, such as DNA replication, transcription, and recombination, are carried out by multiprotein complexes (MPCs). The yeast origin recognition complex contains six individual subunits, of which five are needed for sequence-specific DNA binding.[3] Transcription of most genes coding for proteins in eukaryotic cells is carried out by RNA polymerase II holoenzyme, a complex of 32 individual subunits with a molecular mass of about 1.2 MDa.[4] Recently, several chromatin-modeling machines have been identified as MPCs with 10 to 20 subunits.[5] In spite of the great biological interest in many of these, comparatively little is known about their structure and function. There are several reasons for this. MPCs are often difficult to study in recombinant form, because the large number of interacting polypeptides makes them extremely difficult to reconstitute *in vitro*. In many cases, the individual MPC subunits also display low solubility when expressed in recombinant form, probably owing to the existence of sticky hydrophobic surfaces normally involved in protein–protein interactions with other subunits in the individual complexes. In almost all cases, active MPCs must, therefore, be purified directly from the eukaryotic cell. This fact often puts restraints on the amount of protein available and makes functional and structural studies problematic.

As an alternative to the pure biochemical approach, MPCs can be studied with a mix of genetic and biochemical techniques. *S. cerevisiae* has several important advantages that makes it an excellent system for this kind of scientific strategy. It is well established as a genetic system and refined to a point where researchers with no previous training in genetics can accomplish simple genetic manipulations in a short period of time (e.g., disrupting genes of interest or creating fusion proteins with specific sequence tags). With the sequencing of the entire *S. cerevisiae* genome complete, such genetic manipulations are even easier. From a protein biochemist's point of view, *S. cerevisiae* is very attractive. Large quantities of this simple eukaryote can be obtained quickly and at a low cost. The low cost is an important feature, because many of the most interesting MPCs are present at a comparatively low level in the eukaryotic cell, and protein preparations routinely start with 0.5 to 1.0 kg of cells (see Protocol 1).

One plan for a combined biochemical and genetic approach to study MPCs begins with the introduction of epitope tags, which are recognized by highly specific monoclonal antibodies (see Figure 7.1). These tags make purification and *in vivo* monitoring easier. Protein domains with specific affinities can also be fused to individual subunits and used for high-affinity protein purification of an entire complex. Purification of MPCs and subunit identification frequently identify products of previously uncharacterized open reading frames. Novel nonessential yeast genes can simply be deleted from the genome, and the resulting strain characterized. To analyze the functional organization of the gene, mutant alleles can be constructed *in vitro* and introduced into the deletion strain on replicating plasmid vectors.

Genetic analysis can give important information about the functional role of a specific gene, but it is often difficult to make definitive conclusions about the molecular basis for an observed phenotype. This is especially true for MPCs in which many proteins make up a functional unit and the deletion of one subunit can have significant consequences for the integrity of the whole or at least parts of the analyzed complex. To address this problem, MPCs from mutant strains can be purified and compared to the wildtype complex for biochemical activity and subunit composition. A careful comparison of genetic and biochemistry data will enable the

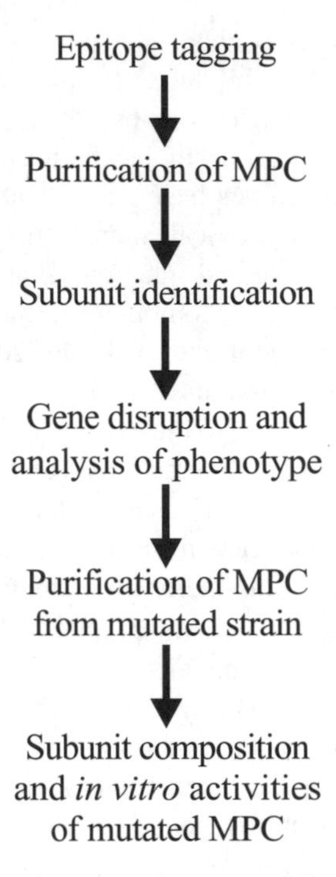

FIGURE 7.1 A combined biochemistry–genetics approach for studying multiprotein complexes.

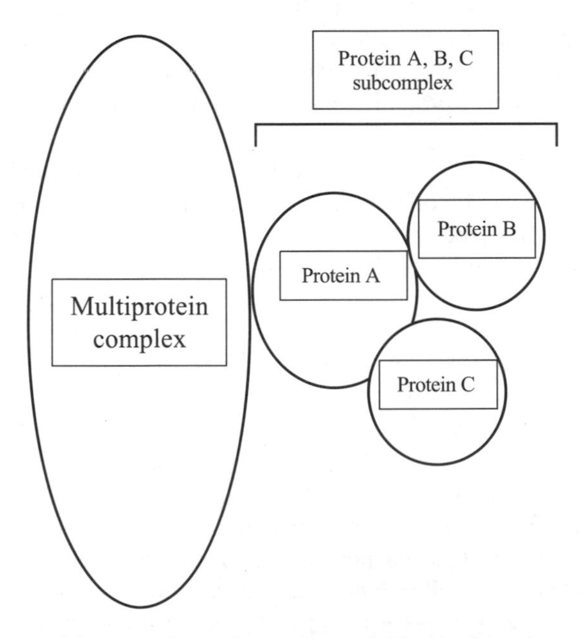

FIGURE 7.2 A careful comparison of genetics and biochemistry is needed for making predictions about which functional effects are direct or indirect. In this model, deletion of protein A will also lead to loss of proteins B and C from the multiprotein complex.

identification of subcomplexes within the MPC and make predictions about which functional effects are direct or indirect (see Figure 7.2). The existence of subcomplexes within a MPC can be substantiated with various forms of the two-hybrid method which allows monitoring of protein–protein interactions.[6] The DNA microarray technology provides another attractive possibility for the study of the protein function in yeast.[7] Recently, this technology has been used to study effects on global gene expression of specific mutations in various MPCs linked to transcription.[8, 9]

There are no strict rules for calling something an MPC. The definition is arbitrary and often based on genetic as well as biochemical observations. During the purification of the complex, the individual subunits should co-migrate over several different column types (e.g., ion-exchange and gel filtration). Additional evidence for a stable complex can be obtained by co-immunoprecipitation of the individual subunits under stringent conditions. Genetic data often provide important evidence as well, because the genes coding for the individual subunits of a specific MPC generally behave in a coherent way.

Gene Disruption in Yeast

Frequently, novel proteins identified as subunits of MPCs are products of previously uncharacterized ORFs. With reverse genetics, it is possible to start with such a cloned gene and through various genetic methods deduce its function. A cornerstone in reverse genetics is techniques for replacing the wildtype gene with various mutant forms generated *in vitro*. The mutant strains thus generated can be examined for interesting phenotypes (e.g., temperature sensitivity, ability to grow on various carbon sources) that can be correlated with the specific mutations introduced. Generation of mutant versions of a cloned nonessential gene is relatively easy. The wildtype gene can be deleted in its entirety from the yeast genome and mutant versions can be introduced on replicating plasmids. To study essential genes is more problematic (see below).

Homologous recombination is an extremely efficient process in *S. cerevisiae*. In fact, homologies of 30 bp on each side of a selectable marker are sufficient to obtain a large fraction of targeted integration events.[10] This fundamental property of budding yeast makes it easy to create deletion mutants for individual genes (see Protocol 2). The procedure begins with PCR amplification of a selective marker gene and oligonucleotide primers with homology to the gene of interest. This gives rise to a linear disruption construct in which the marker gene is flanked by short segments of the target gene. The construct is used to transform yeast cells, which are subsequently selected on appropriate drop-out medium (see Protocol 3).

Essential genes are more difficult to characterize because disruption of the wildtype gene leads to a nonviable phenotype. To avoid this obstacle, techniques have been developed for the generation of temperature-sensitive mutants in essential genes. These techniques work well in most cases and are described in manuals of general methods.

Recently, Moqtaderi and co-workers developed an alternative approach, the double-shutoff technique.[11] They used this method to study holo-TFIID, a MPC found in higher eukaryotes containing the TATA-binding protein (TBP) and TBP-associated

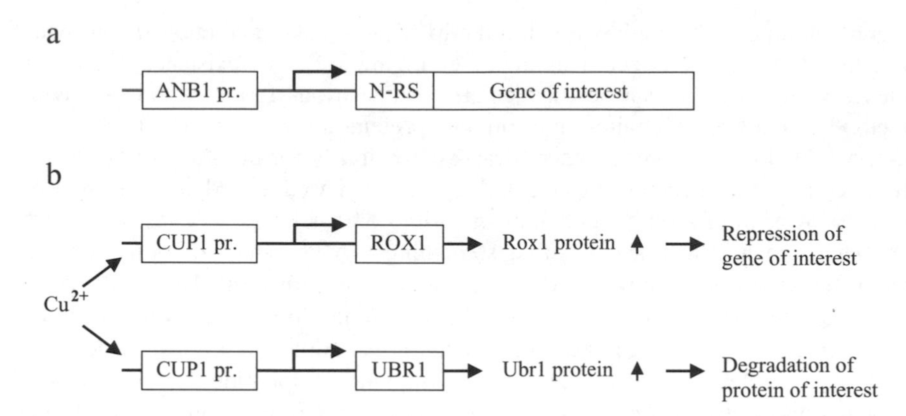

FIGURE 7.3 Creation of conditional mutants through the double-shutoff technique. (a) The gene of interest is fused to the N-end recognition signal (N-RS) for the rapid ubiquitin degradation pathway and is placed under the control of the *ANB1* promoter. (b) In the presence of copper, the *UBR1* and *ROX1* genes are expressed. *ROX1* represses transcription from the *ANB1* promoter and *UBR1* is needed for the rapid ubiquitin degradation pathway.

factors (TAFs). Holo-TFIID has been proposed to mediate the effects of activators on the general transcription machinery at almost all genes. Moqtaderi and co-workers could selectively deplete the cells for a number of these TAFs and provide evidence for a role for TAFs in transcription of only a subset of genes.[11]

In the double-shutoff technique, the yeast strain used has the two genes, *ROX1* and *UBR1*, under the control of a tightly regulated copper-inducible promoter (see Figure 7.3). *ROX1* encodes for a transcriptional repressor,[12] and the *UBR1* product is the N-terminal recognition protein involved in the ubiquitin degradation pathway.[13] The gene of interest is placed under the control of the *ANB1* promoter, which is effectively shut off in the presence of *ROX1*. This means that in the presence of copper, the *ROX1* will be expressed, and transcription of the gene of interest will be repressed. In addition, the protein of interest is fused to a N-end recognition signal for rapid ubiquitin-dependent degradation. In the absence of *UBR1,* this pathway does not work. This means that in presence of copper, *UBR1* will be expressed, and our protein degraded. In short, this means that, in the absence of the copper ion, neither *ROX1* nor *UBR1* are expressed, but the gene of interest is expressed, and its product is stable. In the presence of copper, however, both *ROX1* and *UBR1* are expressed. This will, in turn, lead to transcriptional repression of the gene of interest and an immediate degradation of its product.

Tagging Proteins in Yeast

Purification and characterization of an individual MPC can be simplified by the introduction of various forms of sequence tags and functional protein domains into the subunits of the complex. This is easily accomplished in *S. cerevisiae* with the use of genetic recombination. These fusions have primarily three different purposes: to introduce epitopes that can be specifically detected with monoclonal antibodies, to aid in purification, and to contribute a new function.

Epitope tags, such as the c-myc and FLAG tags, introduce a highly antigenic sequence into the protein, which eliminates the need for protein-specific antibodies. These tags can be used to localize gene products in living cells as well as tracking proteins during purification. However, one should remember that even highly purified monoclonal antibodies often cross-react with yeast proteins. If the epitope is to be used for monitoring an MPC during purification, more than one protein should be tagged. This facilitates the interpretation of immuno-blots and limits the risk of chasing a cross-reacting protein during purification. The small sizes of the FLAG (eight amino acids) and c-myc tags (ten amino acids) make interference with the architectural integrity of an MPC less likely.

Fusion proteins can also be used to introduce protein domains with specific affinities, such as the 6 × His tag or glutathione-S transferase (GST) domain. When fused to individual subunits of an MPC, these affinity tags allow a gentle way of specifically enriching for the complex. The GST domain binds with high affinity to a glutathione resin, while the 6 × His tag interacts specifically with a Ni^{2+}-resin. It is conceivable that, by introducing such tags into every single yeast protein, it would be possible to isolate protein complexes and identify their components with peptide mass fingerprinting. Such an analysis would generate a map for all the interactions taking place between yeast proteins. However, even if these tags present an attractive short cut for protein purification, practical experience has shown that MPCs with such tags often fail to bind to the specific resins. A method of improving the chances of success for this approach is to first use a couple of purification steps with conventional chromatography. A fusion protein that binds poorly in a crude extract may work perfectly after a 50- to 100-fold enrichment. The 26 kDa GST-domain presents an extra problem in MPCs because its large size can have severe effects on the subunit organization. The 6 × His tag has the benefit of being small (0.84 kDa) and, therefore, seldom has consequences for complex formation.

Green fluorescent protein (Gfp) adds a novel function to the fusion protein. It generates a highly visible internal fluorophore and has, in recent years, become an established marker of gene expression and protein targeting in intact cells and organisms.[14]

One elegant method for introducing epitope tags without the need for any cloning steps was developed by Schneider and co-workers.[15] They designed vectors in which the *URA3* gene was flanked by direct repeats of epitope tags. The tag-*URA3*-tag cassette was amplified by PCR with primers containing 40 to 50 bp of sequence homology to the genomic location of interest. The resulting PCR fragment was used to transform yeast and integrants were selected on *URA3* omission plates. The two tags flanking the *URA3* gene represent a direct repeat. In yeast, sequences between such repeats are frequently lost as a result of homologous recombination. When recombination takes place between the flanking tags, the intervening *URA3* gene and generation of one single epitope tag occurs at the desired location. It is easy to select the cells that have lost the *URA3* gene because they are unable to grow on plates containing 5-fluoro-orotic acid (5-FOA). The 5-FOA is converted to a toxic product (5-fluorouracil) in the presence of the *URA3* gene product, which prevents cell growth.

The correct insertion of the DNA fragment encoding the tag can be checked with PCR genotyping of the total yeast DNA (see Protocol 4). Total cell proteins should be prepared (see Protocol 5), and immunoblot analysis performed to verify the presence of the tag in the expressed fusion protein.

Measurement of Global Transcription in Yeast

Cellular development, differentiation, and function are precisely regulated by a very large number of genes. To understand the complex molecular mechanisms underlying these events, it is essential to study the expression of many genes simultaneously. By studying changes in global expression patterns in cells with specific mutations under different physiological and developmental conditions, it is also possible to make conclusions about the function of an individual gene. Studies of global genome transcription in yeast have been made possible with the development of DNA micro-array technology. This new technology has recently allowed a functional dissection of a number of MPCs involved in different aspects of transcriptional regulation.[8, 9]

Measurements of gene expression are achieved mainly by various techniques involving DNA–RNA hybridization but can also involve a cDNA synthetic step and subsequent DNA–DNA hybridization. The most common method for this is Northern blotting, which is described in a number of manuals of standard methods. With the generation of more and more DNA sequence data from various sequencing projects, DNA micro-array technologies have become important alternatives to Northern blotting. These technologies take advantage of all available sequence data and monitor the expression of thousands of genes in parallel.

The *S. cerevisiae* sequencing project was completed in 1996, leading to the identification of about 6000 individual genes.[1] Of these genes, 70% could be classified by sequence similarities (cyclin-dependent kinases, helicases), whereas no less than 30% lacked significant homologies with any proteins present in the protein databases. A key element in our understanding of all these novel genes is, of course, to know the pattern of their expression. Advances in micro-array technology have enabled hybridization-based monitoring of the steady state mRNA levels of every gene in the *S. cerevisiae* genome.[16] Because gene expression in eukaryotic cells is mainly controlled at the level of transcription, the technology allows us to essentially monitor the expression of each gene in the yeast genome simultaneously.

Genome-wide expression analysis has been used to study the functions of various forms of MPCs involved in transcription and chromatin remodeling in *S. cerevisiae,* demonstrating the importance of individual subunits in these complexes.[8, 9] It is also obvious that a combined biochemical and genetic approach is necessary because the functional impact of a certain mutation often can be ascribed to effects on other subunits in the MPC. Disruption of one subunit often leads to the loss of other subunits from the complex.

Micro-arrays can be manufactured in two different ways, by synthesis *in situ* or by delivery.[17] In the *in situ* synthesis approach, nucleic acids are synthesized directly on a chip surface. This is made possible by a light-directed chemical synthesis technology developed by scientists at Affymetrix.[18] By using different sets of photolithographic masks, scientitsts can define the chip exposure sites and, thereby, the sequence of the oligonucleotides. In this way, arrays can be synthesized with up to 400,000 different oligonucleotides in an area of 1.6 cm^2. Every spot contains about

10 million copies of an oligonucleotide with a defined sequence. One of the advantages of this technique is that it is easy to get started. All the necessary components are commercially available (Affymetrix). Another important advantage is that a specific chip can be constructed directly with data obtained from sequence databases. No cumbersome steps with PCR amplification and sample handling are needed.

The delivery approach for construction of the DNA micro-array was developed by Brown and co-workers at Stanford University.[19] The principle is straightforward. Each gene to be studied is amplified by PCR, and the PCR products are then printed onto glass microscope slides with an arraying machine. The slides subsequently are processed with chemical and heat treatment to attach the DNA sequences to the glass surface and denature them. Polyadenylated mRNA is then prepared from the cell type that will be studied (see Protocol 6), and cDNA is synthesized from this material in the presence of fluorescein-labeled dCTP. This cDNA is then being hybridized to the array under very stringent conditions. Finally, the array is scanned with a laser. The amount of fluorescence in each spot gives a value for the level of gene expression of the corresponding gene. A refined form of this technique can be used to study differential gene expression. In this case, mRNA is prepared from two different sources (e.g., a cancer cell and a wildtype cell from the same type of tissue). cDNA is then synthesized in the presence of fluorescein-12-dCTP in one case and lissamine-5-dCTP in the other. The two probes are mixed together in equal proportions and hybridized to a single array, after which the amount of lissamine and fluorescein in each spot is determined by laser scanning. The relative amount of lissamine to fluorescein in a spot gives a value for the relative value for the expression of that gene.

Studies of the Med2 Subunit of the Yeast Mediator Complex

A number of powerful techniques can be employed to study multiprotein complex function in yeast (e.g., gene disruption, two-hybrid assays, and DNA micro-array technology). To explain how some of these techniques can be used to study a specific complex, the characterization of a subunit in the Mediator complex of *S. cerevisiae* will be reviewed.

Basal transcription can be regulated positively by transcription activators that bind to regulatory sequences around the promoter sequence. According to a current model for transcriptional activation, these factors interact directly or indirectly with general transcription factors via their activating domains.[20] This stimulates the assembly of the basal transcription factors on the promoter and transcription can commence.

Activators can only elicit a very low level of activation in a transcription system reconstituted from pure RNA polymerase II (pol II) and basal factors, some types of co-factor are needed in addition. One such factor was found and termed Mediator.[21] In an *in vitro* transcription system reconstituted from essentially pure factors, the mediator complex was shown to stimulate basal transcription 50-fold and to support transcriptional activation from the Gal4-VP16 and Gcn4 activating protein. The Mediator complex could also be purified in a complex with RNA polymerase II (pol II), thus forming a holo-enzyme form of pol II.

Purified to homogeneity, the Mediator complex was shown to be a complex consisting of 20 polypeptides.[22] One of the new subunits identified was Med2, the product of a previously uncharacterized open-reading frame. Before further studies of the Med2 function were initiated, it was essential to confirm that this protein was a true member of the Mediator complex. To this end, polyclonal antibodies were obtained against the Med2 protein, and the stoichiometric association of Med2 with the Mediator was shown by co-purification and co-immunoprecipitation.[22]

As a first step in a functional characterization, the Med2 gene was deleted from the genome with PCR-mediated gene disruption. Tetrad dissection of the Med2 deletion strain yielded four viable spores, meaning that the gene was not needed for vegetative growth on a rich medium.[22] To assay for the functional importance of the Med2 protein, the haploid Med2 deletion strain was tested for growth on different carbon sources.[9] The deletion strain grew well on all carbon sources tested except galactose. This specific inability to grow on galactose was further examined by whole-genome DNA microarrays.[9] Differences in specific transcript levels between Med2 and wildtype strains were determined under galactose growth conditions. Approximately 200 of the 6000 genes analyzed showed a 75% decrease in expression in the Med2 strain under these conditions. This experiment clearly demonstrated that galactose induction of several genes needed for galactose metabolism was defective in the mutant strain. The effect of the Med2 deletion on galactose induction was selective: transcript levels of several genes shown by previous micro-array analysis to be induced by galactose were essentially unchanged.

Because the Med2 deletion strain was viable, large amounts of this strain could be grown and Mediator lacking Med2 purified to near homogeneity. The subunit composition of this mutant Mediator was determined by SDS/PAGE, immunoblotting, and silver staining. The results demonstrated that loss of Med2 also led to the specific loss of another Mediator subunit, Pgd1. Conversely, Med2 was not retained in the Mediator purified from a Pgd1 deletion strain. It could, therefore, be concluded that Med2 and Pgd1 must interact, directly or indirectly, to stabilize their mutual association with the holo-enzyme. Thus, by comparing the Mediator purified from wildtype and mutant strains, it was possible to make conclusions about the structural organization of the Mediator.

The Mediator isolated from the Med2 deletion strain could be further tested for *in vitro* activities.[9] In an *in vitro* transcription system, this mutant mediator was unresponsive to the activator Gal4-VP16. By contrast, both basal transcription and stimulation of CTD phosphorylation were unaffected. The yeast Gcn4 protein is, along with VP-16, a member of the family of acidic activator proteins. Like VP-16, Gcn4 requires the Mediator for activation in a fully reconstituted transcription system. Surprisingly, Gcn4 did not require the Med2–Pgd1 proteins in the reconstituted system. Evidently, the two acidic activators contact different members of the Mediator complex or function through Mediator by different mechanisms. The characteristics of Gal4-VP16- and Gcn4-dependent activation in the *in vitro* system reconstituted with Mediator lacking the Med2–Pgd1 module were consistent with those *in vivo* providing a parallel between effects of Mediator mutations *in vivo* and *in vitro*.

GENETICS TO STUDY SINGLE PROTEINS IN TRANSGENIC MICE

The transgenic technology is today an integrated part of biomedical research, and the use of transgenic mice is one of the major success stories of modern molecular biology. Expression of mammalian genes in transgenic mice is frequently undertaken to study the functional consequences of overexpression of a gene, to study protein structure–function relationships, and to identify the DNA sequences that control the tissue-specific expression of a gene. Analysis of these mouse models has increased our knowledge about basic mechanisms that control biological systems and the pathological processes in human genetic disorder.

The mouse is fairly inexpensive, easy to feed, raise, and handle, and is available world wide. It has a short gestation period (18 to 21 days) and a long period of reproductive activity (from 2 to about 14 months of age) during which it can produce over 10 litters and 100 offspring.[23] Furthermore, strains can be intercrossed with genetically modified animals. The drawback of generating transgenic mice, however, is that it requires expensive equipment for oocyte micro-injection, is time-consuming, and requires elaborate animal-care facilities.

DESIGNING TRANSGENES

Two different approaches can be used to design transgenes. The most common approach to express a foreign gene relies on insertion of a cDNA copy encoding the desired protein into an appropriate expression plasmid. The cDNA minigenes are easy to manipulate *in vitro* due to their smaller size (<20 kb). In addition, the appropriate coding region can be inserted behind an efficient promotor that also includes all expression elements to ensure efficient expression.

The second approach uses genomic DNA and has several advantages. In general, genomic clones are better expressed than the cDNA clones. Second, many genes are greater than 20 kb in length and cannot be isolated intact from a single bacteriophage λ clone. Furthermore, reconstructing genes from overlapping λ or cosmid clones is often difficult owing to the paucity of unique cloning sites and the limitations of cloning large DNA fragments in plasmid vectors. Third, the gene of interest is flanked by its natural 5′ and 3′ sequences. Furthermore, P1 bacteriophage clones, bacterial artificial chromosomes (BAC), and yeast artificial chromosomes (YAC) are commercially available. These vectors permit the cloning of large genomic DNA fragments from approximately100 kb to several Mb in length.[23] However, the handling of these large vectors is more difficult than handling small ones, and the low yield of DNA from yeast artificial chromosomes is a major drawback.

MODIFICATION OF TRANSGENES

The transgene is often modified to custom design the final protein product. Furthermore, many expression vectors are designed to allow various tags to be linked to the expressed recombinant protein. In this manner, the recombinant hybrid protein containing the tag and the desired expressed gene protein may be detected and/or isolated on the basis of the unique properties of the tag. Several of these tags have been described previously. Manipulation of small DNAs has become a routine, and several very efficient commercial kits are available. However, manipulation of large-insert bacterial clones is a nontrivial task given the size of these clones and the lack of convenient unique restriction sites.

One approach to modify P1 bacteriophage and BAC clones is to use RecA-assisted restriction endonuclease (RARE) cleavage, a technique developed by Ferrin et al.[24] RARE cleavage consists of protecting a specific restriction endonuclease site with a complementary oligonucleotide (see Figure 7.4). In the presence of RecA, a triplex DNA complex is formed that prevents methylation at the protected site(s), while unprotected sites are methylated by the corresponding methylase. After dissociation of the complex, the protected sites can be cleaved with the specific restriction endonuclease. This procedure makes it possible to introduce point-specific mutations and deletions in P1 bacteriophage and BAC clones of up to approximately 150 kb in length.[24–29] The most common approach has been to remove a specific *Eco*RI fragment from a P1 bacteriophage or BAC clone, which typically could contain 10 to 20 internal sites for these restriction endonucleases. The specific *Eco*RI fragment could then be mutated and ligated back into the P1 clone. However, other methylases (e.g., *Alu*I that protects cleavage sites recognized by *Hind*III) are also available for use with the RARE cleavage system, so the strategy need not be confined to the use of the *Eco*RI fragments.[25, 29] A protocol for RARE cleavage is shown (see Protocol 7). To assess the integrity of the mutated P1 bacteriophage or BAC clone, it is advisable to perform restriction endonuclease mapping of the manipulated clones. A protocol for miniprep and mapping of P1 bacteriophage and BAC DNA is shown in Protocol 8.

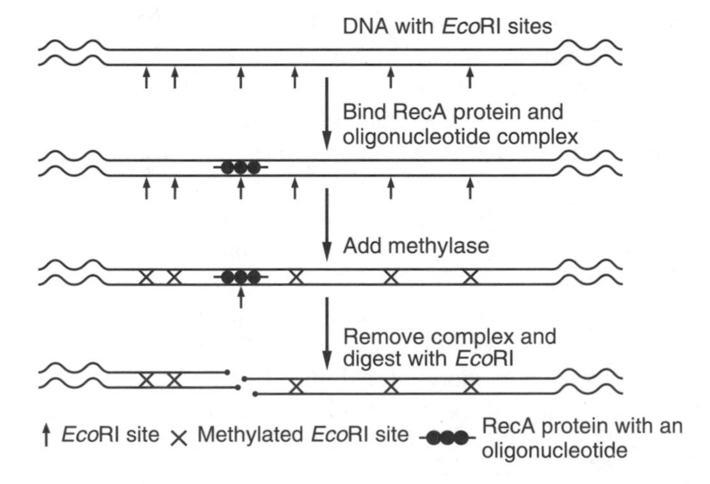

FIGURE 7.4 In RARE cleavage, specific restriction endonuclease sites are protected with a complementary oligonucleotide. In the presence of RecA, a triplex DNA complex is formed that prevents methylation at the protected site(s), whereas unprotected sites are methylated by the corresponding methylase. After dissociation of the complex, the protected sites can be cleaved with the specific restriction endonuclease.

Another excellent approach that introduces small insertions, deletions, or point mutations in large DNA is the YAC gene–targeting approach.[30–32] This technique exploits the efficient recombination between homologous segments of DNA that is a hallmark genetic characteristic of *S. cerevisiae*. This feature, along with the availability of a variety of selectable markers, facilitates gene targeting in yeast. As discussed previously gene targeting has been widely used to introduce mutations into endogenous yeast genes.[33, 34] More recently, the same gene-targeting techniques have been utilized to introduce mutations into human genes within YACs, for the purpose of generating transgenic mice.[30, 32]

However, the YAC-based technique is more labor intensive. The yield of pure YAC DNA is only a small fraction of the amount of the P1 DNA that can be obtained from bacteria,[35] and the protocol uses a two-step pop-in, pop-out strategy.[30, 31] Furthermore, the efficient machinery for homologous recombination in the yeast can, unfortunately, give rise to spontaneous rearrangement of the DNA.

A disadvantage of the RARE cleavage technique, which does not apply to the YAC gene-targeting approach, is that the appropriate restriction sites must be conveniently located with respect to the region of interest before this technique can be utilized. Moreover, it is necessary to have rather extensive DNA sequence information around the sites of interest before applying this technique. This drawback could, however, be overcome using the quantitative DNA fiber mapping (QDFM) technique for mapping of multiple restriction sites and precise localization of restriction fragments in large genomic clones as shown by Duell et al.[36]

Preparation of DNA for Micro-injection into the Pronucleus

The quality of the DNA used for micro-injection is one of the most important determinants of success in transgenic experiments. Thus, it is essential that the DNA for micro-injection is absolutely free from all contaminants that are toxic to the embryo, including traces of agarose, phenol, or ethanol.

Although prokaryotic cloning vector sequences have no apparent effect on the integration efficiency of the injected gene, they should be removed to the greatest extent possible.[23] Tissue specificity and the level of expression of the transgene are greatly influenced by the presence of the vector sequences. A 100- to 1000-fold increase in the level of expression was observed when the vector sequences were deleted from the human β-globulin gene.[37] Furthermore, linear molecules integrate with at least five times higher frequency than circular molecules.[38, 39] There seems to be a little effect of structure at the ends of DNA. In general, DNA with single-stranded ends has higher efficiency of integration.[38, 39]

Small-sized plasmid DNA (less than 20 kb) easily can be purified with standard techniques, and an efficient protocol is shown in Protocol 9. High-quality P1 bacteriophage, BAC, and even YAC DNA for micro-injection can be purified from preparative agarose pulsed-field gels.[35] This procedure is somewhat more complicated than handling small-sized plasmid DNA but can easily be performed with some experience. A protocol for purification of P1 bacteriophage and BAC DNA is shown in Protocol 10.

Micro-Injection of DNA into the Zygote

There are three different methods to prepare transgenic animal. The first method is the micro-injection of DNA into the pronucleus of the zygote. Though the DNA can be micro-injected into either of the two pronuclei in the fertilized egg, the micro-injection into the male pronucleus is more convenient because it is larger than the female pronucleus and not obstructed by polar bodies. The second method is the retroviral vectors and retroviral infection of embryo. The final method is the Embryonic stem cell injection into blastocysts. The most commonly and successfully employed method is the micro-injection of DNA directly into the pronuclei of fertilized mouse eggs. The overall strategy is shown in Figure 7.5. The techniques for injecting small-sized DNA from plasmids have become routine, and several

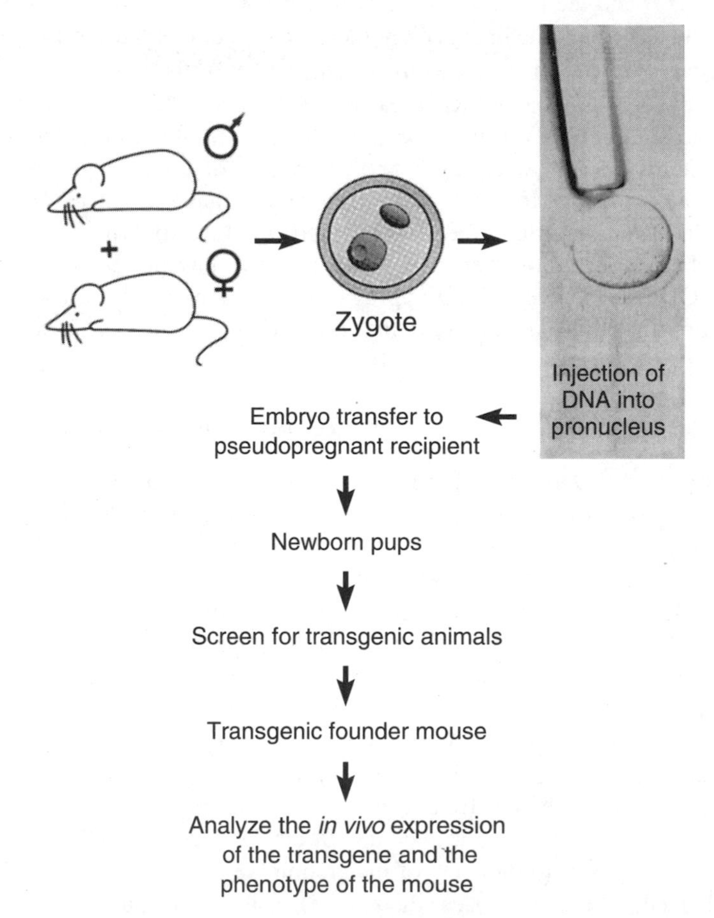

FIGURE 7.5 The flow chart for generation of transgenic mice by the pronuclear injection technique. A purified DNA construct is injected into fertilized mouse eggs at the one-cell stage. After injection, eggs are transferred into oviducts of pseudopregnant females, where they develop further, and offspring is born approximately 20 days later. Mice that have the transgene integrated in their genome are identified by DNA analysis, and the transgenic mice are analyzed.

useful books have been written on this topic.[40] In general, no modifications of standard transgenic procedures are required when micro-injecting P1 bacteriophage, BAC, or even YAC DNA except that 100 m*M* NaCl must be included in the micro-injection buffers to avoid shearing of the large DNA.

Concentration of DNA for pronuclear injection plays a significant role in the viability of embryos. Usually, 1 to 2 μg/ml is the optimum concentration of DNA. This is equivalent to about 600 to 800 copies of a plasmid with a length of 5 kb.[23] At lower concentrations, the frequency of integrations is also low, whereas at concentrations of 10 μg/ml or more, the survival rate of embryos is drastically reduced.

The pronuclear injection method results in the stable chromosomal integration of the foreign DNA in 5 to 40% of the resulting mice. In most cases, integration appears to occur at the one-cell stage because the foreign DNA is present in every cell of the transgenic animal, including all primordial germ cells. In approximately 20 to 30% of transgenic mice, the foreign DNA apparently integrates at a later stage, resulting in mice that are mosaic for the presence of foreign DNA. The number of copies of the foreign DNA sequence that integrate into the host genome ranges from one to several hundred and bears little relation to the number of DNA molecules injected into the egg. When multiple copies are present, they are usually found at a single chromosomal locus. However, there may occasionally be separate integration sites on two different chromosomes. The copies of foreign DNA at each integration site are arranged primarily in a head-to-tail array, although some integration events may be more complicated.

Analysis of Transgenic Animals

Potentially transgenic mice are usually screened for the presence of the injected gene by performing PCR or Southern analysis with a DNA sample extracted from the tail or ear. Several protocols are available to extract DNA for these analyses; and Protocols 11 and 12 show two excellent examples.

Mice that develop from injected eggs are often termed founder mice. To establish a transgenic line, allow the founder mice to breed with nontransgenic mice. Males can mate with females within 24 hours of the birth of a litter, and females can produce litters at 3-week intervals. The mice should be allowed to breed again for up to four to five generations. If the introduced gene is stably expressed in several generations, a transgenic mouse line has been established. Although most transgenic founders will transmit the foreign integrated DNA to 50% of their offspring, approximately 20 to 30% of transgenic founders are mosaic and transmit the gene at a lower frequency.

A founder mouse derived from an inbred strain is normally mated with mice of the same inbred strain to maintain the gene on a defined genetic background. If the mouse is an F_2 hybrid or if it not important to maintain the gene on an inbred background, it is often more efficient to mate it with F_1 hybrid mice, which are healthier and make better mothers than inbred mice. Homozygous transgenic mice are produced by setting up heterozygous intercrosses. As expected from the Mendelian transmission of the gene, about onequarter of the progency from this cross will be homozygous for the gene. Homozygous animals can be identified by dot blot or Southern blot analysis in which the homozygous animals will display a stronger signal. Homozygosity is further confirmed by mating. All progeny from this cross should be positive for the transgene.

Approximately 5 to 15% of the random DNA integration events in transgenic mice produce recessive lethal mutations; consequently, not all transgenic lines can be maintained in the homozygous state.

Analysis of the *In Vivo* Expression of the Transgene and the Phenotype of the Mouse

The analysis of the transgenic mouse can involve purification of the recombinant protein, analysis of its tissue expression and gene regulation, and its biological/metabolic consequences. To illustrate how transgenic mice can be used to study different aspects of the same protein, the characterization of mice overexpressing apolipoprotein B (apoB) will be reviewed.

All mammalian triacylglycerol-rich lipoproteins contain apoB that has been shown to be an obligatory factor for their formation and degradation.[41] From the clinical point of view, much interest has focused on apoB because elevated levels of the apoB-containing lipoproteins have been shown to have a central role in the development of the most common circulatory disease in the industrialized world, atherosclerosis.[42]

Structure–Function Studies of Human ApoB

Even though the ApoB cDNA and gene were cloned and sequenced more than a decade ago, progress in structure–function studies of apoB100 has been slow. There are several reasons for the slow progress. ApoB100, one of the largest proteins known (4536 amino acids), is highly hydrophobic and aggregates when not associated with lipids. Thus, it is impossible to perform structure–function studies of apoB in cell-free systems. Instead, the protein must be synthesized as part of a complex lipoprotein particle in liver cells. Moreover, recombinant LDL secreted by hepatic cells in culture transfected with apoB minigenes have been shown to display defective LDL receptor-binding activity, and heterologous minigenes yield extremely poor expression in transgenic mice.[43]

Fortunately, about 6 years ago, the laboratories of Drs. S. Young and E. Rubin identified and mapped a P1 bacteriophage clone that spanned the entire human apoB gene. They used the p158 clone to generate human apoB transgenic mice.[44, 45] In transgenic lines with more than 10 copies of the transgene, the plasma levels of human apoB in chow-fed hemizygous mice were 60 to 80 mg/dl—similar to those in normolipidemic humans.[46] The development of the p158-human apoB transgenic mice represented a breakthrough in our understanding of apoB biology.

The ability to express high levels of apoB was essential for apoB structure–function studies, and several studies have now been performed using the human apoB transgenic mice. The overall approach for these studies has been to insert mutations into clone p158, with the YAC-gene targeting approach or RARE cleavage protocol, then express the mutant construct in transgenic mice, and finally analyze the properties of the mutant human apoB protein (see Figure 7.6). This strategy made it feasible to generate a series of apoB mutants to dissect the

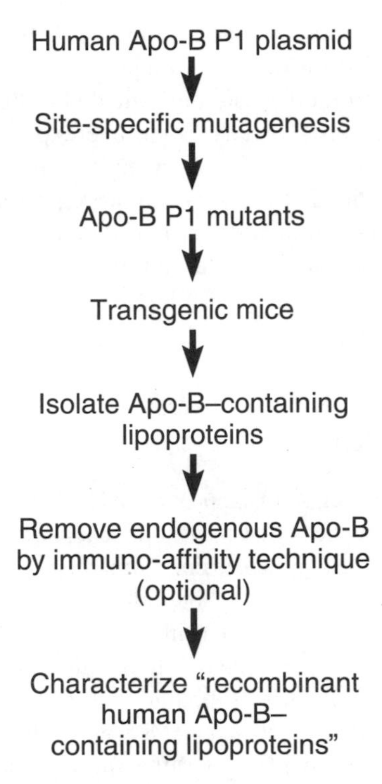

FIGURE 7.6 A schematic of the generation, purification, and characterization of recombinant human apo-B-containing lipoproteins in transgenic mice.

molecular interaction between apo-B100 and the LDL receptor, identifying the sequence in apoB that mediates the interaction with proteoglycans and defining the features of the apoB molecule that are important for its interaction with apo(a) in the assembly of Lp(a).[26, 27, 29, 30, 47]

Studies of the Atherogenicity of ApoB

Human apoB transgenic mice have also been a major asset for studies of atherosclerosis. Characterization of the human apoB transgenic mice revealed that they did not develop atherosclerotic lesions when fed a chow diet.[45, 46] When fed a high-fat diet, however, the human apoB transgenic mice developed extensive atherosclerotic lesions in the proximal aortic root.[46] After 6 months of the high-fat diet, lesions in the human apoB transgenic mice contained fibrous caps, necrotic cores with large pools of extracellular lipid, and abundant cholesterol clefts—the hallmarks of an atherosclerotic lesion.[46] Furthermore, the mice have been mated with mice overexpressing human apolipoprotein(a) to investigate the atherogenicity of Lp(a).[44, 45]

The mice have also been used to investigate if the interaction between apoB and proteoglycans in the extracellular matrix is important for atherogenesis. For this reason, a series of mutations in the human apoB gene were made and expressed in transgenic mice. Recombinant human LDL were isolated, and their abilities to bind to smooth muscle cell proteoglycans were determined. The results showed that by mutating all of the basic amino acids between residues 3359 and 3369 of apoB100, the binding of recombinant apoB to both proteoglycans and the LDL receptor was disrupted.[27] The biological significance of this finding was then tested in an atherosclerosis study. Mice overexpressing wildtype human or proteoglycan-binding-defective were fed a high-cholesterol diet for 17 weeks. The mice were then sacrificed, and the aortas were perfusion fixed and analyzed. The results from this study indicated that proteoglycan-binding-defective LDL have a greatly reduced atherogenic potential and provide for the first time direct experimental evidence that binding of LDL to artery wall proteoglycans is an early step in atherogenesis.[27]

An Approach for Studying Gene Regulation by Distant DNA Elements

The transgenic mice expressing human apoB have also been used to define distant *cis*-acting regulatory DNA sequences that control the expression of the apoB gene. When RNA samples from the mice overexpressing human apoB were analyzed, two unexpected findings were made.[44, 45] First, although the transgene was highly expressed in the liver, transgene expression was absent in the intestine. Second, the human apoB transgene was expressed in the heart, an organ thought to lack apoB gene expression.

In interpreting these results, it was thought that the peculiar pattern of transgene expression was because of the fact that p158 was simply too short to contain the *cis*-acting DNA sequence elements governing the correct spatial pattern of apoB gene expression. Thus, it was hypothesized that the human apoB transgene lacked an enhancer element governing apo-B gene expression in the intestine. However, the absence of transgene expression in the intestine was surprising because both P1 clones contained more than 11 kb of flanking sequences both 5′ and 3′ to the gene. To investigate if the p158 plasmid lacked an enhancer element, Nielsen et al.[48] isolated and characterized 145- and 207-kb bacterial artificial chromosome (BAC) clones that spanned the human apoB gene. Each of these BACs contained extensive 5′ and 3′ flanking sequences, and each directed spatially and physiologically appropriate apoB gene expression in the intestines of transgenic mice. To define the location of the sequences that control intestinal expression of the apoB gene, they generated transgenic mice by co-micro-injecting the p158 plasmid (which did not confer intestinal expression of apoB) with neither the 5′ sequences nor the 3′ sequences from the 145-kb BAC. Analysis of the apoB expression pattern in those mice revealed that the DNA sequences controlling intestinal expression were located 5′ to the apoB gene. Next, Nielsen et al. used RARE cleavage to truncate specific segments of the 5′ and 3′ flanking sequences from the 145-kb BAC.[49] A series of the truncated BACs containing different lengths of 5′ and 3′ sequences were used to generate more than 40 additional lines of human apoB transgenic mice. Analysis of human apoB gene expression in those mice demonstrated that the sequences controlling the expression of the apoB gene in the intestine are located more than 50 kb 5′ to the apoB gene. These studies show that the RARE cleavage/transgenic expression is a powerful approach for structure–function studies and for the study of gene regulation by distant gene-regulatory elements.[49]

APOB-CONTAINING LIPOPROTEINS SECRETED BY THE HEART

During the initial characterization of the p158-human apoB transgenic mice, it was observed that the human apoB transgene was expressed in the heart. This unexpected observation was confirmed by both RNA slot blot studies and by RNase protection assays in several independent lines of the p158 transgenic mice. The amount of apoB mRNA in the heart was approximately 4% of that in the liver.[44, 45] This finding was initially considered as a transgenic artifact. However, subsequent studies showed that the apoB gene could be detected in both human heart tissue and in nontransgenic mouse hearts. These findings could indicate that the heart could secrete apoB-containing lipoproteins.

To test this hypothesis, heart tissue from several different lines of human apoB transgenic mice was metabolically labeled with [^{35}S]methionine/cysteine, and the medium was fractionated by sucrose density gradient ultracentrifugation. ApoB was immunoprecipitated from each fraction and examined by SDS-PAGE and autoradiography. The human apoB transgenic mouse hearts secreted apoB-containing lipoproteins.[50] Metabolic labeling experiments were also performed on fresh human heart tissue (obtained from the explanted diseased hearts at the time of cardiac transplantation) and on the hearts of nontransgenic mice.[50] In both cases, the synthesis and secretion of apoB100-containing lipoproteins could be detected.

The finding that the heart is an apoB-secreting organ was a completely unexpected result from the characterization of the human apoB transgenic mice. It is still unclear if the apoB secretion from the heart is important. The next steps in understanding apoB biology will be to characterize heart lipoprotein assembly and secretion and to assess its physiologic significance. The human apoB transgenic mice also will play a very important role in these studies.

SAMPLE PROTOCOLS

PROTOCOL 1 LARGE-SCALE PROTEIN PREPARATION FROM YEAST

1. A single colony of the yeast strain to be used is picked and added to 100 ml of autoclaved YPD. The preculture is incubated in a shaker at 30°C for 48 h.
2. Inoculate 20 L of YPD with the preculture. Grow the yeast to an absorbance of 3.8 to 4.0 at 600 nm. For most wildtype strains, this will take approximately 12 h.
3. Collect the cells by centrifugation in a clinical centrifuge (7 min, 2500 rpm, 4°C).
4. Resuspend the cells in 500 ml of ultrapure water and centrifuge as before.
5. Resuspend in the yeast in 3× lysis buffer (1.0 g of cells corresponds to approximately 1.0 ml) and freeze them in liquid nitrogen and store them at –80°C.
6. Equal volumes of thawed cells and glass beads are put in the metal container of a bead beater (Mini Bead Beater, Microspec, U.S.A.). The

container is covered with ice and NaCl to keep the cells cool during the lysis process. Lysis is performed with a program (30 s of beating alternated with 90 s of rest) that is repeated 25 times. The resting step is necessary to prevent heating of the cells.

7. Collect the lysed cells by centrifugation (Beckman JA-10 rotor, 9000 rpm, 4°C, 25 min).
8. Add one ninth cell volume of 5 *M* KAc to the supernatant and stir the mixture for 15 min.
9. Add polyethylenimin to a concentration of 0.2% (w/v) and stir for an additional 30 min.
10. Collect the polyethylenimim precipitate with centrifugation (Beckman Ti-45 rotor, 42,000 rpm, 4°C, 90 min).
11. Freeze the supernatant in liquid nitrogen and store it in a –80°C freezer.

Protocol 2 PCR-Mediated Gene Disruption in Yeast

1. Two primers are constructed for amplification of the selectable marker. PCR primer 1 is constructed with a region of 50 bp at the 5′ end homologous to the flanking sequence of the gene of interest and 20 bases of homology at the 3′ end (5′-CTGTGCGGTATTTCACACCG-3′) to the flanking sequence of the marker genes in the pRS303-306 series of vectors.[51] PCR primer 2 is constructed with a region of 50 bp at the 5′ end homologous to the other side of the targeted region, followed by 20 bases of homology at the 3′ end: 5′-AGATTGTACTGAGAGTGCAC-3′ to the flanking sequence of the marker genes in the pRS303-306 series of vectors. By using primers 1 and 2, the selective marker genes are amplified by PCR from the plasmids pRS303-306 and pRS313-316.
2. Perform PCR under the following conditions: 5 min 94°C; 30 cycles of 1 min 94°C, 2 min 50°C, 3 min 72°C; 1× 8 min 72°C.
3. The resulting PCR product is purified by gel electrophoresis.
4. Transform yeast cells (W303) (Protocol 3).
5. After transformation, cells are plated on the appropriate selective medium.
6. Individual colonies are picked and grown on YPD agar plates overnight.
7. Yeast genomic DNA is isolated and PCR analysis is performed to confirm that the marker gene has recombined correctly into the target gene. The diploid yeast cells carrying the marker gene integrated correctly are sporulated and tetrads are dissected on YPD agar.

Protocol 3 Transformation of Yeast with Lithium Acetate

Solutions

1. 1.0 *M* lithium acetate. Dissolve lithium acetate in distilled de-ionized water and filter sterilize.
2. Polyethylene glycol (PEG-3350 50% w/v). Dissolve PEG3350 in distilled de-ionized water and filter sterilize.

Method

1. Inoculate the yeast strain to be transformed into 2 ml of YPAD medium and grow with shaking at 30°C overnight.
2. Dilute the cells in 50 ml of prewarmed YPAD to a density of 5×10^6 cells/ml.
3. Incubate at 30°C with shaking (200 rpm) until a density of 2×10^7 cells/ml has been reached. This will take 3 to 4 h.
4. Pellet the cells by centrifugation (3000× *g*, 5 min).
5. Discard the supernatant and resuspend the cells in 25 ml sterile water and recentrifuge.
6. Resuspend the cell pellet in 1 ml of 100 m*M* sterile LiAc and transfer to a centrifuge tube.
7. Pellet the cells by centrifugation (10,000× *g*, 15 s) and remove the LiAc with a micropipette.
8. Resuspend the cell pellet in 100 m*M* LiAc to a final volume of 500 µl.
9. Vortex the cell suspension and transfer 50-µl samples to a centrifuge tube.
10. Pellet the cells by centrifugation (10,000× *g*, 15 s) and carefully remove the LiAc.
11. Add 240 µl of 50% PEG-3350, 36 µl 1.0 *M* LiAc, 50 µl of herring testes carrier DNA (2 mg/ml) × µl library plasmid DNA (0.1 to 10 µl), and sterile water to a final volume of 360 µl. It is important that the components are added in the exact sequence listed above. Vortex vigorously until the cell pellet has been completely mixed. This usually takes about 1 min.
12. Incubate at 30°C for 30 min.
13. Heat shock for 30 min at 42°C. Mix by inversion for 15 s after every 10 min.
14. Centrifuge the cells at 3000× *g* for 5 min at room temperature.
15. Resuspend the cells in 1 ml of sterile water. This should be done with great care because this step significantly influences the transformation efficiency. We normally cut off the tip of a 1-ml plastic micropipette tip and carefully pipette up and down a couple of times.
16. Plate 2 to 200 µl of the transformation mix onto the appropriate drop out plates. Incubate for 2 to 4 days at 30°C.

Protocol 4 Rapid Preparation of DNA from Yeast

1. Grow a yeast culture overnight, with shaking, in 5 ml of YPD medium at 30°C. Harvest when cell density is about 10^8 cells/ml.
2. Pellet 1.2 ml of the culture in a bench-top centrifuge at 6000 rpm and wash once with 1 ml sterile water.
3. Resuspend the pellet in 0.2 ml protoplasting buffer (TE buffer [100 m*M* Tris-HCl, pH 7.5, 10 m*M* EDTA, 150 m*M* β-mercaptoethanal] with 0.2 mg/ml Zymolase (20,000 units/mg).
4. Incubate at 37°C for 90 min with occasional inversion of the tube.
5. Add 0.2 ml of lysis buffer (200 m*M* NaOH, 10 g/l SDS) and mix gently by inversion.

6. Incubate at 65°C for 20 min and then cool it rapidly on ice.
7. Add 0.2 ml of 5 *M* KAc (pH 5.3) and mix gently by inversion. Incubate on ice for 15 min.
8. Collect the pellet by centrifugation for 5 min, 13,000 rpm, at room temperature.
9. Add 0.6 volumes of isopropanol, mix gently by inversion, and incubate for 5 min at room temperature.
10. Pellet the DNA by centrifugation for 1 min at 13,000 rpm at room temperature.
11. Add 1 ml of 70% ethanol to the pellet. Incubate at room temperature for 10 min.
12. Centrifuge as in step 10.
13. Dry the pellet and dissolve the DNA in 30 μl of TE buffer.
14. Store the DNA at 4°C.

Protocol 5 Analysis of Total Cell Proteins from Yeast

Required Solutions

1. Sample buffer: 60 m*M* Tris-HCl, pH 6.8, 10% (v/v) glycerol, 2% (w/v) SDS, 5% (v/v) β-mercaptoethanal, 0.0025% (w/v) bromophenol blue
2. 100 m*M* phenylmethylsulfonyl fluoride (PMSF)
3. 0.5 *M* benzamidine

Method

1. Grow 25 ml of cells to early log phase.
2. Pellet the cells by centrifugation (2500 rpm for 5 min). Wash once with water and spin again.
3. Resuspend cells in 1 ml of water and transfer to a 1.5-ml microfuge tube. Pellet by centrifugation for 20 s and pour off water.
4. Resuspend in 0.5 ml of ice cold sample buffer with freshly added PMSF (0.5 m*M*) and benzamidine (0.5 m*M*).
5. Add glass beads (approximately 0.5 ml).
6. Vortex vigorously for four periods of 45 s each. Cool the suspension on ice for 30 s between each cycle of vortexing.
7. Pellet by centrifugation for 5 min in microfuge at 4°C.
8. Transfer supernatant to a new tube and boil for 5 min.
9. Load 10 μl on a SDS-PAGE.

Protocol 6 Isolation of Undegraded RNA from Yeast

Several methods exist for the isolation and purification of RNA from yeast cells. The method described here is based on simple lysis in phenol. The method chosen, however, depends on the yeast strain used. Some yeast strains are easier to lyse than others, and sometimes, it might prove necessary to protoplast the cells partially before commencing RNA isolation. The protocol of RNA isolation described here has been used satisfactorily for a couple of years in the laboratory of one of the authors (Gustafsson) to provide RNA for Northern blotting as well as genome-wide expression analysis with DNA micro-arrays.

Precautions to Minimize RNase Contamination

Great care should be taken to ensure that all apparatus and chemicals are sterile and free from RNase contamination. Glassware should be incubated at above 150°C for at least 6 h. All solutions not containing amines should be treated with diethyl-pyrocarbonate (DEPC) and autoclaved. If solutions contain amines (Tris), RNase-free chemicals should be added to DEPC-treated distilled water. Gloves should be worn at all times and changed frequently.

Required Solutions

1. 50 m*M* sodium acetate (pH 5.3), 1 m*M* EDTA
2. 10% w/v SDS
3. Phenol equilibrated with 0.05 *M* sodium acetate (pH 5.3), 1 m*M* EDTA
4. 5 *M* sodium acetate (pH 5.3)
5. Phenol-chloroform (1 : 1)
6. Ethanol

Method

1. Grow cells until the OD_{600} = 3.
2. Harvest cells by centrifugation at 2000× *g* for 5 min at 4°C, and wash the cell pellets twice in ice-cold sterile distilled water.
3. Resuspend the cells in 10 ml of 50 m*M* sodium acetate, pH 5.3, 1 m*M* EDTA in a plastic Falcon tube.
4. Vortex the cell suspension for 3 min in the cold.
5. Add an equal volume of hot (65°C) phenol and mix thoroughly for 5 min.
6. Rapidly chill the mixture in an ethanol/dry ice bath until phenol crystals appear.
7. Centrifuge at 4000× *g* to separate the phases and decant the aqueous phase to a clean sterile centrifuge tube.
8. Repeat steps 4 through 7.
9. Extract the aqueous phase with a half volume of phenol/chloroform for 5 min at room temperature and transfer the aqueous layer to a clean sterile centrifuge tube.
10. Bring the aqueous phase to 0.3 *M* final concentration of sodium acetate by addition of 5 *M* sodium acetate solution and add 2.5 volumes of ethanol (99.5%) to precipitate the RNA.
11. Dissolve the RNA in an appropriate volume of DEPC-treated water.

Protocol 7 RARE Cleavage of Large DNA

1. Add the following solutions in the order described to a 1.5-ml Eppendorf tube: 32 μl of RARE buffer (125 m*M* Tris-acetate, pH 7.85, 20 m*M* $MgCl_2$, 2.5 m*M* spermidine hydrochloride, 2 m*M* DTT), 40 μg (approximately equal to 2.0 mg/ml)) of RecA protein, 16 μl of ADP/ATPgS (11 m*M* ADP, 3 m*M* ATPgS), 0.38 μg (0.16 μg/μl) each of the two 60-mer oligomers (or 0.72 μg of one oligomer if the RARE cleavage is performed with only one oligomer to linearize the P1 plasmid), 4 μg of DNA, and 8 μl of acetylated bovine serum albumin (2 mg/ml), and H_2O to achieve a final volume of 160 μl.
2. Mix gently but carefully and incubate at 37°C for 10 min to allow formation of a triplex DNA complex.
3. Add 8 μl *Eco*RI-methylase (32 U freshly diluted in water) and 8 μl of S-adenosylmethionine (2.4 m*M* freshly diluted in water) to methylate all unprotected *Eco*RI sites.*
4. Incubate the mixture at 37°C for 30 min.*
5. Heat-inactivate the methylase at 65°C for 15 min.
6. Spot-dialyze the mixture for 30 min on a 25-mm filter (Millipore VSWP0 2500) against 50 ml of 0.5× TE buffer. Wet the filter for 1 min before applying the sample.
7. Add one ninth volume of the tenfold concentrated restriction endonuclease buffer provided by the manufacturer, and 80 U of *Eco*RI restriction endonuclease enzyme.
8. Incubate at 37°C for 60 min.
9. Add DNA agarose loading solution and examine the products by agarose gel electrophoresis or pulsed-field gel electrophoresis.

Protocol 8 P1 and BAC Miniprep Protocol with the Modified Alkaline Lysis Method

This protocol was developed by Shizuya et al.[52]

1. Inoculate 5 ml liquid broth (LB) medium containing 25 μg/ml of kanamycin (for P1 plasmids) or 12.5 μg/ml chloramphenicol (for BAC plasmids) with a single bacterial colony and incubated at 37°C overnight with vigorous shaking.
2. Concentrate the bacteria in a 2.0-ml microcentrifuge tube (Eppendorf) and resuspend in 100 μl of an ice-cold solution of 50 m*M* glucose, 10 m*M* EDTA and 25 m*M* Tris-HCl, pH 8.0.
3. Place the tubes on ice and add 200 μl freshly prepared lysis buffer (0.2 *M* NaOH, 1% SDS).

---

* It is possible to use *Alu*I methylase to methylate the sites for the restriction endonuclease *Hind*III instead of *Eco*RI methylase. The same concentration of *Alu*I methylase is used, but the incubation with the methylase is carried out for 45 min instead of 30 min.

4. Invert the tubes carefully five times and add 150 μl of ice-cold potassium acetate (pH 4.8).
5. Invert the tubes carefully five times and centrifuge for 5 min at full speed in a microfuge at room temperature.
6. The supernatant is transferred to a fresh tube and 2 vol of cold 100% ethanol is added, followed by a 5-min centrifugation.
7. Pour off supernatant, add 500 μl of 70% ethanol at room temperature to rinse pellet, and dry the pellet briefly at room temperature for 10 min.
8. Add 40 μl TE (1 m*M* EDTA, 10 m*M* Tris-HCl, pH 7.5) containing DNase-free pancreatic RNases (20 μg/ml), and resuspend by tapping.
9. Perform restriction digestion of 8 μl DNA solution in a reaction volume of 25 μl.

Protocol 9 Isolation of DNA Fragments for Pronuclear Injection

This protocol has been described by The Karolinska Institute MouseCamp (*http://www.ki.se/core/*).

1. Prepare plasmid DNA by commercial preparation kits or by CsCl purification.
2. Digest approximately 30 μg of plasmid DNA with the appropriate restriction enzymes (do not take less DNA or the final concentration will be too low). Heat inactivate the restriction enzyme (5 to 10 min at 70°C) and separate the DNA fragments by horizontal agarose gel electrophoresis in TAE buffer (important). For most purposes, a 0.7% agarose gel works well. Run the sample in several lanes to avoid overloading of the gel.
3. Visualize the EtBr-stained gel by long-wave UV light and cut out the fragment of interest in a minimal amount of agarose. Avoid excessive exposure of the gel (and yourself!) to UV light.
4. Place the fragments in six to eight 1.5-ml microcentrifuge tubes and determine the weight of the fragments.
5. To each tube, add 2.5 volume (i.e., 250 μl/100 mg of agarose slice) of the NaI solution from the GeneClean kit (GeneClean, Bio 101, 1070 Joshua Way, Vista CA 92083, e-mail: bio101@bio101.com).
6. Incubate at 50°C for 1 to 2 min, vortex briefly, and incubate 5 min at 50°C.
7. Check that all agarose is in solution. To each tube, add 5 μl GlassMilk (GeneClean kit) for up to 5 μg DNA; then add 1 extra μl GlassMilk for each additional μg DNA.
8. Vortex briefly, incubate on ice for 5 min, and centrifuge at max speed in a microcentrifuge for 5 s (to spin the glass beads down).
9. Pour off the supernatant, add 500 μl of New Wash (GeneClean kit) and redissolve the pellet (with the DNA on the glass beads). Make sure that the New Wash solution is very cold (take directly from –20°C). Then centrifuge the tubes for 5 s at max speed, as described above.
10. Repeat step 9 two more times (i.e., a total of three washes). After the last centrifugation step, remove all New Wash with a micropipette and leave

the tubes with open lid for 4 min at room temperature (for the last New Wash solution to evaporate).

11. Resuspend the pellet in 30 μl of sterile injection buffer (10 m*M* Tris, 0.2 m*M* EDTA, pH 7.4) for each tube. Incubate for 3 min at 50°C. Spin down the pellet for 5 s at max speed in a microcentrifuge.
12. Transfer the supernatants from all the microcentrifuge tubes to one new 1.5-ml microcentrifuge tube. Add another 30-μl injection buffer to each pellet, redissolve the pellet and repeat the centrifugation. Transfer these supernatants to the new tube.
13. To remove all residual glass milk, centrifuge the tube containing the pooled supernatants for 5 s and transfer to a new tube. Repeat until no pellet is visible (four to six times). The volume should now be approximately 300 μl.
14. Solubilize the DNA in TE buffer (10 m*M* Tris-HCl, pH 7.4, 1 m*M* EDTA) and assess the concentration of the DNA by spectrophotometry (260 to 280 nm), fluorometry, by comparison to DNA markers of known concentration after electrophoresis on agarose gels, or by the DNA Dipstick kit (Invitrogen, San Diego, CA).
15. Adjust the DNA concentration to approximately 2 μg/ml in TE immediately prior to micro-injection.
16. Optional: filter the DNA through a 0.22-μm Acrodisk filter (Gelman, Ann Arbor, MI) immediately prior to injection.

PROTOCOL 10 PREPARATION OF P1 BACTERIOPHAGE AND BAC DNA FOR PRONUCLEAR INJECTION

Preparation of P1 Bacteriophage and BAC DNA*

1. Innoculate 1 l LB medium containing 25 μg/ml of kanamycin (for P1) or 12.5 μg/ml chloramphenicol (for BAC) with 10 ml overnight of an *E. coli* strain DH10B (Gibco BRL) harboring the P1 or BAC.
2. Incubate in a shaking incubator at 37°C.
3. Add 10 ml of 100 m*M* isopropyl-β-D-thiogalactopyranoside (i.e., 0.25 g in water) to the culture after 1 h incubation and continue to grow the bacteria in a shaking incubator for 4 h (for P1) or 6 h (BAC).
4. Cool on ice and pellet the bacteria in a 1000-ml bottle (4000 rpm for 15 min).
5. Prepare the P1 and BAC DNAs by alkaline lysis using the Qiagen Plasmid Maxi kit (Qiagen). Resuspend cells in 16 ml of P1 solution. Add 2 ml of 50 mg/ml lysozyme *freshly made* in P1 solution. Mix and leave at room temperature for 15 min.

* Care should be taken during the whole procedure to avoid overdrying pellets or vortexing at any step. From a 1-l culture, the typical yield was 600 to 800 μg P1 DNA and 100 to 200 μg of BAC DNA.

6. Purify the P1 or BAC DNA on three Qiagen-tip 500 columns, using the washing and elution buffers contained in the kit except for buffer P2, which must be freshly prepared before use (0.2 *M* NaOH, 1% SDS).
7. Resuspend the DNA pellet in 600 μl of TE. Let stand for >30 min at room temperature until dissolved.
8. Add 70 μl of 3 *M* sodium acetate.
9. Extract once with 700 μl of phenol/chloroform and once with 700 μl of chloroform. Carefully invert tube by hand for 1 min in each step. Do not vortex.
10. Precipitate by adding 400 μl of isopropanol, and centrifuge for 10 min in a microcentrifuge at room temperature.
11. Wash the DNA precipitate with 1 ml 70% ethanol, and centrifuge for 5 min in a microcentrifuge at room temperature.
12. Assess the concentration of the DNA by spectrophotometry (260 to 280 nm), after solubilization of the DNA in 300 μl of TE buffer (10 m*M* Tris-HCl, pH 7.4, 1 m*M* EDTA).

Preparative Pulsed-Field Gel Electrophoresis

1. For a generation of transgenic mice, it is not desirable to micro-inject vector sequences.[23] The P1 and BAC vectors are equipped with several rare restriction endonuclease cleavage sites flanking the DNA insert. Thus, the DNA insert frequently can be cleaved from the vector with these enzymes.
2. Digest a total of 10 to 20 μg P1 or BAC DNA with the appropriate restriction endonuclease and load it onto a 1% preparative low-melt agarose gel (Sea Plaque® GTG®, FMC BioProducts, Rockland, ME).
3. Run the preparative pulsed-field gel electrophoresis (PFGE) for 12 h using a linear ramp from 2 to 6 s with 6 V/cm in 0.5× TAE buffer at 14°C.
4. Following electrophoresis, marker lanes on both ends of the gel were removed, stained with ethidium bromide (0.5 μg/ml), and photographed to visualize the position of DNA insert.
5. The large DNA band containing the insert of the P1 or BAC clone is excised from the unstained portion of the gel and equilibrated for 30 min at room temperature in 25 ml Gelase buffer (Epicentre Technologies, Madison, WI) supplemented with 100 m*M* NaCl, 30 μ*M* spermine, and 70 μ*M* spermidine.
6. Melt the agarose (approximately 100 to 200 μl) at 68°C for 10 min, equilibrated it for 10 min at 45°C, and digest it for 60 min with 1 U of Gelase.
7. Dialyze the DNA by placing the DNA solution on a microdialysis membrane (type VS, 0.025 μm, Millipore, Bedford, MA) floating in TE buffer containing 100 m*M* NaC1, 30 μ*M* spermine, and 70 μ*M* spermidine.
8. Centrifuge the dialyzed DNA-containing solution was at 12,500× *g* for 15 min.

9. The concentration of the DNA is determined by fluorometry by comparison to DNA markers of known concentration after electrophoresis on 1% agarose gels, or by the DNA Dipstick kit (Invitrogen, San Diego, CA).
10. To assess the integrity of the DNA, approximately 0.5 μg is subjected to pulsed-field electrophoresis on a 1% agarose gel (Sea Kem GTG from FMC BioProducts, Rockland, ME).*
11. Adjust the DNA concentration to approximately 2 μg/ml in TE buffer containing 100 m*M* NaCl, 30 μ*M* spermine, and 70 μ*M* spermidine prior to micro-injection. Inclusion of spermine or spermidine in the micro-injection buffers does not adversely affect the generation of transgenic mice.
12. Optional: filter the DNA through a 0.22-μm Acrodisk filter (Gelman, Ann Arbor, MI) immediately prior to injection.

Protocol 11 Mini DNA Preparation for PCR and/or Southern Analysis

[As described by the Mouse Molecular Genetics Groups at Göteborg University (http://cbz.medkem.gu.se/)]

1. Take a ≤5 mm biopsy from a mouse tail. Preferably, take the biopsy from a mouse less than 16 d of age. Mice under this age bleed little, and they do not have to be subjected to the double stress of biopsing and weaning at the same time. From a younger mouse, a smaller biopsy might be taken, as cell density is higher. From an older mouse, more than 5 mm never has to be taken as the thicker tail makes the volume of the biopsy larger.
2. Put the tissue in a 1.5-ml Eppendorf tube. It is not necessary to cut the biopsy into smaller pieces.
3. Add to each sample 500 μl sample tail-lyse buffer (50 m*M* Tris, pH 8.0, 100 m*M* EDTA, 100 m*M* NaCl; store at room temperature), 25 μl Proteinase K (10 mg/ml, dissolved in dH_2O, stored at –20°C), 25 μl 20% SDS.
4. Incubate at 56°C overnight. Take the samples out after some hours and mix the samples by pipetting or vortexing. At this stage, samples may be put in refrigerator for a few days or may be frozen.
5. Add 1 volume (550 μl) Phenol : Chisam (Chisam = 24 : 25 chloroform, 1 : 25 isoamylalcohol). Vortex for 20 s or shake until an emulsion has formed. Separate phases by spinning in a table centrifuge at full speed for 10 min. Transfer 500 μl of upper phase to a new 1.5-ml tube using a cut 100- to 1000-μl pipette tip to make the opening larger.
6. If your restriction enzyme is sensitive, repeat step 5.
7. Precipitate the DNA by adding 2 volumes of 99% ethanol. Turn the tubes several times until a nest of DNA is formed. (If only small pieces of DNA show, do not worry, go ahead, but you might want to use the longer times for spinning.) Do not add more than 2 volumes ethanol because other

* This procedure consistently yields P1 and BAC DNA that is completely free of sheared fragments, as judged by ethidium bromide-stained pulsed-field agarose gels.

substances then might precipitate and the chloroform might form an emulsion.

8. Spin the nest in a table centrifuge at full speed for 3 min. Pour off the 99% ethanol and immediately add approximately 1 ml of 70% ethanol for washing. Leave for some minutes (even overnight) at room temperature.
9. Pour on ethanol and watch the nest so it is not poured out as well. Spin briefly in a table centrifuge. Remove the remaining ethanol with a pipette. Do not let the DNA dry completely. Add 50 to 200 µl dH_2O. Leave the tubes for several hours or overnight at room temperature for the DNA to dissolve, or at 37°C several hours. The DNA may now be stored at –20°C for several months.
10. Pipette DNA to make it less viscous.
11. Mix 5 µl DNA lysate with 995 µl dH_2O and measure OD^{260}. Calculate DNA concentration. OD 1 = 50 µg/ml. The preceding dilution gives 10 µg = 1/OD = x µl of DNA lysate.
12. Run a PCR or digest 10 to 20 µg DNA with a specific restriction enzyme for Southern blot. Digest at 37°C overnight. Run one third to one half of the restriction digest on an agarose gel. Voltage and concentration of the gel depend on the size of the bands to be separated.

Protocol 12 Quick Preparation of DNA for PCR Analysis

This protocol has been described by the Mouse Molecular Genetics Groups at Göteborg University; see *http://cbz.medkem.gu.se/*.

1. A biopsy of approximately 1 mm in length is taken from the outer tip of the tail. Try to cut only in the skin avoiding the bone. Try to biopsy the mice at 10 to 16 d of age. Because young mice have higher density of cells a smaller biopsy can be used for genotyping.
2. Biopsies are lysed in 25 to 35 µl of lysisbuffer each (a standard PCR buffer that comes with the Taq DNA-polymerase, or Gittschier buffer). Lysis is done in a 56°C-incubator overnight. Tap the tubes to separate cells after a few hours. Alternatively, tap the tubes in the morning and then return them to the incubator for 1 h.
3. Inactivate the Proteinase K at 85°C for 10 min before setting up PCR reactions. It is usually sufficient to use 4 to 5 µl of lysate for one PCR reaction.

Buffers for Lysis before PCR

1× Gittschier buffer without gelatin (stored as 10× stock at –20°C)
0.5% Triton X-100 (stored as 20× stock at RT).
500 ug/ml proteinase K(10 mg/ml in dH_2O at –20°C).
10× Gittschier buffer (166 m*M* ammonium sulfate, 670 m*M* Tris pH 8.8, 67 m*M* $MgCl_2$, 5 m*M* β-mercaptoethanol, 67 µ*M* EDTA, for PCR buffer include 0.01% gelatin); store at –20°C in aliquots

1× PCR buffer (with or without $MgCl_2$) (included with Taq DNA-polymerase)
0.045% NP40 (stored as 10× stock at RT)
0.045% Tween 20 (stored as 10× stock at RT)
500 μg/ml proteinase K

REFERENCES

1. Goffeau, A., Barrell, B. G., Bussey, H., Davis, R. W., Dujon, B., Feldmann, H., Galibert, F., Hoheisel, J. D., Jacq, C., Johnston, M., Louis, E. J., Mewes, H. W., Murakami, Y., Philippsen, P., Tettelin, H., and Oliver, S. G. 1996. Life with 6000 genes, *Science*, 274, 536–537.
2. The *C. elegans* Sequencing Consortium. 1998. Genome sequence of the nematode *C. elegans*: a platform for investigating biology, *Science*, 282, 2012–2018.
3. Dutta, A. and Bell, S. P. 1997. Initiation of DNA replication in eukaryotic cells, *Ann. Rev. Cell Dev. Biol.*, 13, 293–332.
4. Björklund, S. and Kim, Y. J. 1996. Mediator of transcriptional regulation, *Trends Biochem. Sci.*, 21, 335–337.
5. Cairns, B. R. 1998. Chromatin remodeling machines: similar motors, ulterior motives, *Trends Biochem. Sci.*, 23, 20–25.
6. Brent, R. and Finley, J. R. I. 1997. Understanding gene and allele function with two-hybrid methods, *Ann. Rev. Genet.*, 31, 663–704.
7. Winzeler, E. A. and Davis, R. W. 1997. Functional analysis of the yeast genome, *Curr. Opin. Genet. Dev.*, 7, 771–776.
8. Holstege, F. G., Jennings, E. G., Wyrick, J. J., Lee, T. I., Hengartner, C. J., Green, M. R., Golub, T. R., Lander, E. S., and Young, R. A. 1998. Dissecting the regulatory circuitry of a eukaryotic genome, *Cell*, 95, 717–728.
9. Myers, L. C., Gustafsson, C. M., Hayashibara, K. C., Brown, P. O., and Kornberg, R. D. 1999. Mediator protein mutations that selectively abolish activated transcription, *Proc. Natl. Acad. Sci. U.S.A.*, 96, 67–72.
10. Lorenz, M. C., Muir, R. S., Lim, E., McElver, J., Weber, S. C., and Heitman, J. 1995. Gene disruption with PCR products in *Saccharomyces cerevisiae*, *Gene*, 158, 113–117.
11. Moqtaderi, Z., Bai, Y., Poon, D., Weil, P. A., and Struhl, K. 1996. TBP-associated factors are not generally required for transcriptional activation in yeast, *Nature*, 383, 188–191.
12. Lowry, C. V. and Zitomer, R. S. 1988. ROX1 encodes a heme-induced repression factor regulating ANB1 and CYC7 of *Saccharomyces cerevisiae*, *Mol. Cell. Biol.*, 8, 4651–4658.
13. Varshavsky, A. 1997. The N-end rule pathway of protein degradation, *Gene Cells*, 2, 13–28.
14. Tsien, R. Y. 1998. The green fluorescent protein, *Ann. Rev. Biochem.*, 67, 509–544.
15. Schneider, B. L., Seufert, W., Steiner, B., Yang, Q. H., and Futcher, A. B. 1995. Use of polymerase chain reaction epitope tagging for protein tagging in *Saccharomyces cerevisiae*, *Yeast*, 11, 1265–1274.
16. DeRisi, J. L., Iyer, V. R., and Brown, P. O. 1997. Exploring the metabolic and genetic control of gene expression on a genomic scale, *Science*, 278, 680–686.
17. Schena, M., Heller, R. A., Theriault, T. P., Konrad, K., Lachenmeier, E., and Davis, R. W. 1998. Microarrays: biotechnology's discovery platform for functional genomics, *Trends Biotechnol.*, 16, 301–306.

18. Lipshutz, R. J., Morris, D., Chee, M., Hubbell, E., Kozal, M. J., Shah, N., Shen, N., Yang, R., and Fodor, S. P. 1995. Using oligonucleotide probe arrays to access genetic diversity, *Biotechniques*, 19, 442–447.
19. Schena, M., Shalon, D., Davis, R. W., and Brown, P. O. 1995. Quantitative monitoring of gene expression patterns with a complementary DNA microarray, *Science*, 270, 467–470.
20. Ptashne, M. and Gann, A. 1997. Transcriptional activation by recruitment, *Nature*, 386, 569–577.
21. Kim, Y. J., Björklund, S., Li, Y., Sayre, M. H., and Kornberg, R. D. 1994. A multiprotein mediator of transcriptional activation and its interaction with the C-terminal repeat domain of RNA polymerase II, *Cell*, 77, 599–608.
22. Myers, L. C., Gustafsson, C. M., Bushnell, D. A., Lui, M., Erdjument-Bromage, H., Tempst, P., and Kornberg, R. D. 1998. The Med proteins of yeast and their function through the RNA polymerase II carboxy-terminal domain, *Genes Dev.*, 12, 45–54.
23. Hogan, B., Beddington, R., Costantini, F., and Lacy, E. 1994. *Manipulating the Mouse Embryo. A Laboratory Manual*, Cold Spring Harbor Laboratory Press, Plainview, NY.
24. Ferrin, L. J. and Camerini-Otero, R. D. 1991. Selective cleavage of human DNA: RecA-assisted restriction endonuclease (RARE) cleavage, *Science*, 254, 1494–1497.
25. Borén, J., Lee, I., Callow, M. J., Rubin, E. M., and Innerarity, T. L. 1996. A simple and efficient method for making site-directed mutants, deletions, and fusions of large DNA such as P1 and BAC clones, *Genome Res.*, 6, 1123–1130.
26. Borén, J., Lee, I., Zhu, Arnold, W. K., Taylor, S., and Innerarity, T. L. 1998. Identification of the low density lipoprotein receptor-binding site in apolipoprotein B100 and the modulation of its binding activity by the carboxyl terminus in familial defective apoB100, *J. Clin. Invest.*, 101, 1084–1093.
27. Borén, J., Olin, K., Lee, I., Chait, A., Wight, T., and Innerarity, T. 1998. Identification of the principal proteoglycan binding site in LDL: a single point mutation in apolipoprotein B100 severely affects proteoglycan interaction without affecting LDL receptor binding, *J. Clin. Invest.*, 101, 2658–2664.
28. Callow, M. J., Ferrin, L. J., and Rubin, E. M. 1994. Single base, site-directed mutagenesis of a 90 kilobase-pair P1 clone, *Nucleic Acids Res.*, 22, 4348–4349.
29. Callow, M. J. and Rubin, E. M. 1995. Site-specific mutagenesis demonstrates that cysteine 4326 of apolipoprotein B is required for covalent linkage with apolipoprotein(a) *in vivo*, *J. Biol. Chem.*, 270, 23914–23917.
30. McCormick, S. P. A., Ng, J. K., Taylor, S., Flynn, L. M., Hammer, R. E., and Young, S. G. 1995. Mutagenesis of the human apolipoprotein B gene in a yeast artificial chromosome reveals the site of attachment for apolipoprotein(a), *Proc. Natl. Acad. Sci. U.S.A.*, 92, 10147–10151.
31. McCormick, S. P. A., Peterson, K. R., Hammer, R. E., Clegg, C. H., and Young, S. G. 1996. Generation of transgenic mice from yeast artificial chromosome DNA that has been modified by gene targeting, *Trends Cardiovasc. Med.*, 6, 16–24.
32. Peterson, K. R., Li, Q.-L., Clegg, C. H., Furukawa, T., Navas, P. A., Norton, E. J., Kimbrough, T. G., and Stamatoyannopoulos, G. 1995. Use of yeast artificial chromosomes (YACs) in studies of mammalian development: production of β-globin locus YAC mice carrying human globin developmental mutants, *Proc. Natl. Acad. Sci. U.S.A.*, 92, 5655–5659.

33. Rothstein, R. 1991. Targeting, disruption, replacement, and allele rescue: integrative DNA transformation in yeast, *Methods Enzymol.*, 194, 281–301.
34. Scherer, S. and Davis, R. W. 1979. Replacement of chromosome segments with altered DNA sequences constructed *in vitro*, *Proc. Natl. Acad. Sci. U.S.A.*, 76, 4951–4955.
35. McCormick, S. P. A., Linton, M. F., and Young, S. G. 1994. Expression of P1 DNA in mammalian cells and transgenic mice, *Genet. Anal. Tech. Appl.*, 11, 158–164.
36. Duell, T., Nielsen, L. B., Jones, A., Young, S. G., and Weier, H. U. 1997. Construction of two near-kilobase resolution restriction maps of the 5′ regulatory region of the human apolipoprotein B gene by quantitative DNA fiber mapping (QDFM), *Cytogenet. Cell Genet.*, 79, 64–70.
37. Townes, T. M., Chen, H. Y., Lingrel, J. B., Palmiter, R. D., and Brinster, R. L. 1983. Expression of human β-globin genes in transgenic mice: effects of flanking metallothionein-human growth hormone fusion gene, *Mol. Cell. Biol.*, 5, 1977–1983.
38. Khillan, J. S. 1997. Transgenic animals as bioreactors for expression of recombinant proteins, in *Recombinant Protein Protocols*, Vol. 63, Tuan, R., Ed., Humana Press, Totowa, NJ, 327–342.
39. Brinster, R. L., Chen, H. Y., Trumbauer, M. E., Yagle, M. K., and Palmiter, R. D. 1985. Factors affecting the efficiency of introducing foreign DNA into mice by microinjecting eggs, *Proc. Natl. Acad. Sci. U.S.A.*, 82, 4438–4442.
40. Cid-Arregui, A. and Garcia-Carranca, A. 1998. *Micro-Injection and Transgenesis: Strategies and Protocols*, Springer-Verlag, Berlin.
41. Havel, R. J. and Kane, J. P. 1995. Introduction: structure and metabolism of plasma lipoproteins, in *The metabolic and molecular bases of inherited disease*, Vol. 2, 7th ed., Scriver, C. R., Beaudet, A. L., Sly, W. S., and Valle, D., Eds., McGraw-Hill, New York, 1841–1851.
42. Ross, R. 1995. Cell biology of atherosclerosis, *Ann. Rev. Physiol.*, 57, 791–804.
43. Chiesa, G., Johnson, D. F., Yao, Z., Innerarity, T. L., Mahley, R. W., Young, S. G., Hammer, R. H., and Hobbs, H. H. 1993. Expression of human apolipoprotein B100 in transgenic mice. Editing of human apolipoprotein B100 mRNA, *J. Biol. Chem.*, 268, 23747–23750.
44. Linton, M. F., Farese, Jr., R. V., Chiesa, G., Grass, D. S., Chin, P., Hammer, R. E., Hobbs, H. H., and Young, S. G. 1993. Transgenic mice expressing high plasma concentrations of human apolipoprotein B100 and lipoprotein(a), *J. Clin. Invest.*, 92, 3029–3037.
45. Callow, M. J., Stoltzfus, L. J., Lawn, R. M., and Rubin, E. M. 1994. Expression of human apolipoprotein B and assembly of lipoprotein(a) in transgenic mice, *Proc. Natl. Acad. Sci. U.S.A.*, 91, 2130–2134.
46. Purcell-Huynh, D. A., Farese, Jr., R. V., Johnson, D. F., Flynn, L. M., Pierotti, V., Newland, D. L., Linton, M. F., Sanan, D. A., and Young, S. G. 1995. Transgenic mice expressing high levels of human apolipoprotein B develop severe atherosclerotic lesions in response to a high-fat diet, *J. Clin. Invest.*, 95, 2246–2257.
47. McCormick, S. P. A., Ng, J. K., Cham, C. M., Taylor, S., Marcovina, S. M., Segrest, J. P., Hammer, R. E., and Young, S. G. 1997. Transgenic mice expressing human apoB95 and apoB97. Evidence that sequences within the carboxyl-terminal portion of human apoB100 are important for the assembly of lipoprotein(a), *J. Biol. Chem.*, 272, 23616–23622.

48. Nielsen, L. B., McCormick, S. P. A., Pierotti, V., Tam, C., Gunn, M. D., Shizuya, H., and Young, S. G. 1998. Human apolipoprotein B transgenic mice generated with 207-kb and 145-kb bacterial artificial chromosomes: evidence that a distant 5′ element confers appropriate transgene expression in the intestine, *J. Biol. Chem.*, 272, 29752–29758.
49. Nielsen, L. B., Kahn, D., Duell, T., Weier, H. U., Taylor, S., and Young, S. G. 1998. Apolipoprotein B gene expression in a series of human apolipoprotein B transgenic mice generated with recA-assisted restriction endonuclease cleavage-modified bacterial artificial chromosomes. An intestine-specific enhancer element is located between 54 and 62 kilobases 5′ to the structural gene, *J. Biol. Chem.*, 273, 21800–21807.
50. Borén, J., Véniant, M. M., and Young, S. G. 1998. Apo-B100-containing lipoproteins are secreted by the heart, *J. Clin. Invest.*, 101, 1197–1202.
51. Sikorski, R. S. and Hieter, P. 1989. A system of shuttle vectors and yeast host strains designed for efficient manipulation of DNA in *Saccharomyces cerevisiae*, *Genetics*, 122, 19–27.
52. Shizuya, H., Birren, B., Kim, U. J., Mancino, V., Slepak, T., Tachiiri, Y., and Simon, M. 1992. Cloning and stable maintenance of 300-kilobase-pair fragments of human DNA in *Escherichia coli* using an *F*-factor-based vector, *Proc. Natl. Acad. Sci. U.S.A.*, 89, 8794–8797.

8 How to Employ Proteins in Nonaqueous Environments

Douglas G. Hayes

CONTENTS

Introduction 180
Selection of Biocatalyst 181
Preparation of Biocatalyst 181
    Is Further Purification Needed? 181
    Lyophilization 182
        Lyoprotectants 185
        pH Buffer Salts 185
        Salts 185
        Ligands (Bio-Imprinting) 186
        Surfactants 186
        Crown Ethers 186
        Co-factors 186
        Polymers 186
    Immobolization 187
    Formation of Noncovalent, Solvent-Soluble Protein-Lipid Aggregates 190
    Covalent Modification of Enzymes 193
Preparation of Substrates 194
Choice of Continuous Phase 194
Control of Water Activity of Proteins, Solvents, and Substrates 198
Use of Microemulsions (Reversed Micelles) 201
Operation of Enzymatic Reactions 202
    Screening Experiments 202
    Temperature 203
    Dispersion of Biocatalyst and Reactor Design 203
    Continuous Control of Water Activity 204
    Use of Water Mimics 208
    Use of Organic Phase Buffers to Control pH 208
    Continuous Recovery of Product 209
    Engineering of Conditions during the Course of Reaction 209
    Monitoring Progress of Biocatalytic Reactions 210

0-8493-9453-8/02/$0.00+$1.50

Conclusions 211
References 211

INTRODUCTION

The employment of proteins in nonaqueous environments (e.g., organic solvents, supercritical fluids, and vapors), once believed to be a fallacy, is becoming common practice in biotechnology. In particular, the research field of enzymology in nonaqueous media has grown increasingly in the past 15 years, owing to numerous investigations undertaken by laboratories throughout the world in the academic, governmental, and industrial sectors. Major driving forces behind the increased attention include the use of enzymes in producing chiral drugs,[1–4] and the development of more active and thermostable enzymes through micro-organism screening,[5–7] protein engineering (site-directed mutagenesis),[7–9] and random mutagenesis.[7, 8, 10, 11] Some novel applications of nonaqueous enzymology are listed in Table 8.1.[1, 8, 12–30] Several reviews have been written on nonaqueous enzymology.[31–38] These reviews suggest the following advantages for hosting enzymatic reactions in nonaqueous media:

- Enhanced solubilization of lipophilic substrates
- Reduction of enzymatic hydrolysis
- Alteration of enzymatic substrate, regio-, and stereo-selectivity
- Enhanced thermostability (because of the reduced water activity)
- Simplified post-reaction recovery of protein and products
- Reduction of microbial contamination

The challenge for more widespread and industrial acceptance is to increase the productivity rates of many nonaqueous biocatalytic reactions.[36] It is believed for many situations that this goal can be achieved through proper methodology and design of reaction medium and conditions.[39] In particular, these specific and related design issues must be addressed:

- Protein preparation method (e.g., immobilization, lyophilization, derivatization, and encapsulation)
- Pre-treatment of reaction media (e.g., organic solvent or other continuous phase, substrates, and co-factors)
- Selection of reaction operating conditions (e.g., temperature and solvent system programming, water activity control, procedure used to form the medium)

This chapter reviews and examines the methodology employed in nonaqueous enzymology, and presents the current state of the art. The material is specifically directed at the novice in this field. It is also desired that this review will serve as a source of origin for study of specific aspects of nonaqueous enzymatic methods.

TABLE 8.1
Examples of Reactions Catalyzed by Enzymes in Nonaqueous Media

Application	Enzyme
Aldehyde and ketone modification	Alcohol dehydrogenase[12,13]
Biosurfactants (sugar–fatty acid esters)	Lipase[14] and Subtilisin[15]
Chiral drugs (Ibuprofen and Naproxen)	Lipase[16,17]
(–)-Lactam (HIV drug intermediate)	Acylase[1]
Nerve gas degradation	Organophosphorus hydrolase[18]
Optically pure polyesters	Lipase[19,20]
Peptide synthesis	Penicillin amidase,[21] Proteases,[8,22] and Thermolysin[23]
Phenol polymerization	Horseradish peroxidase[24,25]
Steroid modification	Cholesterol oxidase,[264] Dehydrogenase,[26] and Lipase[27,28]
Structured lipids (infant formula)	Lipase[29,30]

SELECTION OF BIOCATALYST

Enzymes and proteins are available commercially from a wide variety of sources (Table 8.2). The extensive technical literature related to nonaqueous enzymology will assist the user in selecting enzymes to fulfill a specific application. A series of screening reactions (discussed below) can be employed to further narrow the selection. In addition, a relatively new tool, molecular modeling, may be valuable in determining whether a given substrate is accessible to a given enzyme's active site. Modeling is reported to be very effective for predicting enzyme stereoselectivity for a substrate of interest.[40–42] Several software packages are available that import protein crystal structures from the Protein Data Bank[43] contained at a Web site (*www.rcsb.org*) maintained by a consortium led by Rutgers University and the U.S. National Institutes of Standards. In addition to molecular modeling, several groups, through extensive screening experiments performed on substrate libraries, have been able to predict the spacial positions and structure of active and substrate binding sites, hence the development of stereoselectivity models.[44–46] As additional protein crystal structures are resolved and combinatorial chemistry continues its growth, the use of molecular modeling and substrate libraries in nonaqueous enzymology will most probably expand.

PREPARATION OF BIOCATALYST

IS FURTHER PURIFICATION NEEDED?

In many cases, commercially available enzyme preparations (Table 8.2), often provided in lyophilized form, are highly impure, containing small quantities of salts, carbohydrates, and lyoprotectants, such as sorbitol (discussed below). For example, *Candida rugosa* lipase, a product from Sigma-Aldrich, contains stabilizers, such as lactose, and hydrolases, which possess contaminating biocatalytic activity.[47, 48] Purification requires significant time and effort and, therefore, is not recommended unless high-level optimization is being pursued. Whether protein purification greatly improves enzyme activity in organic solvent is a debatable issue. Poor activity, owing to ineffective

TABLE 8.2
Suppliers of Enzyme and Enzyme Technology for Use in Nonaqueous Media

Company	Address	Enzymes
Albany Molecular Research, Inc.	Albany, NY www.albmolecular.com	Enzymes in drug discovery and combinatorial analysis
Altus Biologics	Cambridge, MA *www.altus.com*	Enzyme cross-linked crystals
Bio-Cat, Inc.	Troy, VA *www.bio-cat.com*	Several
Biocatalysts Ltd.	Mid Glamorgan, U.K.	Lipases
Biosun Biochemicals	Tampa, FL *www.biosunbc.com*	Amylases and food-processing enzymes
Biozyme	San Diego, CA *www.biozyme.com*	Dehydrogenases, peroxidase, glucosidase, catalase
Genecor, Intl.	San Francisco, CA *www.genencor.com*	Detergent, starch enzymes
Gist-Brocades	Delft, Netherlands www.dsm.nl	Food and dairy enzymes
Novo-Nordisk	Bagsvaerd, Denmark www.enzymes.novo.dk	Several
Roche Molecular Biochemicals	Indianapolis, IN *www.biochem.roche.com*	Amidases, lipases, proteases
Sigma-Aldrich-Fluka	St. Louis, MO www.sigma-aldrich.com	Several
U.S. Biological	San Antonio, TX *www.nefind.com*	Several
Valley Research	South Bend, IN www.valleyenzymes.com	Proteases, lipases, amylases
Worthington Biochemical	Lakewood, NJ www.worthington-biochem.com	Several

dispersion in nonaqueous media, has been reported for several purified enzymes.[49, 50] Recently, Dordick and co-workers demonstrated increased activity occurs for purified *C. rugosa* lipase relative to the crude form when the procedure of the enzyme's incorporation into nonaqueous medium is carefully performed (discussed below).[48]

Enzyme purification may be required for selective reactions, particularly those that require high stereoselectivity. Moreover, the removal of contaminents has been reported to increase the enantioselectivity significantly.[47, 48] Methods for protein purification include various low-pressure column chromatographies (e.g., size exclusion, affinity, or hydrophobic interaction) and dialysis to remove added salts.

Lyophilization

Lyophilization, or freeze-drying, has been commonly practiced in the food and pharmaceutical industries for several decades. Most enzyme products (e.g., Table 8.2) are supplied as lyophilized powders. The technique is also commonly used for the stabilization and long-term storage of microbial cells, proteins, nucleotides, and peptides. Lyophilization consists of the removal of solvent (typically, water) from solute through the sublimation of the former. For sublimation to occur, subzero temperatures and vacuum-level pressures are required.

The science of lyophilization will be summarized here. The literature contains more detail (e.g.,[51]). Lyophilization of proteins from aqueous buffer involves three basic steps.

- The rapid cooling of the buffer solution to subzero temperatures (e.g., –20 to –40°C)
- Removal of bulk water referred to as primary drying
- Removal of strongly bound water referred to as secondary drying

The first step must occur as quickly as possible to reduce denaturation.[52] Typically, it is accomplished by immersing a vessel containing the aqueous solution into liquid nitrogen or cold acetone at a temperature between –35 and –80°C. Recently, the development of a new spray freezing technique has led to reduced protein denaturation, presumably because of its more rapid freezing rate.[52] This method, which is expensive and requires more sophisticated equipment, has not yet been applied to nonaqueous enzymology. On a theoretical level, the purpose of the freezing step is to lower the temperature of the mixture below a critical value needed to completely solidify all of its components. The freezing process may consist of the actual solidification of the solute–water eutectic mixture, or the formation of a solute glass phase, depending on the composition and thermodynamics of the aqueous mixture. A quick solidification process is required to limit the period of time the protein will have to unfold, or denature.[52] The mechanism of the unfolding appears to be the adsorption of protein molecules on the surface of the solid ice phase.[53]

The second step, primary drying, consists of the removal of water through sublimation. For sublimination to occur, according to the phase diagram for water, the pressure must be lowered at subzero temperatures to below 50 μm Hg (7×10^{-5} atm) by a vacuum pump. In addition, sublimation requires the transfer of energy equal in amount to the latent heat of sublimation to the solid phase. Furthermore, the water vapor that forms must be trapped. This latter objective is achieved using a condenser. Water condenses to solid on the condenser walls when the tube-side fluid is cooler than the subliming water vapor. The entire removal of the bulk water (i.e., the end of the primary drying stage) is noted by the sharp decrease in the partial pressure of water and by visual observation of the disappearance of ice.

Upon completion of primary drying, the water content of the freeze-dried protein may still be as high as 7 to 8%. Secondary drying is then applied to remove tightly bound water molecules, yielding a final water content of 0 to 2%. The process typically occurs at warmer conditions than primary drying, often at refrigerator or even ambient temperatures. It is not uncommon for secondary drying to occur as a series of constant temperature steps increasing in temperature from the primary drying temperature up to ambient.[52]

Freeze-drying equipment is available from several sources on several scales of operation. Manufacturers of laboratory-scale (e.g., bench, scale) lyophilizers include Labconco (Kansas City, MO), VirTis (Gardiner, NY), FTS Systems (New York), and

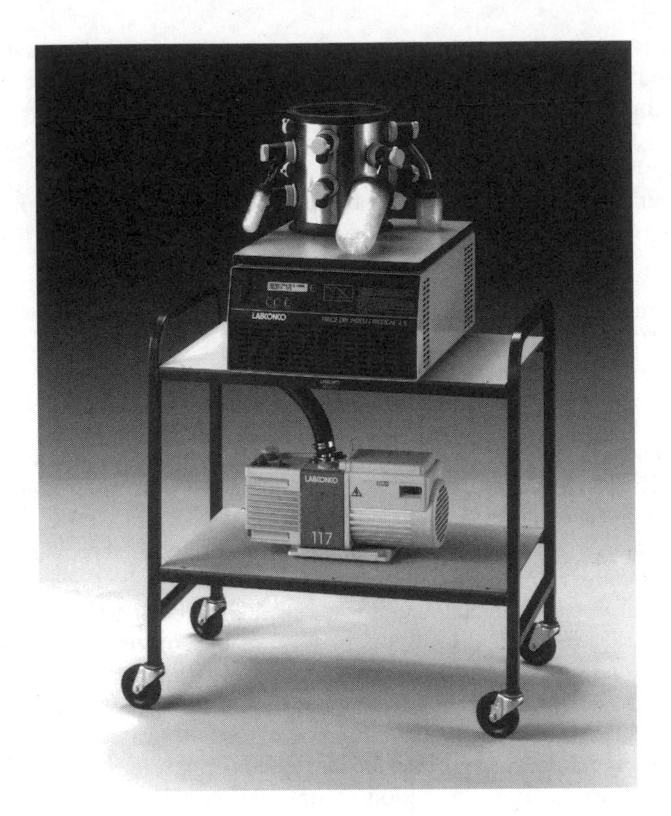

FIGURE 8.1 Illustration of a bench-scale freeze dryer. (Reprinted with permission from Labconco, Kansas City, MO.)

Wheaton Scientific (Millville, NJ). An example of a bench-scale freeze dryer is depicted in Figure 8.1. The freeze-drying apparatus is used for carrying out only the primary and secondary drying steps. Typically, the frozen sample is placed in the freeze dryer, which is preset to the primary drying temperature. Then vacuum pressure is applied to induce sublimation. There are two categories of freeze dryer design. In the first category, referred to as the manifold design (Figure 8.1), frozen samples are placed in vials that attach externally to ports of a drying chamber, the latter of which is subjected to subzero temperature and vacuum pressure. The placement of the vials external to the chamber allows the ambient air to supply the latent heat of sublimation/evaporation through the vial walls during primary drying, a process named evaporative cooling. The second category, known as a batch design, consists of a drying chamber containing several shelves for sample placement. Latent heat is supplied to the solid phase through resistive heating of the shelves. Most freeze dryers include gauges to monitor temperature and pressure. Further information on the operation of freeze dryers can be obtained from the manufacturers and several review articles (e.g.,[54]). The key operational parameters to be chosen are the temperature, pressure, and duration of primary drying.

It is well known that lyophilization induces denaturation and inactivation of proteins to varying degrees depending on the lyophilization conditions. In other words, lyophilization can lock biomolecules into inactive conformations, which are often fully reversible upon rehydration.[39, 55, 56] It is believed that improper lyophilization methodology is a primary cause of the low activity often exhibited by enzymes in nonaqueous media. Moreover, contact of lyophilized enzymes with most organic solvents does not induce further denaturation beyond that caused by lyophilization.[57, 58] Significant progress has been achieved in the past decade on the improvement of catalytic activity and binding efficiency of proteins in nonaqueous media by the inclusion of additives in the aqueous phase prior to lyophilization. These include the following.

Lyoprotectants

Lyoprotectants refer to sugars, sugar alcohols, and glycols that reduce denaturation during lyophilization.[56] Examples include sorbitol, polyethylene glycol (PEG), sucrose, and trehalose.[59, 60] The washing away of lyoprotectant from the lyophilized enzyme by polar solvent or water reverses the activity enhancement.[60] Lyoprotectants function during primary drying. Proposed mechanisms of their activity enhancement include the elevation of the minimum or glass transition temperature for the aqueous mixture, the prevention of protein adsorption on the surface of ice crystals, the reduction of protein flexibility owing to entrapment, and their ability to induce formation of a protective molecular water shell surrounding the protein.[53–55, 61] In addition, lyoprotectants reduce the occurrence of hydrolysis during enzyme-catalyzed alcoholysis.[60] The optimal level of lyoprotectant (e.g., 10 mmol/g protein for sorbitol) is relatively low in comparison to other additives, such as salts.[59, 62] Also, improvement of nonaqueous biocatalytic rates by lyoprotectants is often modest (within an order of magnitude) by comparison.[60]

pH Buffer Salts

It has been established that lyophilized proteins "remember" the pH history of their prelyophilization aqueous phase.[63] In the absence of buffer salts, lyophilization can induce pH changes in the protein environment.[61] Therefore, it is now common practice to lyophilize a given enzyme from a buffered aqueous phase set at its optimal activity pH value.

Salts

The presence of salt in the prelyophilization buffer solution at salt/enzyme ratios of 0.5 to 1 w/w can reduce denaturation and dramatically enhance enzyme activity.[58, 64] Specifically, salts increase the Michaelis-Menten turnover number, k_{cat}, and only slightly decrease the Michaelis constant, K_M.[64] Salt inclusion may also reduce substrate inhibition.[65] Underlying mechanisms of nonaqueous enzymatic activity include their buffering capabilities in the solid biocatalyst phase,[62] their rigidification of the solid phase,[64, 66] their provision of a polar environment for the solid-phase enzyme,[64] and their prevention of protein mobility during lyophilization.[64] Despite the large amount of salt present in the solid phase, substrate diffusion in the solid phase is not drastically reduced.[67] On the negative side, buffer salts can reduce thermostability.[62] Potassium salts are preferred over sodium salts because the latter offer less protection against pH changes during lyophilization,[65] with KCl being the most effective.[54, 61, 64]

Ligands (Bio-Imprinting)

The research groups of Klibanov and Mosbach in the late 1980s demonstrated that when lyophilization occurred in the presence of ligands, nonaqueous enzymatic activity was significantly improved. Upon completion of lyophilization, removal of the ligand through washing with anhydrous organic solvents is required to increase bioactivity and binding capacity.[68] Bio-imprinting can induce novel substrate specificity or binding affinity not present with native protein, and can create biospecific binding in polymers.[69, 70] It is believed that the ligand helps lock the biomolecule (or polymer) into an active conformation during lyophilization. The presence of water unlocks the bio-imprinting effect.[59, 71] The bio-imprinted proteins remain stable for several weeks.[68] In addition to ligands, transition state analogs are reported to be effective bio-imprinting agents.[72]

Surfactants

For lipolytic enzymes—lipases and phospholipases—Braco and co-workers demonstrated that the presence of surfactants (e.g., octyl β-D-glucopyranosides) and phospholipids in the prelyophilization aqueous phase at concentrations above their critical micellar concentrations (i.e., in the form of micelles or vesicles), increased enzymatic activity by several orders of magnitude.[73–75] It is believed for lipases that the formation of a small bulk aqueous interface induces an open conformation of the enzyme, where a lid that covers the active site recedes. Treatment of crude lipase with 2-propanol may also promote an open conformation of lipases.[76] Similar to bio-imprinting, the activation agent, the surfactant, must be rinsed away from the lyophilized enzyme by anhydrous solvent prior to use. The activity enhancement decreases with increasing water content in the reaction medium.[73]

Crown Ethers

The presence of crown ethers in aqueous medium before lyophilization can greatly enhance the activity of proteases (e.g., chymotrypsin, subtilisin, trypsin) for peptide synthesis in organic solvents.[77, 78] The crown ether must also be rinsed away from the lyophilized enzyme before use. The activity enhancement is solvent dependent.[78] It is speculated that crown ethers accelerate enzymatic rates either by preventing a salt bridge from forming in the enzyme's secondary structure[78] or inducing microscopic changes in the structure of the solid phase.[66]

Co-factors

Dehydrogenases are quite active when their co-factors, $NAD(P)^+$ or NAD(P)H, are co-lyophilized.[12, 13]

Polymers

In addition to PEG, often employed as a lyoprotectant, other polymers, when present in the aqueous mixture, can yield more active lyophilized enzymes.[12, 79]

Based on the technical literature[59, 60, 65, 80] and the preceding discussion, the following overall procedure is suggested for lyophilization.

- Prepare an aqueous buffer: ~10 mg/ml of protein, 7 to 10 mg/ml of KCl in a 10 m*M* buffer solution (at the optimal pH) in the presence of a bio-imprinting agent, surfactant, or crown ether if applicable.
- Freeze the buffer rapidly in liquid nitrogen (–40 to –80°C).
- Induce primary drying at –40°C and 5 to 300 μm Hg pressure for 24 to 48 h; maintain condenser temperature at –50 to –60°C.
- Stop primary drying when ice crystals disappear (visual observation) or when sublimation slows, noted by a sharp decrease in the partial pressure of water, or in the water content of the solid phase determined gravimetrically, or by a novel solid-phase Karl Fischer titration procedure.[65]
- Perform secondary drying by adjusting the water content of the lyophilized protein over saturated salt solutions or dessicant at refrigerator or ambient temperature (discussed below).

Immobilization

Immobilization, defined as the physical confinement or localization of an enzyme into a specific micro-environment,[81] has been a very common approach to prepare enzymes for aqueous as well as nonaqueous applications. For nonaqueous enzymology, immobilization improves storage and thermal stability, facilitates enzyme recovery, and enhances enzyme dispersion. In addition, immobilized enzymes are readily incorporated in packed bed bioreactors, allowing for continuous operation of reactions. Moreover, lyophilized enzyme powders often aggregate and attach to reactor walls, particularly when the water activity is moderately high. The major disadvantage of immobilization is low activity, induced by pore diffusion mass transfer limitations[82, 83] and by alteration of protein structure. For enzymes in nonaqueous media, the following broad categories of immobilization exist:

- Cross-linking of enzyme molecules (e.g., with glutaraldehyde)
- Entrapment into gels, membranes, liposomes, foams, liquid crystals, and other porous media
- Covalent attachment (e.g., through bioconjugate chemistry) between free amino groups of enzyme, particularly for lysine amino acid residues, and charged groups on the immobilization matrix
- Use of mycelia containing enzymes
- Adsorption onto matrices

Research and development is ongoing in all of these areas, with numerous preparations, methodologies, and applications reported. Coverage of each category listed previously is well beyond the scope of this report. The reader is referred to the numerous review articles and textbooks available (e.g., [81, 84–90]).

In this chapter, immobilization via adsorption will be briefly expanded because it is quickly, simply, and easily performed, and in the author's opinion, provides the highest precision for enzymatic reactions among the immobilized enzymes. Thus,

TABLE 8.3
Commonly Employed Matrices for Immobilization of Proteases and Hydrolases by Adsorption for Use in Nonaqueous Enzymology[a]

Matrix	Source	Notes
Porous silica	Amicon[b]	20–130 mesh, 50 nm pore size
DEAE Sephadex A50[c]	Sigma	
Celite 535 or 545[d]	The Celite Corp.[e]	130–180 mesh
Amberlite XAD-4	Sigma	4.0 nm pore size; 20–60 mesh
Amberlite XAD-7	Sigma	9.0 nm pore size; 20–60 mesh
Porous glass	Sigma	120–200 mesh (150 A)
Duolite ES562 or A-568[f]	Sigma	190 nm pore size
Dowex MWA-1	Sigma	16–50 mesh
Accurel[g] EP-100 or EG 700	Akzo-Nobel[h]	

a Data listed in their approximate order of hydrophobicity, with Accurel being the most hydrophobic
b Division of Millipore Corp., Bedford, MA
c Dextran
d Diatomaceous earth
e Lompoc, CA
f Immobilization matrix used for Lipozyme IM (immobilized *Rhizomucor miehei* lipase produced by Nordisk Bioindustrials (Bagsvaerd, Denmark)
g Macroporous polypropylene powder
h Odenburg, Germany

for a researcher beginning in the field of nonaqueous enzymology, adsorption may be of interest to increase enzyme stability and facilitate enzyme re-use in screening experiments. Adsorption has been applied successfully for several proteases and hydrolases, particularly lipases.

The most important factor for adsorption is the choice of immobilization matrix. Several commercially available matrices commonly employed for nonaqueous enzymology are listed in Table 8.3. They consist of macroporous millimeter-sized particulates that are hydrophilic with the exception of Accurel (polypropylene). Celite (diatomaceous earth) is probably, the most commonly employed matrix of those listed in Table 8.3. In addition, a few enzymes are commercially available in immobilized forms, including Lipozyme IM and Novozyme from Novo-Nordisk, the Chirazyme product line from Roche Molecular Biochemicals, and *Pseudomonas cepacia* lipase immobilized in Sol-Gel AK (Fluka).

Adlercreutz and co-workers suggest that the matrices be pretreated as follows.[83]

- Wash with ethanol to remove organic impurities.
- Wash with deionized water.
- Remove small particulates or fines through decanting.
- Repeat steps 2 and 3 several times.
- Place particulates in 10% (w/v) nitric acid overnight.
- Wash with deionized water to remove acid.
- Dry overnight at 80°C.

For Accurel EP-100, granular polypropylene, it is recommended that it be prewet with 0.5 to 6 ml of absolute ethanol per gram of matrix prior to immobilization.[91–93]

The choice of immobilization matrix material must be made carefully because its polarity, or aquaphilicity (water-adsorption capability), and the chemistry of its charged surface groups can affect the observed enantioselectivity[94, 95] and the occurrence of side reactions, particularly hydrolysis[96] and isomerization (e.g., of mono- and diglycerides during lipase-catalyzed synthesis of structured triglycerides).[97] In the situations alluded to above, it may be worthwhile to screen the various immobilization matrices.

There are two basic procedures for enzyme immobilization by adsorption. Both share the same first steps: preparation of the immobilization matrices (discussed above) and the aqueous enzyme solution. Typically, the aqueous solution is fairly concentrated in enzyme (approximately 5 to 40 mg/ml) and buffered (approximately 10 m*M*) to the optimal pH value of the enzyme. In addition, necessary co-factors, such as $NAD(P)^+$ or NAD(P)H, should be included.[98, 99] Also, the presence of albumin or high-molecular-weight PEG at concentrations of ~2 mg per g of matrix have been shown effective in protecting enzymes during the water removal stage of immobilization.[100] Furthermore, the presence of sorbitol can increase activity and reduce side reactions, such as hydrolysis.[60, 101] It is recommended that the aqueous solutions be centrifuged prior to use to remove any nonsolubilized matter.

The first procedure, referred to as the acetone precipitation method, is frequently employed for common adsorbents, such as Celite.[100, 102, 103]

- Add immobilization matrix at proportion of 0.5 to 2 g per ml of aqueous enzyme solution. Stir slurry for about 30 to 60 min at refrigerator temperature (0 to 5°C).
- Slowly add cold acetone (–20°C or cooler) under stirring for several minutes at refrigerator temperature until its concentration is 7 to 10 *M* in the solution (acetone:buffer, 2:3, v:v).
- Recover immobilized enzyme by filtration; wash twice with acetone.
- Dry *in vacuo* overnight.

The second procedure is referred to as the water evaporation method.[60, 83, 100]

- Add immobilization matrix at proportion of 0.5 to 2 g per ml of aqueous enzyme solution. Stir slurry for 30 to 60 min at refrigerator temperature (0 to 5°C).
- Water is evaporated slowly *in vacuo* overnight (room temperature).

As an additional step, glutaraldehyde, a common cross-linking agent, can be included in the aqueous protein solution. The formation of covalent intra- or inter-enzyme cross linkages will help prevent leakage of enzyme from support and improve stability, but will also reduce the enzyme activity by increasing the resistance to pore diffusion.[104]

Regarding the enzyme loading, a minimum ratio of crude enzyme to matrix of 100 to 500 mg/g is recommended. In other words, the enzyme activity per mass of matrix is not affected by enzyme loading, given that a critical minimum loading is

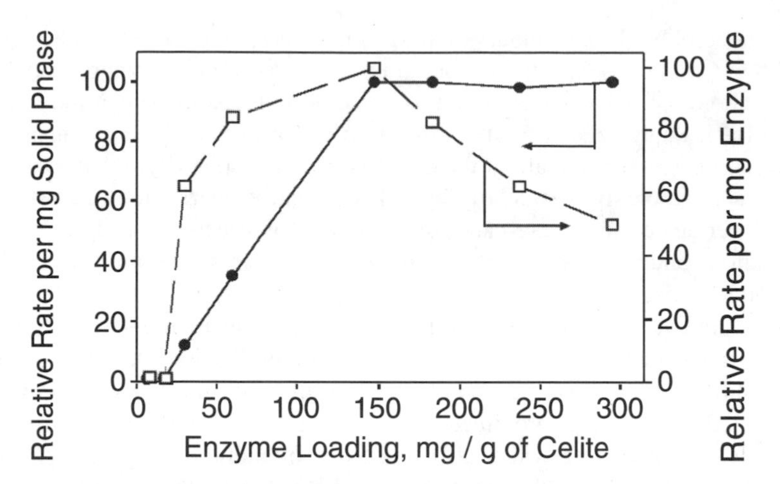

FIGURE 8.2. Transesterification rate of *Pseudomonas* sp. lipase immobilized by adsorption onto Celite as a function of enzyme loading. (Redrawn from Bovara, R. et al., *Biotechnol. Lett.*, 15, 169–174, 1993. With permission.)[105]

exceeded (Figure 8.2).[83, 105] Thus, a high loading rate is usually not detrimental, but can waste costly enzyme. On a *per gram of enzyme* basis, the relative activity reaches a maximum, then decreases as the enzyme loading increases (Figure 8.2). Immobilization preparations containing small enzyme loadings often exhibit poor activity because of strong protein adsorption.[100] The loading is optimized when an initial layer of protein (either the enzyme of use, or other protein, in native or denatured form) is first deposited, to reduce inactivation of additional layers during adsorption.[100] For initial screening experiments (discussed above), it is not worthwhile to examine the enzyme loading, but this variable should be considered for optimization and process scale-up.

A series of novel immobilization products have recently been developed, cross-linked crystals, or CLECs, which are commercially available from Altus Biologics (Cambridge, MA). CLECs consist of microcrystals of purified enzyme (100 μm in length) connected together via glutaraldehyde cross-linkages.[106] The crystals are typically washed with anhydrous solvent containing surfactants to reduce water content, and often contain additives, such as PEG, to improve contact between crystals and substrates.[47, 107] CLECs are reported to be highly active, stable, re-usable, and enantioselective.[47, 108] Altus manufactures CLECs of several different lipases and proteases.

Formation of Noncovalent, Solvent-Soluble Protein-Lipid Aggregates

A recent and exciting area of research is the *solubilization* of enzymes in nonaqueous solvents. One way solubilization is achieved is through noncovalent complexes of lipid (surfactant) and protein, to be referred to here as enzyme–lipid aggregates, or ELAs. Such complexes are reported to be highly active and stable. Moreover, the activity of ELAs can be significantly higher than free, suspended enzyme (in the absence or presence of surfactant), enzymes solubilized in aqueous-organic biphasic systems, or reverse micellar solutions,[109–112] and can approach catalytic rates in aqueous media.[113] In addition, greater thermal and operational enzyme stability is often achieved for enzymes solubilized in organic solvent compared to aqueous solution.[112, 114–118] Furthermore, owing to the actual *solubilization* of enzymes in organic solvent, the substrate specificity of ELAs, compared to water-solubilized enzymes, is often shifted toward more hydrophobic substrates.[119, 120]

ELAs are typically insoluble in water but can solubilize in many organic solvents: long-chain hydrocarbons (hexane and isooctane), aromatics (benzene and toluene), chlorinated solvents (e.g., carbon tetrachloride), and even triglycerides,[121] but are completely insoluble in water and are poorly soluble in polar solvents (e.g., methanol, acetonitrile, and 1,4-dioxane).

Several enzymes and proteins (insulin,[122, 123] peroxidase,[124] phospholipase D,[114] proteases,[112, 113, 116, 125, 126] glucosidase,[127, 128] soybean tryposin inhibitor,[123] and antibodies[129]) have been incorporated into ELAs, but the majority of the reports involve lipases. Lipase-ELAs have been used to catalyze esterification,[109, 118, 120, 130–134] acidolysis,[110, 121, 135–137] acyl exchange between triglycerides,[138] and separation of racemic alcohols[115, 139–141] and carboxylic acids.[142]

The structure of ELAs is not well understood and can vary significantly between preparation methods and surfactant/protein systems. It is believed that typically an enzyme is coated with a monolayer of strongly associated surfactant molecules (approximately 30 to 600 surfactant molecules per enzyme molecule), and additional layers of less strongly bound surfactant, resulting in a preparation that typically contains 10 to 50 wt % protein.[110, 135, 139, 143, 144] Typically, the ratio of tightly bound surfactants per protein for ELAs is similar in magnitude to surfactant aggregation number for aqueous micelles, perhaps indicating that aqueous micelles are involved in ELA formation. ELAs typically contain 1 to 8 wt % water.[120] Beyond these general observations, very little work has been conducted to examine the structure of ELAs, and their structural integrity and stability in organic solvent.

Several different surfactants have been employed in the derivation of ELAs, including nonionics, anionics, cationics, and zwitterionics (phospholipids). Several of the surfactants are commercially available: Aerosol-OT [or AOT: sodium bis (2-ethylhexyl sulfosuccinate)], sodium dodecyl sulfate (SDS), and sorbitan and sucrose fatty acid (mono-) esters; however, the research groups of Okahata and Goto employed unique surfactants synthesized in their laboratories.[145–147] In general, surfactants with two hydrocarbon tails were more successful in forming ELAs compared to single-tail surfactants.[109, 135] Long-to-intermediate surfactant tail length (C_{12} to C_{18}; also, oleyl ($18 : 1^{9cis}$) groups), and branched tail groups yielded ELAs with high enzyme activity.[109, 134] These trends correlate with the lowering of the *cmc* of surfactants in aqueous solution with increasing surfactant tail length, branching, and unsaturation.[148] However, the surfactant head groups more strongly affect the ELA properties than the tail length. Furthermore, when nondenaturing surfactants were employed, Goto et al. report that all reactions proceed to completion without regard to the surfactant's head and tail structure.[109]

Next, the three basic preparation procedures will be presented. However, the reader must realize that the synthesis of ELAs can often be more of an art rather than a science. Moreover, there are many operating variables that can strongly affect the success of complex formation (i.e, the amount of complex formed and its catalytic properties):

- Surfactant type (e.g., HLB value)
- The pH and ion type/ionic strength of the aqueous solution
- The ratio of surfactant to enzyme

The optimum levels or values of each of these parameters will vary from case to case. A series of screening experiments may be necessary to optimize conditions.

The first of the three methods involves the formation of ELAs in aqueous media, and applies primarily to lipases. The method is based on procedures developed by Okahata et al.,[115] Goto et al.,[136] and Nakajima et al.[133]

- Form an aqueous solution of enzyme at fairly high concentrations (approximately 1 to 4 g/l); store at refrigerator temperature, approximately 4°C. The solution should be buffered to an appropriate pH value (discussed below).
- If surfactant is soluble in aqueous media, add surfactant directly to obtain a concentration of 0.75 to 10 g/l. If surfactant is poorly soluble, dissolve surfactant in a polar solvent (e.g., ethanol or acetone [~1 to 5 m*M*]), and add ethanolic solution dropwise and slowly with stirring at refrigerator temperature. In a few cases, a hydrophilic organic solvent e.g., tetrahydrofuran or ethyl alcohol, was added directly to the aqueous solution prior to adding the surfactant.[110, 144]
- Allow solution to stand for at least 2 h, if not for a 1-day period, at refrigerator temperature, in order to form the complex. The ELA gradually precipitates from solution during this period.
- Recover solid precipitate by centrifugation.
- The resultant solid is usually lyophilized or dried *in vacuo* before its use as a biocatalyst in the organic solvent.

In a second procedure, derived by Goto and co-workers,[117, 118] ELAs form from a water-in-oil emulsion. This method applies to a wider variety of proteins than the first method.

- Approximately 10 ml of buffer containing enzyme at ~1 mg/ml is contacted with 90 ml of organic solvent (e.g., isooctane) containing 10 m*M* surfactant.
- The two-phase system is homogenized at 8000 rpm for about 2 min to form a stable water-in-oil emulsion.
- The emulsion is dried at room temperature and pressure for approximately 4 days to produce the complex.

The third method, used to form ELAs of proteases, involves the extraction of protein from aqueous solution to organic solvent containing surfactant through ion pairing.[113, 116, 125] This method requires use of a charged surfactant (e.g., AOT). Recent results indicate a similar ion-pairing mechanism will help solubilize proteins into supercritical carbon dioxide.[149]

- A buffered aqueous solution containing protein at ~1 mg/ml and, importantly, 2 to 20 m*M* of $CaCl_2$ is contacted with an equal volume of organic solvent (e.g., isooctane) containing surfactant (e.g., AOT) *at very low concentration* (e.g., 1 to 2 m*M*), at moderate stirring of ~250 rpm for 5 min.
- Agitation is discontinued to permit phase separation.
- If phase separation is slow, centrifugation can be employed.
- The recovered ELA-containing organic phase can be used directly for biocatalysis.

The selection of pH buffer is based on the following two criteria:

- The charges on the protein's surface groups must be opposite of the surfactant head groups.
- The pH value should be near the optimal value for activity.

Salt is needed in the aqueous phase to prevent the formation of emulsions at the interface.[116] The best solvents for solubilization of ELAs in Method 3 are reported to be hydrophobic (isooctane, decalin, tert-amyl alcohol), while polar solvents (chloroform, methylene chloride, tetrahydrofuran, and tert-amyl alcohol) promote loss of catalytic activity due to denaturation.[113, 116, 125]

Researchers, primarily from Moscow State University, Russia, have also formed noncovalent aggregates between enzymes and polymers [palmitoylated poly (sucrose acrylate)],[79] polyelectrolytes (polybrene),[150, 151] and block copolymers.[152] Reported enzymatic activities were high. For the former, activity decreased as the molecular weight and degree of palmitoylation increased.[79] Chymotrypsin-polyelectrolyte and -block copolymer complexes were active in aqueous-polar organic co-solvent mixtures when the ionic strength was increased.[150–152]

Other polyelectrolytes (heparin, DNA, and cyclodextrins) are reported to enhance the stability of enzymes in aqueous media via complexation.[153, 154]

Covalent Modification of Enzymes

It is reported that the covalent attachment of PEG or other polymers to enzymes allows them to be soluble in polar organic solvents (e.g., dioxane, carbon tetrachloride, and benzene), which in turn enhances their catalytic rates.[155–160] The methodology involves tedious bioconjugation chemistry steps to attach activated forms of polymers (e.g., PEG monomethyl ether) to the free amino (lysine) groups of enzymes. In addition, isolation of the modified protein from the bioconjugate chemical reaction mixture is required. Proteins can also be modified using fatty acid chlorides, to create hydrophobic derivatives for cell membrane insertion.[161] Modification is not recommended by the author for most situations due to the modest increases in reaction rate[160] and the great experimental effort required. The references cited in this section will provide further detail and information.

PREPARATION OF SUBSTRATES

In most cases, substrates can be used directly out of the reagent bottle, or after proper control of water activity (discussed below), given high reagent purity and the absence of any contaminants that may harm the biocatalyst (e.g., acids and bases, denaturing solvents). However, there are two cases known to the author where further work-up of the substrate is recommended. The first case involves acylation (ester or amide bond formation) of bulky acyl acceptors, such as saccharides, cholesterol derivatives, polyesters, and phenol derivatives. Nonaqueous biocatalysis is a very desirable route for these reactions owing to the high regio- and stereo-selectivity exhibited and mild reaction conditions. Due to steric hindrance of the bulky substrate in the active site, the rate is extremely slow. Acceleration of the reaction rate by 1 to 2 orders of magnitude occurs upon the *activation* of the acyl substrate,[162] where acyl groups are covalently attached to good leaving groups through simple (but often expensive) organic chemical reactions. An example of a commonly employed good leaving group is vinyl alcohol. Upon the transfer of an acyl group from vinyl ester to an enzyme-binding site, the released vinyl alcohol-leaving group tautomerizes irreversibly into an aldehyde. Because the aldehydes cannot participate as nucleophiles, the overall reaction is driven strongly in the forward direction and essentially becomes irreversible. A thorough list of activated fatty acyl substrates and organic synthetic methods for their synthesis are summarized in Reference 162.

The second case involves poor substrate solubility. An example is the regioselective esterification of polyols by lipases or proteases. Most polyols, e.g., saccharides and glycols, are difficult to co-solubilize with fatty acyl donors, particularly because of the solubilization limitations of polyols in most organic solvents. Thus, several techniques have been applied to improve solubility. The covalent attachment of isopropylidene groups to certain target hydroxyls through the chemical formation of an acetal bond[163–165] is one technique. In addition to improving the solubility, derivitization of polyols with isopropylidene helps control the reaction's regioselectivity. Upon the completion of the reaction, the isopropylidene group is removed via acidification.[164,165] Other approaches to improve polyol solubilization in nonaqueous media include the formation of a soluble, noncovalent complex between polyol and phenyl boronic acid[166,167] and a nonsoluble complex between polyol and silica gel.[168,169] The latter is suspended in nonaqueous media, acting as a substrate reservoir during the course of reaction.[169]

CHOICE OF CONTINUOUS PHASE

Enzyme reactions have been successfully operated in a variety of organic solvents (Table 8.4) as well as in supercritical fluids (e.g., carbon dioxide and fluoroform)[170] and gases.[171,172] The latter two categories offer some intriguing possibilities and potential advantages relative to solvents, including enhanced substrate diffusivity, tunable solvent phase properties (via temperature and pressure), reduced solvent disposal costs, and simplified downstream separation and recovery processes. Enzymology in gas and supercritical fluids is an ongoing area of technical research. This review will focus primarily on the use of organic solvents owing to their simplicity to the novice in nonaqueous enzymology and their more widespread use.

In some enzymatic reactions and, particularly, in those involving lipases, solvent is not necessary (i.e., the substrates and products perform as a solvent). However, in cases where glycerides or fatty acyl substrates are present at significant concentration, the temperature must be elevated (approximately 50 to 80°C) in order to reduce the medium viscosity to allow effective dispersion of a solid-phase catalyst.

In general terms, the rate of enzymatic reactions decreases with solvent polarity. The underlying explanation for this trend is believed to be the detrimental stripping away of essential water molecules by solvent from the enzyme's micro-environment.[31] There are several possible measurements of solvent polarity. The most commonly employed are the solvent permitivity (dielectric constant), water solubility, and the logarithm of the octanol–water partition coefficient (log P value).[31, 33, 173] Values of these three polarity indices for several solvents commonly employed in nonaqueous enzymology are listed in Table 8.4.[173–179] Generally, a plot of activity vs. solvent log P value exhibits a reasonably high correlation in most (but not all) cases; thus, it is the most commonly applied polarity index.[31, 33, 173] Values of log P for solvent, substrates, and products are readily calculated from hydrophobic fragmental constants.[173] Log P values for mixtures, log P_{mix}, are calculated from the component mole fractions and log P values, X_i and log P_i, respectively.

$$\log P_{mix} = \Sigma X_i \log P_i \tag{8.1}$$

For supercritical fluid continuous phase, enzyme activity correlates well with fluid pressure, more precisely, with fluid density.[170]

Thus, when selecting a solvent to host an enzymatic reaction, it is generally best to employ the most lipophilic solvent that will solubilize all substrates. The solvents listed in Table 8.4 are commercially available and should be purchased in a high purity form. An important operational issue is the water content of the solvent (discussed below). It is least expensive and simplest to dry solvents (i.e., to obtain anhydrous solvents) by treating with 3A or 4A molecular sieves. Alternate techniques for water removal include solvent storage over saturated salt solutions (discussed below) or their circulation through a chromatographic column of basic alumina (type WB-2; commercially available). (Binary mixtures of water and polar organic solvents are known to denature proteins, often irreversibly;[177, 180] thus, excess water should be avoided.) In addition, chlorinated solvents frequently contain alcohols (e.g., ethanol) as a stabilizer. Molecular sieves and basic alumina can also be used to remove the stabilizer.

Most of the solvents listed in Table 8.4 are volatile at ambient temperature and above. If nonvolatility is required for either safety reasons or for high temperature reactions, veratrole (boiling point temperature, T_{bp}, of 206°C), diphenyl ether (T_{bp} = 259°C), and decalin (T_{bp} = 196°C) are recommended.[116, 178] Another alternative recently demonstrated is the use of ionic liquids. Ionic liquids, typically consisting of nitrogen-containing organic cations and inorganic anions, are readily synthesized,

TABLE 8.4
Common Solvents Used for Biocatalysis and Measurements of Their Polarities[a]

Solvent	log P	log $S_{w/o}$	ε
N,N-dimethylformamide	−1.0	Miscible	36.7
Acetonitrile	−0.33	Miscible	36.2
1,4-dioxane	−0.27	2.21[b]	
Acetone	−0.23	Miscible	20.7[b]
2-butanone	0.28	−0.44	18.5
Tetrahydrofuran	0.46	Miscible	7.32
Dichloromethane	0.60	−2.15	9.80
Ethyl acetate	0.70	0.21	10.34
Pyridine	0.71		12.3[b]
Tert-butanol	0.79	Miscible	10.9[c]
2-pentanone	0.80	−0.93	15.4
3-pentanone	0.80	−1.28	17.0
Ethyl ether	0.85	−0.24	4.34
Methylene chloride	1.5	−0.84	8.93
2-methyl-2-butanol	1.3	−0.36	5.82[b]
Methyl isobutyl ketone	1.3	−0.08	13.1
Trichloroethylene	1.5	−1.59	3.42
Isopropyl ether	1.9	−0.64	3.88
Benzene	2.0	−1.51	2.27
Chloroform	2.0	−1.12	4.81
Veratrole	2.2		
Toluene	2.5	−1.80	2.38
1,1,1-trichloroethane	2.8	−1.60	7.25
Butyl ether	2.9		3.10
Carbon tetrachloride	3.0	−1.93	2.24
Cyclohexane	0.2	−2.25	2.02
1-hexane	3.5	−2.39	1.88
1-heptane	4.0	−2.30	1.92
Diphenyl ether	4.3		3.65[c]
Isooctane	4.5	−2.67	1.94
1-decane	5.6	−3.03	1.99
1-dodecane	6.2	−2.69	2.02
1-hexadecane	8.8	−2.72	2.10

[a] All values taken at 20°C. Column labels: log P refers to the logarithm of the octanol–water partition coefficient, log $S_{w/o}$ to the logarithm of the water solubility, mol^{-1}, ε to the solvent dielectric constant. Data from References 173 through 179.

[b] 25°C

[c] 30°C

inexpensive, recylcable, and most importantly are non-volatile. Their properties (e.g., polarity) can be controlled by changing the anion or substituents on the cations. Some ionic liquids remain in the lliquid state up to 300°C. Initial reports indicate that enzymes are very stable in ionic liquids; and, reaction rates are comparible to values obtained using organic solvents.[179b,179c]

Reactions can also be operated with solvents undergoing reflux (e.g., for improved substrate solubility and water removal, as discussed below). Binary solvent mixtures of lipophilic and hydrogen bonding species (e.g., tert-butanol and hexane) are recommended for reflux operation because of their ability to form azeotropes, which usually translates to lower operating, or boiling point, temperature than pure solvent.[181] Solvents to avoid for nonaqueous enzymology are those capable of dissolving proteins, including dimethylsulfoxide (DMSO), formamide, and methanol, which promote significant denaturation.[58, 59, 180] However, remarkably, the inclusion of a few percent of the denaturing solvents into anhydrous, slightly polar, solvents, such as acetonitrile, dioxane, and carbon tetrachloride, can enhance catalytic activity, particularly at lower temperatures.[182] It is speculated that the activation is due to enhanced protein flexibility.[182] In addition, a polar/nonpolar solvent mixture effectively adsorbs water generated during enzymatic esterification, which is beneficial for continuous operation (discussed below).[183]

Another issue to consider when selecting solvents is their effects on equilibrium positions in enzymatic reactions. For example, use of more polar solvents induces the formation of monoesters relative to di- and higher esters for polyol ester formation.[176, 184] In addition, as the water solubility of a solvent increases, the equilibrium constant for an esterification reaction decreases, meaning that the extent of hydrolysis is greater.[174, 175] Values of water solubility in various solvents are contained in Table 8.4. The effect of solvent type on equilibrium is accurately modeled using activities rather than concentrations.[175] Activity coefficients can be calculated using a thermodynamic model, such as UNIFAC. This model is available in several software packages.[40, 175, 185]

Solvent polarity has a major impact on the observed selectivity, particularly regio-[186] and enantio-selectivity, of enzyme-catalyzed reactions.[187–189] This has led to the use of the term solvent engineering. Mechanistically, the solvent affects selectivity owing to its effect on substrate and product solvation. Also, the solvent type can have differing effects on stability of different enzymes in the same preparation.[48] Recently, Klibanov and co-workers[40] have developed procedures for predicting the effect of solvent on enantioselectivity that involve Raoult's law of activity coefficients for a given substrate solubilized in a given solvent. The activity coefficients are readily calculated using the UNIFAC model. To illustrate, the racemic separation of two enantiomers, R and S, will be considered. Through a thermodynamic model, the following relationship is derived which relates the enzyme activity, k_{cat}/K_M, for each substrate enantiomer:

$$\log\frac{(k_{cat}/K_M)_S}{(k_{cat}/K_M)_R} = \log\frac{\gamma'^S_i}{\gamma'^R_i} + \text{constant} \tag{8.2}$$

The parameter $\gamma'i$ refers to the activity coefficient of the *desolvated portion* of substrate i and is a strong function of solvent type. Molecular modeling of the enzyme-substrate complex was employed to determine the unsolvated portion of the substrates.[40] Next, molecules similar in structure to the unsolvated portions were used for calculating activity coefficients via UNIFAC.[40] The preceding equation predicts a linear relationship between the logarithm of activity vs. the logarithm of the activity coefficient for a given enzymatic reaction hosted in a series of solvent systems. This model was quite effective for reactions utilizing CLECs, which have relatively unperturbed molecular structure in organic solvents relative to aqueous media.[40]

However, because lyophilized enzymes contain altered structure relative to aqueous phase enzymes, protein structures obtained from molecular modeling with the Brookhaven database (aqueous-derived) are not relevant. For cases involving lyophilized enzymes, and when the desolvated portion of substrates cannot be determined, the same research group discovered that the following equation, similar to that given above, was effective.[189]

$$\log\frac{(k_{cat}/K_M)_S}{(k_{cat}/K_M)_R} = M\log\frac{\gamma_i^S}{\gamma_i^R} + \text{constant} \tag{8.3}$$

The slope, M, is not necessarily equal to unity and varies with enzyme type.[185] Here, γ_i refers to the activity coefficient of the entire substrate, meaning that this model assumes the entire substrate is desolvated.[185] (Desolvation, a critical step in the formation of the enzyme-substrate complex, hence in determining the rate of reaction, is often much slower in organic solvents than in aqueous media.[39] For many cases, desolvation occurs rapidly in aqueous media because of hydrophobic interactions between the substrate and the active site. The presence of organic solvent reduces this driving force.)

The solvent also influences the reaction through its effect on the relative solubility of substrate and product. Ideally, the substrate would be solubilized, while the insoluble product precipitates from solution. Thus, the recovery of product would be improved, thermodynamic equilibrium in the solvent phase would be driven toward ester synthesis, and product inhibition would be prevented. This scenario can occur for the lipase-catalyzed esterification of saccharides and nonpolar, saturated fatty acids. When employing polar solvents such as acetone, saccharides remain solubilized, while the product, saccharide-fatty acid monoester, precipitated.[190]

CONTROL OF WATER ACTIVITY OF PROTEINS, SOLVENTS, AND SUBSTRATES

It is well known that the water content of the reaction medium (i.e., the solvent and solid enzyme-containing phase) has a strong impact on nonaqueous enzymology. Moreover, for a given reaction, enzyme preparation, and medium composition, there is an optimal water content for maximizing the enzyme activity, or the initial rate of reaction. The optimal value is a strong function of the presence and concentration of substrates, and properties of the solid phase. Moreover, the enzyme, immobilization matrix, and continuous phase all compete for adsorption/retention of water molecules. Polar solvents are known to "strip" away water molecules from solid-phase enzymes.[191]

In pioneering research by Halling and co-workers,[194] it was demonstrated that the activity of water is a more representative and useful parameter than water concentration for describing enzymatic rates in nonaqueous enzymology. Water activity, or a_w, is defined as the fugacity of water contained in a mixture divided by the fugacity of pure water at the mixture's temperature. For a typical nonaqueous enzymatic reaction operated in a closed system, the medium will consist of a solvent (or fluid) phase, an enzyme-containing solid phase, and air headspace above the solvent. As a first approximation, the water transport between the three phases is assumed to be at thermodynamic equilibrium. For such a situation, a_w can be defined in terms of the air headspace properties:

$$a_w = P_w/P_w^0 \tag{8.4}$$

where P_w is the partial pressure of water in the air headspace and $P_w{}^0$ is the vapor pressure of (pure) water at the reaction system's temperature. This equation assumes that water behaves ideally. In addition, a_w is equal to the mole fraction of water in the solvent, x_w, multiplied by the activity coefficient γ_w,

$$a_w = \gamma_w x_w \tag{8.5}$$

To demonstrate the utility of a_w, plots of enzyme activity vs. a_w contain similar shapes and optimal a_w values for a variety of different solvents.[38, 192] (However, for a given a_w value, the activity for lipophilic solvent is higher than for polar solvent, as discussed above.) Inconsistent trends exist when enzyme activity is plotted against water concentration in solvent rather than a_w. In addition, a_w-activity profiles for enzyme preparations that differ only in matrix material have similar shapes and maxima.[92] Exceptions to the commonality of a_w-enzyme activity profiles include

- Covalently modified enzymes[50]
- Enzymes from different microbial sources[92, 193]
- Changes in substrate concentration[193]
- Different reactions that employ the same enzyme[193]
- Very polar solvents[194]

Although a_w provides a useful reference scale for predicting the effect of water on enzyme activity, the adsorption (or retention) of water in the solid and fluid phases strongly varies with a_w. The adsorption/retention data is usually plotted in terms of water adsorbed by the solid phase vs. water retained in the solvent, or water adsorbed (retained) vs. a_w. Typically, the solvent water content increases linearly with a_w for lipophilic solvents, sharply at high a_w (approximately 0.6) for slightly polar solvents,[192, 195] and sharply at low a_w for very polar solvents.[194] In addition, water adsorption is strongly influenced by the types and concentration of substrates.[96] Profiles similar to those just described for slightly polar solvents exist for solid-phase (lyophilized or immobilized) enzymes. Typically, water adsorption onto lyophilized or immobilized enzymes occurs within the range of 0 to 0.3 g of water per g of support.[96, 192, 195–197] The adsorption of water by solid phase biocatalyst is strongly affected by the presence of solvent for a_w above 0.6,[195, 196] although the solvent type had little influence.[196] It is believed that the solvent molecules reduce water uptake by disrupting adsorbed water multilayers around the biocatalyst surface, owing to their ability to interact with protein hydrophobic pockets.[196] However, it is important to note that adsorption data can vary between batches of the same protein[196] and that hysteresis can occur.[38] In addition, the composition of the solid phase (enzyme, matrix, and buffer salts) can strongly affect the adsorption of water.[96]

One must be careful to avoid a_w values over approximately 90% owing to the possible formation of a bulk water phase.[38, 196] Moreover, the inactivation of several enzymes can be traced to the formation of a liquid–liquid interface.[198, 199]

One of the major advantages of employing a_w control is the improvement of precision and repeatability of kinetic results because the water content of solvents and protein preparations can vary from batch to batch and with time.[38] In addition, control of water activity is quite simple. It is achieved by storing solvents (of low volatility), substrates, and enzymes over saturated salt solutions. The thermodynamic properties of the salt dictate the observed water activity.[200, 201] Several different salt types used to control a_w for nonaqueous enzymology are listed in Table 8.5.[192, 195, 200, 201] The storage of saturated salt solutions is achieved by the following procedure:[37, 38]

- Select salt type from Table 8.5.
- Prepare a slurry of salt plus saturated aqueous salt solution in a tightly sealable vessel (e.g., a dessicator).
- Place items to be a_w-controlled in a small beaker and place beaker in larger salt-containing vessel. Beakers can be secured by immersing their bottoms into the excess solid phase at the bottom of the vessel.
- Seal vessel and wait approximately 2 days to approach thermodynamic equilibrium.

Although the procedure described above is effective for controlling a_w for most reaction materials, there are some exceptions. First, water-soluble solvents and substrates, particularly saccharides and glycols, continuously adsorb water without reaching an equilibrium condition.[96, 202, 203] In addition, salts may react with substrates.[37] Furthermore, certain salts exhibit either nonideal thermodynamic behavior or too slow a rate of equilibration.[37]

In conclusion, the storage of substrates, solvents, and proteins over saturated salt solutions allows for a more predictable and repeatable control of water concentration in the reaction system. Control of water activity during the course of reaction, required for reactions that consume or generate water, will be discussed later. For measurement of water activity, the following two models of hygrometers are recommended: The HUMIDAT-TH II humidity sensor from Novasina (Pfäffikon, Switzerland)[204, 205] and the Weiss (formerly Phillips) LiCl humidity sensor (Weiss Umwelttechnik GmbH, Reiskirchen, Germany).[206] Sometimes, several hours are required to achieve a consistent hygrometric reading.[206]

TABLE 8.5
Commonly Employed Salts for Controlling Water Activity through Storage over Saturated Salt Solutions[a]

Salt	(a_w) [b]
P_2O_5[c]	<0.01
Molecular Sieves[c]	<0.01
KOH	0.08
LiCl	0.11
CH_3COOK	0.23
$MgCl_2$	0.33
K_2CO_3	0.43
$Mg(NO_3)_2$	0.54
NaBr	0.59
$CuCl_2$	0.68
KI	0.70
$NaNO_3$	0.74
NaCl	0.75
$(NH_4)_2SO_4$	0.79
KBr	0.82
KCl	0.85
$ZnSO_4$	0.90
KNO_3	0.95
K_2SO_4	0.97
Na_2HPO_3	0.98

[a] Data from References 192, 195, 200, 201.
[b] 20°C.
[c] Storage over solid in the absence of water rather than over saturated aqueous salt solution.

USE OF MICROEMULSIONS (REVERSED MICELLES)

Reversed micelles, also known as water-in-oil microemulsions, *w/o-MEs*, consist of nanometer-sized aqueous (or polar) spherical dispersions in lipophilic solvent formed by surfactants (although slight differences exist between the definitions of reversed micelle and micro-emulsion). The dispersions are thermodynamically stable, not requiring agitation, and continuously collide, coalesce, and reform, with the overall process occurring on the scale of microseconds. Use of the w/o-ME solution allows proteins to be solubilized in organic solvents. Reverse micellar enzymology, including methods of protein insertion and spectroscopy of w/o-ME-encapsulated proteins, has been thoroughly reviewed.[207–211] An exciting prospect of encapsulating enzymes into reversed micelles is the *in situ* monitoring of reactions using various spectroscopic techniques, particularly those involving chromophores or fluorophores. The author recommends w/o-MEs for *in situ* assaying of enzymatic reactions or for hosting proteins for the following situations:

- Reactions involving enzymes that require interfaces, such as lipases[212] and phospholipases.[213, 214] Due to the nanometer size of aggregates, the amount of interfacial area is quite large.
- Reactions involving two substrates that differ in solubility, such as fatty acid (lipophilic) and polyols (hydrophilic).[212]
- Reactions where the surfactant interface enhances substrate solubility or induces proper allignment of substrates for catalytic action, such as cholesterol modification,[215] lipase-catalyzed macrocyclic lactone synthesis,[216] and peroxidase-catalyzed phenol polymerization.[217] Of interest, the growing polyphenol chains drop out of solution once a critical molecular weight is achieved, owing to reduced solubility at the of oligomers at the interface, yielding a very monodisperse polymer product.[217] Also, polymer molecular weight can be controlled through selection of the surfactant concentration.[217]
- Solubilization of hydrophobic and membrane proteins, multi-enzyme complexes, and multiple enzyme reactions.[211, 218–220]
- Protein unfolding and refolding/renaturation.[221–223]
- Favorable partitioning of substrates to the interface and product to an excess aqueous phase (Winsor II system[224]) or nonmicellar excess organic phase.[225]

In addition, reversed micelles have been used as scaffolding to immobilize proteins via entrapment into gels. Gel formation is induced through addition of gelatin,[226] phenols,[227] and phospholipid.[210] With the exception of gels, development of large-scale reverse micellar enzyme processes are difficult due to the inherent batch nature of the medium and complications for downstream separations induced by the surfactant. See reviews cited above for further detail.

OPERATION OF ENZYMATIC REACTIONS

SCREENING EXPERIMENTS

Many of the topics discussed in this review involve optimizing nonaqueous enzymatic reactions by proper selection, design, and preparation of enzyme and medium, and control of reaction conditions. For a new reaction or application, a general series of screening experiments should be undertaken to first identify the proper biocatalyst and solvent type (if required). Then, additional screening experiments can be employed to optimize other operational parameters (listed below). Screening experiments should be performed on a small scale (e.g., in small [5- to 20-ml] vials or test or culture tubes, subjected to agitation, as discussed below). The substrate concentration should be fairly dilute. A method of analysis to monitor the reaction should be determined beforehand, particularly one which can be performed quickly and inexpensively, such as thin layer chromatography. The first goal is to screen a series of enzymes. Each of a series of reactions should be operated simultaneously, at identical conditions [e.g., stir rate (sufficiently large to permit dispersion of biocatalyst, determined visually), solvent type (as lipophilic as possible while maintaining substrate solubility), temperature (near room temperature if possible), and reaction time (e.g., 24 to 48 h)]. Commercially available enzymes can be selected from the suppliers listed in Table 8.2. After the best enzyme(s) are identified, the solvent type should be screened using the list given in Table 8.4. Then, additional screening experiments can help determine proper type or levels of the following:

- Substrate type (e.g., for acyl donors, substrate can be a free fatty acid, a fatty acid ester, or an anhydride)
- Substrate concentration, including solventless mode, if applicable
- Substrate ratio (for multiple substrate reactions)
- Initial water activity
- Temperature
- Enzyme preparation method
- Enzyme operational stability and re-usability

It cannot be overemphasized that knowledge of the nonaqueous enzymology literature will reduce the screening and optimization experimentation. The following sections detail specific operational issues that must be considered.

Temperature

In most cases, it is best to employ a temperature near or below the catalyst's thermostable limit (determined from the technical literature, the manufacturer, or through screening experiments). However, in a few cases, low temperature operation is desired. One example is acyl transfer reactions involving peptide. A common problem associated with this category is the occurrence of a side reaction, hydrolysis. Moreover, a competition exists between the amino acid acyl acceptor and water, both acting as nucleophiles toward the acyl-enzyme intermediate. Adlercreutz and co-workers discovered that the selectivity of chymotrypsin toward acyl exchange over hydrolysis increased as temperature decreased.[204, 228] Hence, operation is best at lower temperatures. A second situation is when low temperature selectively precipitates the desired product (discussed below).

Dispersion of Biocatalyst and Reactor Design

Proper dispersion of biocatalyst is critical to avoid reduced reaction rates due to mass transfer limitations. As mentioned previously, it is highly important that excess water and polyols be absent, and that water (or polyol) be removed if generated by the reaction (discussed below) to avoid the clumping together of solid-phase enzyme particles, the adsorption of particles on reactor walls, and inactivation. In addition, the continuous phase must have sufficiently low viscosity to permit suitable dispersion. Agitation, required to disperse biocatalysts, can be achieved using magnetic stirring, impellers, and shakers (e.g., orbital or end-over-end designs, the latter employed with test tubes). Agitation rates should be sufficiently high to avoid convective mass transfer limitations, but not be too high, because the solid phase can be pulvarized. A range of 250 to 400 rpm is recommended.

Operation in batch mode is strongly suggested for conducting small-scale and initial screening experiments. Continuous reactors (e.g., packed bed, stirred tank, and membrane models) have all been employed, mostly for reactions involving lipases. Further information on continuous reactors can be found from a series of review articles.[85, 86, 229] Glass vessels for all scales of batch and continuous processing are recommended.

Continuous Control of Water Activity

Particularly for reactions where the water concentration changes during its course, such as esterification or hydrolysis, it is important that the biocatalyst microenvironment retain its proper degree of hydration (i.e., to remove or replace water generated or consumed, respectively). During esterification, irreversible inactivation of lipase is reported for cases where water is not removed.[230] (Along these same lines, for lipase-catalyzed alcoholysis of triglycerides, reversible inactivation occurs when the product glycerol is allowed to accumulate on the solid phase.[230,231]) In addition, formation of an excess aqueous phase can reduce the dispersion of the biocatalyst.[232]

Perhaps the simplest method for water removal is to operate the reaction in an open vessel, with a large amount of surface area exposed to the atmosphere, employing as high a temperature as can be withstood by the enzyme (free evaporation),[233] or equivalently, operating the reaction *in vacuo*.[233] However, such methods will also lead to the removal of solvent, unless nonvolatile solvents such as veratrole and diphenyl ether are employed. In addition, when employing rapid evaporation rates (e.g., via a vacuum pump) one must be careful not to remove too much water so that the enzyme becomes inactivated by dehydration.

Free evaporation and *in vacuo* operation both involve the same principle: the transport or diffusion of water from a phase at higher activity to one at lower activity. Other evaporation-based methods include

- Bubbling of inert gas (e.g., N_2 or dry air) through the medium[232, 234]
- Bleeding the vapor headspace above the liquid phase reaction medium, by continuously flowing dry air into the closed vessel through the simultaneous use of a vacuum pump, to remove moist air, and a compressor, to introduce dry air[235]
- Pervaporation[236]

For operation of continuous bioreactors, the solvent type employed has a significant effect on the water retention during esterification. One role of the flowing solvent is to remove the generated water. However, when the solvent is lipophilic, such as hexane, its water capacity is too small to remove the water at a rate equal to water production; hence, water accumulates in the solid enzyme phase and leads to irreversible inactivation.[183, 230] The inclusion of a polar co-solvent, or the use of a solvent of intermediate polarity, can improve water removal.[183] The optimal concentration of polar co-solvent is predetermined by equating water removal rate of the co-solvent mixture's (water solubility multiplied by the flow rate) with the water generation rate.[183] Water accumulation also occurs when the polarity of the fluid phase decreases during the course of the reaction, such as during esterification between acids and mono-alcohols.[183] Additional approaches to remove water from packed beds include the use of strong acid anion exchange resins to selectively adsorb water[199, 237] and the introduction of frequent cycles of dry air flow during operation of the reactor.[183]

In addition, water is effectively removed by using solvents at reflux. A condenser is attached to the reaction vessel to liquify the solvent vapors. The condensed solvent flows through a water trap before returning to the vessel.[181, 238–240] The three designs of a water trap—a soxhlet extractor, a reflux trap, and a Dean-Stark trap—are illustrated in Figure 8.3. The Soxhlet extractor and reflux trap both remove water via a packed bed of molecular sieves (usually type 3A or 4A), which are placed in an extraction thimble. During operation of the Dean-Stark trap, water and solvent condensate flow into a gravity-settling column [item f of Figure 8.3] where the heavier solution, water, forms a lower phase, while the upper phase, solvent, overflows the column and travels downward through the "vapor" arm, [item b of Figure 8.3]. A variation of the Dean-Stark trap design includes a stopcock at the bottom of the gravity-settling column to remove water. The moisture removal devices depicted in of Figure 8.3 are designed for small-scale operations. Bloomer et al. have developed an apparatus, that is a hybrid of the reflux and the Dean-Stark traps for preparative enzymatic esterification.[239] As discussed previously, the use of polar/nonpolar azeotropic liquid mixtures allows operation at reflux temperatures below the reflux temperature of either component.[181]

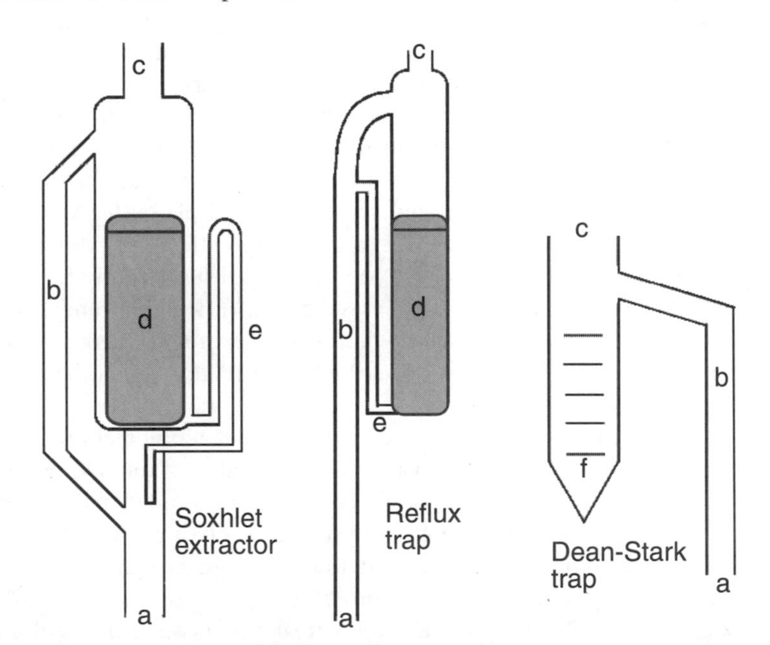

FIGURE 8.3 Columns for water removal used for enzymatic reactions conducted in refluxing solvent. Portions of this figure were adapted from Reference 239. Key to letters: (a) exit from reaction vessel (e.g., round-bottom flask); (b) arm for vapor transport; (c) entrance to condenser; (d) packed bed of molecular sieves (or other drying agent); (e) arm to return dehydrated solvent to reactor; and (f) gravity-settling or phase separation column.

The evaporative water removal methods discussed above can also be employed to remove volatile reaction products which inactivate biocatalysts, such as methanol or aldehydes and ketones, the latter of which form from the vinyl alcohol activated ester leaving group, as discussed above.

Besides evaporation, the moisture level can be controlled through the reactor's water activity, a_w. This approach has been achieved by circulating saturated salt solutions at known a_w (Table 8.5) through silicone tubing[93, 241] or hollow fibers[242] that are suspended in the reaction medium. Of interest, the a_w was successfully programmed using this technique during lipase-catalyzed esterification.[93] A high a_w salt solution was employed during the early stage of the reaction to accelerate the rate and a low a_w salt solution was employed during the latter stages to reduce the extent of hydrolysis.[93] Typical transport rates for such reactors are on the order of 100 μg of water per minute.[241] The use of circulated salt solutions is more effective for polar rather than nonpolar solvents and cannot be used with acidic and basic salts that attack the membrane material or for substrates that can solubilize in the salt solutions.[241] A variation of this design is to circulate a_w controlled air through the reactor.[243] In addition, the reaction can be operated under constant a_w through the use of a dessicator. Lipase-catalyzed hydrolysis was performed successfully by placing a vial containing the reaction medium inside a dessicator containing a saturated salt solution of high a_w (approximately 0.8).[244] The entire apparatus was placed on top of a magnetic stirring/heating plate to provide elevated temperature and stirring to disperse the biocatalyst.[244] The water transport rate was reasonable.[244]

A more recent approach to control water activity during the course of reaction is through salt hydrate pairs. This method involves the suspension of two different hydrate forms of the same salt, such as $Na_2HPO_4 \cdot 12\ H_2O/Na_2HPO_4 \cdot 7\ H_2O$ (abbreviated Na_2HPO_4 12/7), in nonaqueous media. Such hydrate pairs serve as water molecule donors and acceptors and maintain a narrow range of a_w for the reaction media, analogous to proton donation and acceptance by a pH buffer to maintain a pH value near its pKa. The solvent type is known not to affect the exhibited a_w for a given hydrate pair.[206] Salt hydrates have been employed successfully for reactions involving chymotrypsin,[245] lipases,[205, 206, 246] subtilisin,[206] penicillin amidase,[21] and tyrosinase.[247] Table 8.6 lists useful salt hydrates for use in nonaqueous media and their respective a_w values.[205, 206] A list of salt hydrate pairs that should be avoided is contained in Reference 206.

For use of salt hydrates, it is recommended that the reaction medium be stored over a saturated salt solution close in value to the a_w value of the hydrate pair for about 1 h before initiation of the reaction.[206] The substrate is added to the remaining medium components, which are pre-equilibrated as a mixture over saturated salt solution.[206] It is recommended that each hydrate in a given pair be included in equimolar amounts, although employment of only one of the two will also approach the targeted a_w value given a sufficient supply of water molecules in the system.[206] The literature varies on the proper amount of salt hydrate pairs to add to the medium. For lipase-catalyzed esterification, one report recommends employing a sufficient amount of the hydrate pairs to correctly buffer the water activity without harmful effects for large amounts of hydrates,[246] and a second report demonstrates that an optimal salt hydrate concentration exists.[205] Interestingly, for

TABLE 8.6
Salt Hydrate Pairs Used to Effectively Control Water Activity, a_w, in Nonaqueous Enzymology

Salt Hydrate Pair[a]	a_w at 25°C[b]	a_w at 50°C[c]
NaI 2/0	0.07	
Li_2SO_4 1/0	0.09	0.12
Na_2HPO_4 2/0	0.16	
CH_3COONa 3/0	0.34	0.37
$Na_2S_2O_3$ 5/2[d]	0.35	
NaBr 2/0[d]	0.37	
$Ba(OH)_2$ 8/1		0.44
$Na_4P_2O_7$ 10/0	0.47	0.60
Na_2HPO_4 7/2	0.61	
Na_2HPO_4 12/7	0.79	
Na_2SO_4 10/0	0.80	

[a] Nomenclature: Na_2HPO_4 12/7 refers to the salt hydrate pair (e.g., Na_2HPO_4 12 H_2O/Na_2HPO_4 7 H_2O). All salt hydrates are commercially available unless indicated otherwisc.
[b] Data is from Reference 206.
[c] Data is from Reference 205.
[d] Prepared in laboratory ($Na_2S_2O_3$ 2 H_2O and NaBr 2 H_2O).[206]

a series of lipase-catalyzed esterifications of different saccharides, each saccharide possessed a different optimal salt hydrate pair.[205] It is recommended that the user screen for both salt hydrate pair type and concentration before use.

In conclusion, it appears that in many situations salt hydrate pairs may be useful to control a_w during nonaqueous enzymology. Halling and co-workers[194] recommend employment of salt hydrates for systems involving lyophilized enzymes or enzymes immobilized onto hydrophobic matrices, but not for immobilized enzymes on hydrophilic supports.[246] However, the user is warned that hydrate pairs have a limited buffer capacity and require frequent regeneration. Thus, they are not appropriate for industrial use.[242] Also, it must be noted that solid-phase enzymes can receive ions from the salt hydrate pairs, which can be detrimental if multivalent cations are employed in the salt hydrates.[206] A continuous-mode bioreactor that employs salt hydrate pairs has been developed.[248] Its design features the separate compartmentalization of the salt hydrates and immobilized enzyme to reduce the transfer of ions from the hydrates to the enzyme phase.[248]

An extremely low water content must be present during enzymatic polymer synthesis. For example, during lipase-catalyzed ring-opening lactone polymerization, water promotes hydrolysis and cleavage of the initiator from the growing polymer chain.[249] (Water is not produced by this reaction.) In this case, all reaction materials, including the vessel, were thoroughly dried, and the reaction medium assembled quickly under an argon atmosphere, in order to achieve high molecular weight.[249]

Finally, an excess aqueous phase may not lead to inactivation if the method of enzyme and water introduction into nonaqueous media is performed carefully. It is best to introduce enzymes (after their lyophilization at optimal conditions, as discussed above) as highly concentrated aqueous solutions (~60 mg/ml) into organic media rather than to add solid-phase enzymes to micro-aqueous solvents.[48, 80] Such a procedure was particularly effective for incorporating highly purified lipase into nonaqueous media,[48] in contrast to previous results indicating such purified lipases are highly inactive.[50]

Use of Water Mimics

As noted above, small quantities of water are required for an enzyme to function in a nonaqueous media to act as a lubricant to promote the flexibility needed for the enzyme to perform catalysis. However, the presence of water also promotes hydrolysis, which is a particularly undesirable side reaction during enzyme-catalyzed transesterification. In the late 1980s, Klibanov and co-workers demonstrated that polar solvents can successfully substitute for, or mimic, water as a lubricating agent, while suppressing hydrolysis.[23] Useful water mimics include ethylene glycol, formamide, propylene glycol, diethylene glycol, ethanol, and glycerol.[23, 80, 250, 251] It appears that the best mimic depends on the particular application. In all cases, mimics performed optimally when at least a small quantity of water was present. If the use of water mimics is pursued, a series of screening experiments are recommended to determine the best mimic type (including a control experiment with water), mimic concentration, and water content. Two notes of caution must be mentioned. First, excessive amounts of mimic can lead to enzyme inactivation.[251] Second, in at least one instance, mimics have been reported to participate as substrates.[250]

Use of Organic Phase Buffers to Control pH

A new area of research is the control of pH for nonaqueous enzyme reactions using organic pH buffers.[252, 253] These buffer materials, which are soluble in nonaqueous media, strongly control the effect of pH on the reaction, and are able to override the pH memory of the enzyme before lyophilization or immobilization from an aqueous solution.[252, 253] Typically, buffer salts employed are oppositely charged (e.g., R-COOH and $R\text{-COO}^-\text{Na}^+$, where R is hydrophobic in nature). It is the ratio of these two forms that controls the enzymatic rate. The buffers are believed to function by displacing hydrogen atoms of carboxylic acid groups on the surface of the enzyme, E.[252, 253]

$$R\text{-COO}^- \text{Na}^+ + E\text{-COOH} \rightarrow R\text{-COOH} + E\text{-COO}^- \text{Na}^+ \qquad (8.6)$$

Also, the organic buffers can interact with surface amino groups of the enzyme,[252, 253]

$$R\text{-COO}^- \text{Na}^+ + E\text{-NH}_3^+ \text{Cl}^- \rightarrow R\text{-COOH} + E - \text{NH}_2 + \text{NaCl} \qquad (8.7)$$

The development of organic pH buffers is an ongoing area of research and may become of greater significance in the near future.

Continuous Recovery of Product

An ideal situation would exist if the product of interest could be recovered easily during the course of reaction. This would promote use of a continuous reactor scheme, prevent product inhibition, improve product recovery, and control reaction selectivity. A few examples of *in situ* product recovery will be discussed. The first involves the selectivity of lipase-catalyzed esterification of polyols. Monoester solubility decreases strongly with temperature. By using a low temperature, monoester production reaches a certain low level during the course of reaction, at which point additional monoester precipitates from the solution. This approach, used in the esterification of glycerol[254–256] and sorbitol,[257] reduces the formation of diester and yields higher degrees of esterification. The best operation of these reactions occurred by programming the reaction temperature. During the beginning of the reaction, a high temperature was desired to accelerate the rate; however, upon reaching a critical monoester concentration, the operating temperature was lowered to induce precipitation.[254, 255] Alternatives to recovering monoester include the use of polar solvents[190] (discussed above) and the continuous adsorption of monoesters by a column of silica gel.[231, 258]

The second example is the recovery of free hydroxy-fatty acid (H-FFA), a relatively high-melting compound, during lipase-catalyzed hydrolysis of triglyceride. In a membrane bioreactor, Derksen and co-workers selectively recovered H-FFA by including a cold trap in the product recycle stream, which selectively crystallized the H-FFA.[259] In addition, FFA was readily recovered by contacting the product recycle stream with an aqueous stream at basic pH, the two streams being separated by the membrane.[259] Saponification resulted, which allowed the FFA to migrate across the membrane and solubilize in the aqueous phase.[259]

As discussed above, enzymes have been used to synthesize biodegradable polymers. Russell and co-workers demonstrated that the molecular weight and polydispersity of polyesters synthesized by lipases in supercritical solvents can be controlled by selecting the fluid's density, which in turn is controlled by pressure and temperature.[170, 260] The fluid density dictates the solubility of polymer; hence, upon reaching a critical size, polymers precipitate from solution.

Engineering of Conditions during the Course of Reaction

An example was given above for programming the reaction temperature to induce precipitation of monoester, which controls reaction selectivity. Other parameters that can be changed or engineered during the course of reaction include the water activity[93] (discussed previously) and the type or amount of solvent. In addition, if the substrate lowers the enzymatic activity (e.g., by increasing the solvent polarity[173] or viscosity, or by inducing substrate inhibition), the gradual or stepwise addition of substrate may assist. Examples include the introduction of ethanol during lipase-catalyzed esterification;[239] the addition of hydrogen peroxide, a known denaturant, for lipase-catalyzed peroxidation of lipids;[261, 262] and the addition of saccharide for lipase-catalyzed sugar ester formation.[203] In the latter example, it was found the removing the solvent, tert-butanol, during the course of reaction, along with the stepwise addition of fructose, increased the reaction rate twofold (Figure 8.4). Tert-butanol is beneficial for the initial period of the reaction to help co-solubilize the saccharide and fatty acid substrates.[203]

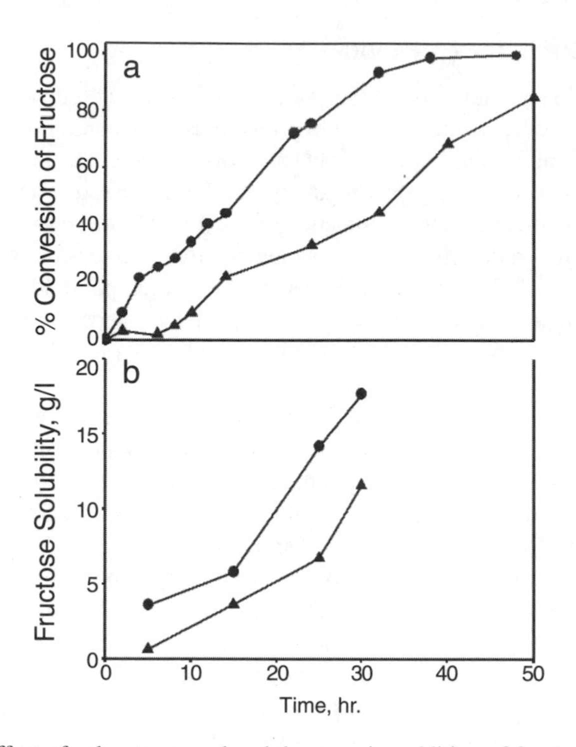

FIGURE 8.4 Effect of solvent removal and the stepwise addition of fructose on the percent conversion of saccharide (a) and on the solubility of fructose (b) for lipase-catalyzed fructose-oleic acid esterification. (•) Fructose was added to 50 mmol oleic acid and 13.4 g tert-butanol in 5 mmol increments for 2-h cycles until the net amount of fructose added was 25 mmol; tert-butanol underwent free evaporation for up to 10 h, at which time, the reaction was stopped, solvent was removed completely by rotary evaporation, then the reaction was continued. (▲) Fructose (25 mmol) was added at time 0 to 50 mmol of oleic acid and 13.4 g (46.9 wt.% fructose-free basis) tert-butanol; solvent freely evaporated away throughout at a rate of 0.47 g per h. Both reactions were operated in stirred batch mode using approximately 0.5 g immobilized *C. antarctica* lipase, (Chirazyme L-2, s.-f, c2, Lyo., Boehringer-Mannheim, Indianapolis, IN), a stir rate of 450 rev min^{-1} and a reaction temperature of 65°C. (From Walde, P. et al., *Chem. Phys. Lipids*, 53, 265–288, 1990. With permission.)

During the course of reaction, the formation of monoester, owing to its hydrophilicity and surfactant activity, leads to a huge increase in fructose solubility; hence, the need for solvent is diminished.[203] In addition, the presence of tert-butanol deters the evaporation of the other esterification product, water.[203]

Monitoring Progress of Biocatalytic Reactions

Enzymatic reactions have been monitored by several procedures. In the case of solid-phase enzymes, analysis is best achieved by periodically withdrawing small aliquots of fluid-phase reaction medium, after solid–fluid separation has occurred via gravity settling (e.g., disabling the agitator in batch reactors), filtration, or centrifugation. The aliquot can then be analyzed via chromatography or spectroscopy. Water content of the fluid phase can be measured using the Karl–Fischer titration. Inexpensive coulometric titrators perform quite adequately unless the water content is above a few percent. For aliquots taken from enzyme-soluble media (e.g., water-in-oil microemulsions), the enzyme must be inactivated (e.g., with high or low temperature or acid–base) to stop the occurrence of further reaction in the aliquot. Reactions catalyzed by solubilized enzymes can also be monitored *in situ* with spectroscopy.

If water is a product of the reaction, its removal by evaporation is induced, and if no other volatile component is present, the progress of the reaction can also be monitored by calculating the mass of water evaporated. The latter is determined from a mass balance, using the overall loss of mass from the system, determined gravimetrically, or using Karl–Fischer titration to monitor water concentration changes.[263]

CONCLUSIONS

This chapter reviewed the methodology required to insert proteins and operate biocatalytic reactions in nonaqueous environments. Nonaqueous enzymology has opened the door to new and exciting avenues in the chemical sciences, such as enantiomerically pure products. Ongoing research will continue to produce more effective biocatalysts and operation procedures for biocatalytic reactions. In conclusion, nonaqueous biotechnology will continue to grow in the upcoming years.

REFERENCES

1. McCoy, M., Biocatalysis grows for drug synthesis, *Chem. Engr. News*, 10–14, 1999.
2. Margolin, A. L., Enzymes in the synthesis of chiral drugs, *Enzyme Microb. Technol.*, 15, 266–280, 1983.
3. Chen, C. S. and Sih, C. J., General aspects and optimization of enantioselective biocatalysis in organic solvents: the use of lipases, *Angew. Chem. Int. Ed. Eng.*, 28, 695–707, 1989.
4. Ferraboschi, P., Santaniello, E., and Grisenti, P., Lipase-catalyzed transesterification in organic solvents: applications to the preparation of enantiomerically pure compounds, *Enzyme Microb. Technol.*, 15, 367–382, 1993.
5. Adams, M. W. W. and Kelley, R. M., Enzymes from microorganisms in extreme environments, *Chem. Eng. Newsl.*, 32–42, 1995.
6. Zamost, B. L., Nielsen, H. K., and Starnes, R. L., Thermostable, enzymes for industrial applications, *J. Ind. Microbiol.*, 8, 71–82, 1991.
7. Govardhan, C. P. and Margolin, A. L., Extremozymes for industry—from nature and by design, *Chem. Ind.*, 680–684, 1995.
8. Sears, P. and Wong, C. H., Engineering enzymes for bioorganic synthesis: peptide bond formation, *Biotechnol. Prog.*, 12, 423–433, 1996.
9. Xu, Z. F., Affleck, R., Wangikar, P., Suzawa, V., Dordick, J. S., and Clark, D. S., Transition state stabilization of subtilisins in organic media, *Biotechnol. Bioeng.*, 43, 515–520, 1994.
10. Moore, J. C. and Arnold, F. H., Directed evolution of a *para*-nitrobenzyl esterase for aqueous-organic solvents, *Natl. Biotechnol.*, 14, 458–467, 1996.

11. Zhao, H., Giver, L., Shao, Z., Affholter, J. A., and Arnold, F. H., Molecular evolution by staggered extension process (StEP) *in vitro* recombination, *Nat. Biotechnol.*, 16, 258–261, 1998.
12. Virto, C., Svensson, I., Adlercreutz, P., and Mattiasson, B., Catalytic activity of noncovalent complexes of horse liver alcohol dehydrogenase, NAD^+, and polymers, dissolved or suspended in organic solvents, *Biotechnol. Lett.*, 17, 877–882, 1995.
13. Yang, F. and Russell, A. J., Optimization of baker's yeast alcohol dehydrogenase activity in an organic solvent, *Biotechnol. Prog.*, 9, 234–241, 1993.
14. Vulfson, E., Enzymatic synthesis of surfactants, in *Novel Surfactants: Preparation, Applications, and Biodegradability*, Holmberg, K., Ed., Surfactant Science Series, Vol. 74, Marcel Dekker, New York, 1998, 279–300.
15. Rich, J. O., Bedell, B. A., and Dordick, J. S., Coontrolling enzyme-catalyzed regioselectivity in sugar ester synthesis, *Biotechnol. Bioeng.*, 45, 426–434, 1995.
16. Mustranta, A., Use of lipases in the resolution of racemic ibuprofen, *Appl. Microbiol. Biotechnol.*, 38, 61–66, 1992.
17. Tsai, S. W. and Wei, H. J., Effect of solvent on enantioselective esterification of naproxen by lipase with trimethylsilyl methanol, *Biotechnol. Bioeng.*, 43, 64–68, 1994.
18. Yang, F., Wild, J. R., and Russell, A. J., Nonaqueous biocatalytic degradation of a nerve gas mimic, *Biotechnol Prog.*, 11, 471–474, 1995.
19. Margolin, A. L., Crenne, J. Y., and Klibanov, A. M., Stereoselective oligomerizations catalyzed by lipases in organic solvents, *Tetrahed. Lett.*, 28, 1607–1610, 1987.
20. Svirkin, Y. Y., Xu, J., Gross, R. A., Kaplan, D. L., and Swift, G., Enzyme-catalyzed stereoelective ring-opening polymerization of α-methyl-β-propiolactone, *Macromolecules*, 29, 4591–4597, 1996.
21. Ebert, C., Gardossi, L., and Linda, P., Control of enzyme hydration in penicillin amidase catalyzed synthesis of amide bond, *Tetrahed. Lett.*, 37, 9377–9380, 1996.
22. Cerovsky, V., Free trypsin-catalyzed peptide synthesis in acetonitrile with low water content, *Biotechnol. Lett.*, 12, 899–904, 1990.
23. Kitaguchi, H. and Klibanov, A. M., Enzymatic peptide synthesis via segment condensation in the presence of water mimics, *J. Am. Chem. Soc.*, 111, 9272–9273, 1989.
24. Dordick, J. S., Marletta, M. A., and Klibanov, A. M., Polymerization of phenols catalyzed by peroxidase in nonaqueous media, *Biotechnol. Bioeng.*, 30, 31–36, 1997.
25. Dordick, J. S., Enzymatic and chemoenzymatic approaches to polymer synthesis and modification, *Ann. N.Y. Acad. Sci.*, 672, 352–362, 1992.
26. Hilhorst, R., Spruijt, R., Laane, C., and Veeger, C., Rules for the regulation of enzyme activity in reversed micelles as illustrated by the conversion of apolar steroids by 20 to β-Hydroxysteroid dehydrogenase, *Eur. J. Biochem.*, 144, 459–466, 1984.
27. Riva, S. and Klibanov, A. M., Enzymochemical regioselective oxidation of steroids without oxidoreductases, *J. Am. Chem. Soc.*, 110, 3291–3295, 1988.
28. Bertinotti, A., Carrea, G., Ottolina, G., and Riva, S., Regioselective esterification of polyhydroxylated steroids by *Candida antarctica* lipase B, *Tetrahedron*, 50, 13165–13172, 1994.
29. Soumanou, M. M., Bornscheuer, U. T., and Schmid, R. D., Two-step enzymatic reaction for the synthesis of pure structured triacylglycerides, *J. Am. Oil Chem. Soc.*, 6, 703–710, 1998.
30. Xu, X., Balchen, S., and Adler-Nissen, J., Pilot batch production of specific-structured lipids by lipase-catalyzed interesterification: preliminary study on incorporation and acyl migration, *J. Am. Oil Chem. Soc.*, 75, 301–308, 1998.
31. Dordick, J. S., Enzymatic catalysis in monophasic organic solvents, *Enzyme Microb. Technol.*, 11, 194–211, 1989.

32. Zaks, A. and Russell, A. J., Enzymes in organic solvents: properties and applications, *J. Biotechnol.*, 8, 259–270, 1988.
33. Margolin, A. L., Enzymes: use them. *CHEMTECH*, 160–167, 1991.
34. Klibanov, A. M., Asymmetric enzymatic catalysis in organic media, *Acc. Chem. Res.*, 23, 114–120, 1990.
35. Wong, C. H., Enzymatic catalysts in organic synthesis, *Science*, 244, 1145–1152.
36. Faber, K. and Franssen, M. C. R., Prospects for the increased application of biocatalysts in organic transformations, *Trends Biotechnol.*, 11, 461–470, 1989.
37. Kvittingen, L., Some aspects of biocatalysis in organic solvents, *Tetrahedron*, 50, 8253–8274, 1994.
38. Halling, P. J., Thermodynamic predictions for biocatalysis in nonconventional media: theory, tests, and recommendations for experimental design and analysis, *Enzyme Microb. Technol.*, 16, 178–206, 1994.
39. Klibanov, A. M., Why are enzymes less active in organic solvents than in water? *Trends Biotechnol.*, 15, 97–101, 1997.
40. Ke, T., Wescott, C.R., and Klibanov, A. M., Prediction of the solvent dependence of enzymatic prochiral selectivity by means of structure-based thermodynamic calculations, *J. Am. Chem. Soc.*, 118, 3366–3374, 1996.
41. Ke, T., Tidor, B., and Klibanov, A. M., Molecular-modeling calculations for enzymatic enantioselectivity taking hydration into account, *Biotechnol. Bioeng.*, 57, 741–745, 1998.
42. Haeffner, F., Norin, T., and Hult, K., Molecular modeling of the enantioselectivity in lipase-catalyzed transesterification reactions, *Biophys. J.*, 74, 1251–1262, 1998.
43. Bernstein, F. C., Koetzle, T. F., Williams, G. J. B., Meyer, E. F. J., Brice, M. D., Rodgers, J. R., Kennard, O., Shimanouchi, T., and Tasumi, M., The protein data bank: a computer-based archival file for macromolecular structures, *J. Mol. Biol.*, 112, 535–542, 1977.
44. Parida, S. and Dordick, J. S., Tailoring lipase specificity by solvent and substrate chemistries, *J. Org. Chem.*, 58, 3238–3244, 1993.
45. Kazlauskas, R. J., Weissfloch, A. N. E., Rappaport, A.T., and Cuccia, L. A., A rule to predict which enantiomer of a secondary alcohol reacts faster in reactions catalyzed by cholesterol esterase, lipase from *Peudomonas cepacia* and lipase from *Candida rugosa, J. Org. Chem.*, 56, 2656–2665, 1991.
46. Franssen, M. C. R., Jongejan, H., Kooijman, H., Spek, A. L., Camacho Mondril, N.L.F.L., Boavida, dos Santos, P.M.A.C., and de Groot A., Resolution of a tetrahydrofuran ester by *Candida rugosa* lipase (CRL) and an examination of CRL's stereochemical preference in organic media, *Tetrahed. Assym.*, 7, 497–510, 1996.
47. Lalonde, J. J., Govardhan, C., Khalaf, N., Martinez, A. G., Visuri, K., and Margolin, A. L., Cross-linked crystals of *Candida rugosa* lipase: highly efficient catalysts for the resolution of chiral esters, *J. Am. Chem. Soc.*, 117, 6845–6852, 1995.
48. Tsai, S. W. and Dordick, J. S., Extraordinary enantiospecificity of lipase catalysis in organic media induced by purification and catalyst engineering, *Biotechnol. Bioeng.*, 52, 296–300, 1996.
49. Yamane, T., Ichiryu, T., Nagata, M., Ueno, A., and Shimizu, S., Intramolecular esterification by lipase powder in microequeous benzene: factors affecting activity of pure enzyme, *Biotechnol. Bioeng.*, 36, 1063–1069, 1990.
50. Bovara, R., Carrea, G., Ottolina, G., and Riva, S., Effects of water activity on V_{max} and K_M of lipase catalyzed transesterification in organic media, *Biotechnol. Lett.*, 15, 937–942, 1993.
51. Mellor, J. D., *Fundamentals of Freeze-Drying*, Academic Press, London, 1978.

52. Heller, M. C., Carpenter, J. F., and Randolph, T. W., Protein formulation and lyophilization cycle design: prevention of damage owing to freeze-concentration induced phase separation, *Biotechnol. Bioeng.*, 63, 166–174, 1999.
53. Strambini, G. B. and Gabellieri, E., Proteins in frozen solutions: evidence of ice-induced partial unfolding, *Biophys. J.*, 70, 971–976, 1996.
54. Fágáin, C. Ó., *Stabilizing Protein Function*, Springer-Verlag, Berlin, 1997.
55. Burke, P. A., Griffin, R. G., and Klibanov, A. M., Solid-state NMR assessment of enzyme active center structure under nonaqueous conditions, *J. Biol. Chem.*, 267, 20057–20064, 1992.
56. Desai, U. R., Osterhout, J. J., and Klibanov, A. M., Protein structure in the lyophilized state: a hydrogen isotope exchange NMR study with bovine pancreatic trypsin inhibitor, *J. Am. Chem. Soc.*, 116, 9420–9422, 1994.
57. Desai, U. R. and Klibanov, A. M., Assessing the structural integrity of a lyophilized protein in organic solvents, *J. Am. Chem. Soc.*, 117, 3940–3945, 1995.
58. Griebenow, K. and Klibanov, A. M., Can conformational changes be responsible for solvent and excipient effects on the catalytic behavior of subtilisin Carlsberg in organic solvent? *Biotechnol. Bioeng.*, 53, 351–362, 1997.
59. Dabulis, K. and Klibanov, A. M., Dramatic enhancement of enzymatic activity in organic solvents, *Biotechnol. Bioeng.*, 41, 566–571, 1993.
60. Triantafyllou, A. O., Wehtje, E., Adlercreutz, P., and Mattiasson, B., Effects of sorbitol addition on the action of free and immobilized hydrolytic enzymes in organic media, *Biotechnol. Bioeng.*, 45, 406–414, 1995.
61. Chang, B. S. and Randall, C. S., Use of subambient thermal analysis to optimize protein lyophilization, *Cryobiology*, 29, 632–656, 1992.
62. Triantafyllou, A. O., Wehtje, E. Adlercreutz, P., and Mattiasson, B., How do additives affect enzyme activity and stability in nonaqueous media?, *Biotechnol. Bioeng.*, 54, 67–76, 1997.
63. Constantino, H. R., Griebenow, K., Langer, R., and Klibanov, A. M., On the pH memory of lyophilized compounds containing protein functional groups, *Biotechnol. Bioeng*, 53, 345–348, 1997.
64. Khmelnitsky, Y. L., Welch, S. H., Clark, D. S., and Dordick, J. S., Salts dramatically enhance activity of enzymes suspended in organic solvents, *J. Am. Chem. Soc.*, 116, 2647–2648, 1994.
65. Ru, M. T., Dordick, J. S., Reimer, J. A., and Clark, D. S., Optimizing the salt-induced activation of enzymes in organic solvents effects of lyophilization time and water content, *Biotechnol. Bioeng.*, 63, 233–241, 1999.
66. Blackwood, A. D., Moore, B. D., and Halling, P. J., Are associated ions important for biocatalysis in organic media? *Biocatalysis*, 9, 269–276, 1994.
67. Bedell, B. A., Mozhaev, V. V., Clark, D. S., and Dordick, J. S., Testing for diffusion limitations in salt-activated enzyme catalysts operating in organic solvents, *Biotechnol. Bioeng.*, 58, 654–657, 1998.
68. Braco, L., Dabulis, K., and Klibanov, A. M., Production of abiotic receptors by molecular imprinting of proteins, *Proc. Natl. Acad. Sci. U.S.A.*, 87, 274–277, 1990.
69. Russell, A. J., Trude, L. J., Skippe, P. Lr., Groopman, J. D., Tannenbaum, S. R., and Klibanov, A. M., Antibody-antigen binding in organic solvents, *Biochem. Biophys. Res. Comm.*, 158, 80–85, 1989.
70. Russell, A. J. and Klibanov, A. M., Inhibitor-induced enzyme activation in organic solvents, *J. Biol. Chem.*, 263, 11624–11626, 1988.
71. Dabulis, K. and Klibanov, A. M., Molecular imprinting of proteins and other macromolecules resulting in new adsorbents, *Biotechnol. Bioeng.*, 39, 176–185, 1992.

72. Slade, C. J. and Vulfson, E. N., Induction of catalytic activity in proteins by lyophilization in the presence of a transition state analogue, *Biotechnol. Bioeng.*, 57, 211–215, 1998.
73. Mingarro, I., Abad, C., and Braco, L., Interfacial activation-based molecular bioimprinting of lipolytic enzymes, *Proc. Natl. Acad. Sci. U.S.A.*, 92, 3308–3312, 1995.
74. Gonzalez-Navarro, H. and Braco, L., Improving lipase activity in solvent-free media by interfacial activation-based molecular bioimprinting, *J. Molec. Catal. B*, 3, 111–119, 1997.
75. Gonzalez-Navarro, H. and Braco, L., Lipase-enhanced activity in flavour ester reactions by trapping enzyme conformers in the presence of interfaces, *Biotechnol. Bioeng.*, 59, 122–127, 1998.
76. Colton, L. J., Ahmed, S. N., and Kazlauskas, R. J., A 2-propanol treatment increases the enantioselectivity of *Candida rugosa* lipase toward esters of chiral carboxylic acids, *J. Org. Chem.*, 60, 212–217, 1995.
77. Broos, J., Sakodinskaya, I. K., Engbersen, J. F. J., and Reinhoudt, D. N., Large activation of serine proteases by pretreatment with crown ethers, *J. Chem. Soc. Chem. Comm.*, 255–256, 1995.
78. van Unen, D. J, Engbersen, J. F. J., and Reinhoudt, D. N., Large acceleration of α-chymotrypsin-catalyzed dipeptide formation by 18-crown-6 in organic solvents, *Biotechnol. Bioeng.*, 59, 553–556, 1998.
79. Blinkovsky, A. M., Khmelnitsky, Y. L., and Dordick, J. S., Organosoluble enzyme-polymer complexes: a novel type of biocatalyst for nonaqueous media, *Biotechnol. Tech.*, 8, 33–38, 1994.
80. Ke, T. and Klibanov, A. M., On enzymatic activity in organic solvents as a function of enzyme history, *Biotechnol. Bioeng.*, 57, 746–750, 1998.
81. Kennedy, J., Melo, E. H. M., and Jumel, K., Immobilized enzymes and cells, *Chem. Engr. Prog.*, 86, 81–89, 1990.
82. Ison, A. P., Macrae, A. R., Smith, C. G., and Bosley, J., Mass transfer effects in solvent-free fat interesterification reactions: influences on catalyst design, *Biotechnol. Bioeng.*, 43, 122–130, 1994.
83. Barros, R. J., Wehtje, E., and Adlercreutz, P., Mass transfer studies on immobilized α-chymotrypsin biocatalysts prepared by deposition for use in organic medium, *Biotechnol. Bioeng.*, 59, 364–373, 1998.
84. Katchalski-Katzi, E., Immobilized enzymes—learning from past successes and failures, *Trends Biotechnol.*, 11, 471–478, 1993.
85. Malcata, F. X., Reyes, H. R. Garcia, H. S., and Hill, C. G. Jr., Immobilized lipase reactors for modification of fats and oils—A review, *J. Am. Oil Chem. Soc.*, 67, 890–910, 1990.
86. Balcao, V. M., Paiva, A. L., and Malcata, F. X., Bioreactors with immobilized lipases: state of the art, *Enzyme Microb. Technol.*, 18, 392–416, 1996.
87. Mosbach, K. Ed., *Immobilized Enzymes: Methods in Enzymology*, Vol. 44, Academic Press, San Diego, 1976.
88. Mosbach, K., Ed., *Immobilized Enzymes and Cells, Part B: Methods in Enzymology*, Vol. 135, Academic Press, San Diego, 1987.
89. Mosbach, K., Ed., *Immobilized Enzymes and Cells, Part C: Methods in Enzymology*, Vol. 136, Academic Press, San Diego, 1987.
90. Mosbach, K., Ed., *Immobilized Enzymes and Cells, Part D: Methods in Enzymology*, Vol. 137, Academic Press, San Diego, 1988.
91. Svensson, I., Adlercreutz, P., and Mattiasson, B., Lipase-catalyzed transesterification of phosphatidylcholine at controlled water activity, *J. Am. Oil Chem. Soc.*, 69, 986–991, 1992.

92. Valivety, R. H., Halling, P. J., Peilow, A. D., and Macrae, A. R., Lipases from different sources vary widely in dependence of catalytic activity on water activity, *Biochim. Biophys. Acta*, 1122, 143–146, 1992.
93. Svensson, L., Wehtje, E., Adlercreutz, P., and Mattiasson, B., Effects of water activity on reaction rates and equilibrium positions in enzymatic esterifications, *Biotechnol. Bioeng.*, 44, 549–556, 1994.
94. Orsat, B., Drtina, G. J., Williams, M. G., and Klibanov, A. M., Effect of support material and enzyme pretreatment on enantioselectivity of immobilized subtilisin in organic solvents, *Biotechnol. Bioeng.*, 44, 1265–1269, 1994.
95. Norin, M., Boutelje, J., Holmberg, E., and Hult, K., Lipase immobilized by adsorption. Effect of support hydrophobicity on the reaction rate of ester synthesis in cyclohexane, *Appl. Microbiol. Biotechnol.*, 28, 527–530, 1988.
96. Adlercreutz, P., On the importance of the support material for enzymatic synthesis in organic media. Support effects at controlled water activity, *Eur. J. Biochem.*, 199, 609–614, 1991.
97. Millqvist-Fureby, A., Virto, C., Adlercreutz, P., and Mattiasson, B., Acyl group migrations in 2-monoolein, *Biocat. Biotrans.*, 14, 89–111, 1996.
98. Johansson, A., Mosbach, K., and Mansson, M. O., Horse liver alcohol dehydrogenase can accept $NADP^+$ as coenzyme in high concentrations of acetonitrile, *Eur. J. Biochem.*, 227, 551–555, 1995.
99. Parida, S., Datta, R., and Dordick, J. S., Supported aqueous-phase enzymatic catalysis in organic media, *Appl. Biochem. Biophys.*, 33, 1–14, 1992.
100. Wehtje, E., Adlercreutz, P., and Mattiasson, B., Improved activity retention of enzymes deposited on solid supports, *Biotechnol. Bioeng.*, 41, 171–178.
101. Adlercreutz, P., Activation of enzymes in organic media at low water activity by polyols and saccharides, *Biochim. Biophys. Acta*, 1163, 144–148, 1993.
102. Wisdom, R. A., Dunnill, P., and Lilly, M. D., Enzymic interesterification of fats: the effect of non-lipase material on immobilized enzyme activity, *Enzyme Microb. Technol.*, 71, 567–572, 1985.
103. Arcos, J. A. and Otero, C., Enzyme, medium, and reaction engineering to design a low-cost, selective production method for mono- and dioleoylglycerols, *J. Am. Oil Chem. Soc.*, 73, 673–682, 1996.
104. Aydemir, T. and Telefoncu, A., Lipase immobilization on different supports for interesterification of fats, *Ind. J. Chem.*, 33B, 387–39, 1994.
105. Bovara, R., Carrea, G., Ottolina, G., and Riva, S., Water activity does not influence the enantioselectivity of lipase and lipoprotein lipase in organic solvents, *Biotechnol. Lett.*, 15, 169–174, 1993.
106. St. Clair, N. L. and Navia, M. A., Cross-linked enzyme crystals as robust biocatalysts, *J. Am. Chem. Soc.*, 114, 7314–7316, 1992.
107. Khalaf, N., Govardhan, C. P., Lalonde, J. J., Persichetti, R. A., Wang, Y. F., and Margolin, A. L., Cross-linked enzyme crystals as highly active catalysts in organic solvents, *J. Am. Chem. Soc.*, 118, 5494–5495, 1996.
108. Margolin, A. L., Novel crystalline catalysts, *Trends Biotechnol.*, 14, 223–230, 1996.
109. Goto, M., Kameyama, H. K., Goto, M., Miyata, M., and Nakashio, F., Design of surfactants suitable for surfactant-coated enzymes as catalysts in organic media, *J. Chem. Eng. Jpn.*, 26, 109–111, 1993.
110. Isono, Y., Nabetani, H., and Nakajima, M., Lipase-surfactant complex as catalyst of interesterification and esterification in organic media, *J. Ferment. Bioeng.*, 80, 170–175, 1995.

111. Paradkar, V. M. and Dordick, J. S., Mechanism of extraction of chymotrypsin into isooctane at very low concentrations of Aerosol OT in the absence of reversed micelles, *Biotechnol. Bioeng.*, 43, 529–540, 1994.
112. Abe, K., Goto, M., and Nakashio, F., Surfactant-chymotrypsin complex as a novel biocatalyst in organic media, *J. Ferment. Bioeng.*, 83, 555–560, 1996.
113. Wangikar, P. P., Michels, P. C., Clark, D. S., and Dordick, J. S., Structure and function of subtilisin BPN′ solubilized in organic solvents, *J. Am. Chem. Soc.*, 119, 70–76, 1997.
114. Okahat, Y., Niikura, K., and Ijiro, K., Simple transphosphatidylation of phospholipids catalysed by a lipid-coated phospholipase D in organic solvents, *J. Chem. Soc. Perkin Trans. I*, 919–925, 1995.
115. Okahata, Y., Hatano, A., and Ijiro, K., Enhancing enantioselectivity of a lipid-coated lipase via imprinting methods for esterification in organic solvents, *Tetrahed. Assym.*, 6, 1311–1322, 1995.
116. Meyer, J. D., Matsuura, J. E., Evans, B. S., Evans, G. J., and Manning, M. C., Solution behavior of *alpha*-chymotrypsin dissolved in nonpolar organic solvents via hydrophobic ion pairing, *Biopolymers*, 35, 451–456, 1995.
117. Goto, M., Sumura, H., Abe, K., and Nakashio, F., Novel preparation method for surfactant-coated enzymes using W/O emulsion, *Biotechnol. Tech.*, 9, 101–104, 1995.
118. Okazaki, S. Y., Kamiya, N., Abe, K., Goto, M., and Nakashio, F., Novel preparation method for surfactant-lipase complexes utilizing water in oil emulsions, *Biotechnol. Bioeng.*, 55, 455–459, 1997.
119. Tsuzuki, W. and Suzuki, T., Reactive properties of the organic solvent-soluble lipase, *Biochim. Biophys. Acta*, 1083, 201–206, 1991.
120. Goto, M., Kamiya, N., Miyata, M., and Nakashio, F., Enzymatic esterification by surfactant-coated lipase in organic media, *Biotechnol. Prog.*, 10, 263–268, 1994.
121. Mogi, K. I. and Nakajima, M., Selection of surfactant-modified lipase for interesterification of triglyceride and fatty acid, *J. Am. Oil Chem. Soc.*, 73, 1505–1512, 1996.
122. Matsuura, J., Powers, M. E., Manning, M. C., and Shefter, E., Structure and stability of insulin dissolved in 1-octanol, *J. Am. Chem. Soc.*, 115, 1261–1264, 1993.
123. Powers, M. E., Matsuura, J., Brassell, J., Manning, M. C., and Shefter, E., Enhanced solubility of proteins and peptides in nonpolar solvents through hydrophobic ion pairing, *Biopolymers*, 33, 927–932, 1993.
124. Kiamiya, N., Okazaki, S. Y., and Goto, M., Surfactant-horseradish peroxidase complex catalytically active in anhydrous benzene, *Biotechnol. Tech.*, 11, 375–378, 1997.
125. Paradkar, V. M. and Dordick, J. S., Aqueous-like activity of α-chymotrypsin dissolved in nearly anhydrous organic solvents, *J. Am. Chem. Soc.*, 116, 5009, 1994.
126. Abe, K., Kawazoe, T., Okazaki, S. Y., Goto, M., and Nakashio, F., Peptide synthesis by surfactant-chymotrypsin complexes in organic media, *Biotechnol. Tech.*, 11, 25–29, 1997.
127. Okahata, Y., Tsuruta, T., Ijiro, K., and Ariga, K., Langmuir-Blodgett films of an enzyme-lipid complex for sensor membranes, *Langmuir*, 4, 1373–1375, 1998.
128. Okahata, Y., Tsuruta, T., Ijiro, K., and Ariga, K., Preparations of Langmuir-Blodgett films of enzyme-lipid complexes: a glucose sensor membrane, *Thin Solid Films*, 180, 65–72, 1989.
129. Okahata, Y., Yamaguchi, M., Tanaka, F. T., and Fujii, I., A lipid-coated catalytic antibody in water-miscible organic solvents, *Tetrahedron*, 51, 7673–7680, 1995.
130. Okahata, Y., Fujimoto, Y., and Ijiro, K., Lipase-lipid complex as a resolution catalyst of racemic alcohols in organic solvents, *Tetrahed. Lett.*, 29, 5133–5134, 1988.
131. Okahata, Y. and Ijiro, K., A lipid-coated lipase as a new catalyst for triglyceride synthesis in organic solvents, *J. Chem. Soc. Chem. Comm.*, 1392–1394, 1988.

132. Isono, Y., Nabetani, H., and Nakajima, M., Wax ester synthesis in a membrane reactor with lipase-surfactant complex in hexane, *J. Am. Oil Chem. Soc.*, 72, 887–890, 1995.
133. Basheer, S., Mogi, K., and Nakajima, M., Surfactant-modified lipase for the catalysis of the interesterification of triglycerides and fatty acids, *Biotechnol. Bioeng.*, 45, 187–195, 1995.
134. Basheer, S., Nakajima, M., and Cogan, U., Sugar ester-modified lipase for the esterification of fatty acids and long-chain alcohols, *J. Am. Oil Chem. Soc.*, 73, 1475–1479, 1996.
135. Okahata, Y. and Ijiro, K., Preparation of a lipid-coated lipase and catalysis of glyceride ester synthesis in homogeneous organic solvents, *Bull. Chem. Soc. Jpn.*, 65, 2411–2420, 1992.
136. Goto, M., Goto, M., Kamiya, N., and Nakashio, F., Enzymatic interesterification of triglyceride with surfactant-coated lipase in organic media, *Biotechnol. Bioeng.*, 45, 27–32, 1995.
137. Basheer, S., Mogi, K., and Nakajima, M., Interesterification kinetics of triglycerides and fatty acids with modified lipase in *n*-hexane, *J. Am. Oil Chem. Soc.*, 72, 511–518, 1995.
138. Basheer, S., Snape, J. B., Mogi, K., and Nakajima, M., Transesterification kinetics of triglycerides for a modified lipase in *n*-hexane, *J. Am. Oil Chem. Soc.*, 72, 231–237, 1995.
139. Okahata, Y., Fujimoto, Y., and Ijiro, K., A lipid-coated lipase as an enantioselective ester synthesis catalyst in homogeneous organic solvents, *J. Org. Chem.*, 60, 2244–2250, 1995.
140. Kamiya, N., Goto, M., and Nakashio, F., Surfactant-coated lipase suitable for the enzymatic resolution of menthol as a biocatalyst in organic media, *Biotechnol. Prog.*, 11, 270–275. 1995.
141. Akita, H., Umezawa, I., Matsukura, H., and Oishi, T., A lipid-lipase aggregate as a new type of immobilized enzyme, *Chem. Pharm. Bull.*, 39, 1632–1633, 1991.
142. Goto, M., Noda, S., Kamiya, N., and Nakashio, F., Enzymatic resolution of racemic ibuprofen by surfactant-coated lipases in organic media, *Biotechnol. Lett.*, 18, 839–844, 1996.
143. Tsuzuki, W., Sasaki, T., and Suzuki, T., Effect of detergent attached to enzyme molecules on the activity of organic-solvent-soluble lipases, *J. Chem. Soc. Perkin Trans. I*, 1851–1854, 1991.
144. Tsuzuki, W., Okahata, Y., Katayama, O., and Suzuki, T., Preparation of organic-solvent-soluble enzyme (lipase B) and characterization by gel permeation chromatography, *J. Chem. Soc. Perkin Trans. I*, 1245–1247, 1991.
145. Goto, M., Kondo, K., and Nakashio, F., Acceleration effect of anionic surfactants on extraction rate of copper with liquid surfactant membrane containing LI65N and nonionic surfactant, *J. Chem. Engr. Jpn.*, 22, 79–83, 1979.
146. Okahata, Y., Tanamachi, S., Nagai, M., and Kunitake, T., Synthetic bilayer membranes prepared from dialkyl amphiphiles with nonionic and zwitterionic head groups, *J. Colloid Interf. Sci.*, 82, 401–417, 1981.
147. Okahata, Y. and Seki, T., pH-sensitive capsule membranes. Reversible permeability control from the dissociative bilayer-coated capsule membrane by an ambient pH change, *J. Am. Chem. Soc.*, 106, 8065–8070, 1984.
148. Jones, M. N. and Chapman, D., *Micelles, Monolayers, and Biomembranes,* Wiley-Liss, New York, 1995.
149. Ghenciu, E. G., Russell, A. J., Beckman, E. J., Steele, L., and Becker, N. T., Solubilization of subtilisin in CO_2 using fluoroether-functional ampliphiles, *Biotechnol. Bioeng.*, 58, 572–580, 1998.

150. Gladilin, A. K., Kudryashova, E. V., Vakurov, A. V., Izumrudov, V. A., Mozhaev, V. V., and Levashov, A. V., Enzyme-polyelectrolyte noncovalent complexes as catalysts for reactions in binary mixtures of polar organic solvents with water, *Biotechnol. Lett.*, 17, 1329–1344, 1995.
151. Vakurov, A. V., Gladilin, A. K., Levashov, A. V., and Khmelmitsky, Y. L., Dry enzyme-polymer complexes: stable organosoluble biocatalysts for nonaqueous enzymology, *Biotechnol. Lett.*, 16, 175–178, 1994.
152. Sakodinskaya, I. K., Sorokina, E. M., Efremova, N. V., and Topchieva, I. N., Polymer-enzyme complexes in organic medium, *Biotechnol. Tech.*, 12, 607–610, 1988.
153. Ghan, L. G. and Roskoski, R. Jr., Thermal stability and CD analysis of rat tyrosine hydroxylase, *Biochemistry*, 34, 252–256, 1995.
154. Burke, C. J., Volkin, D. B., Mach, H., and Middaugh, C. R., Effect of polyanions on the unfolding of acidic fibroblast growth factor, *Biochemistry*, 32, 6419–6426, 1993.
155. Baillargeon, M. W. and Sonnet, P. E., Polyethylene glycol modification of *Candida rugosa* lipase, *J. Am. Oil Chem. Soc.*, 65, 1812–1815, 1998.
156. Inada, Y., Matsushima, A., Takahashi, K., and Saito, Y., Polyethylene glycol-modified lipase soluble and active in organic solvents, *Biocatalysis*, 3, 317–328, 1990.
157. Ito, Y., Fujii, H., and Imanishi, Y., Catalytic peptide synthesis by trypsin modified with polystyrene in chloroform, *Biotechnol. Prog.*, 9, 128–130, 1993.
158. Mabrouk, P. A., The use of nonaqueous media to probe biochemically significant enzyme intermediates: the generation and stabilization of horseradish peroxidase compound II in neat benzene solution at room temperature, *J. Am. Chem. Soc.*, 117, 2141–2146, 1995.
159. Hernaiz, M. J., Sanchez-Montero, J. M., and Sinisterra, J. V., Influence of the nature of modifier in the enzymatic activity of chemical modified semipurified lipase from *Candida rugosa, Biotechnol. Bioeng.*, 55, 252–260, 1997.
160. Bovara, R., Carrea, G., Gioacchini, A. M., and Riva, S., Activity, stability, and conformation of methyxypoly(ethylene glycol)-subtilisin at different concentrations of water in dioxane, *Biotechnol. Bioeng.*, 54, 50–57, 1997.
161. Kabanov, A. V., Levashov, A. V., and Martinek, K., Transformation of water-soluble enzymes into membrane active form by chemical modification, *Ann. N.Y. Acad. Sci.*, 501, 63–66, 1987.
162. Fang, J. M. and Wong, C. H., Enzymes in organic synthesis: alteration of reversible reactions to irreversible processes, *Synlett.* June, 393–402, 1994.
163. Fregapane, G., Sarney, D. B., and Vulfson, E. N., Enzymic solvent-free synthesis of sugar acetal fatty acid esters, *Enzyme Microb. Technol.*, 13, 796–800, 1991.
164. Pecnik, S. and Knez, Z., Enzymatic fatty ester synthesis, *J. Am. Oil Chem. Soc.*, 69, 261–265, 1992.
165. Sarney, D. B., Barnard, M. J., MacManus, D. A., and Vulfson, E. N., Application of lipases to the regioselective synthesis of sucrose fatty acid monoesters, *J. Am. Oil Chem. Soc.*, 73, 1481–1487, 1996.
166. Ikeda, I. and Klibanov, A. M., Lipase-catalyzed acylation of sugars solubilized in hydrophobic solvents by complexation, *Biotechnol. Bioeng.*, 42, 788–791, 1993.
167. Schlotterbeck Lang, S., Wray, V., and Wagner, F., Lipase-catalyzed monoacylation of fructose, *Biotechnol. Lett.*, 15, 61–64, 1993.
168. Berger, M., Laumen, K., and Schneider, M. P., Enzymatic esterification of glycerol. I. Lipase-catalyzed synthesis of regioisomerically pure 1,3-sn-diacylglycerols, *J. Am. Oil Chem. Soc.*, 69, 955–960, 1992.

169. Castillo, E., Dossat, V., Marty, A., Condoret, J. S., and Combes, D., The role of silica gel in lipase-catalyzed esterification reactions of high-polar substrates, *J. Am. Oil Chem. Soc.*, 74, 77–85, 1997.
170. Russell, A. J., Beckman, E. J., and Chaudhary, A. K., Studying enzyme activity in supercritical fluids, *CHEMTECH*, March, 33–37, 1994.
171. Yang, F. and Russell, A. J., The role of hydration in enzyme activity and stability. 2. Alcohol dehydrogenase activity and stability in a continuous gas phase reactor, *Biotechnol. Bioeng.*, 49, 709–716, 1996.
172. Barton, J. W., Reed, E. K., and Davidson, B. H., Gas-phase enzyme catalysis using immobilized lipase for ester production, *Biotechnol. Tech.*, 11, 747–750, 1997.
173. Laane, C., Boeren, S., Vos, K., and Veeger, C., Rules for optimization of biocatalysis in organic solvents, *Biotechnol. Bioeng.*, 30, 81–87, 1987.
174. Valivety, R. H., Johnston, G. A., Suckling, C. J., and Halling, P. J., Solvent effects on biocatalysis in organic systems: equilibrium position and rates of lipase catalyzed esterification, *Biotechnol. Bioeng.*, 38, 1137–1143, 1991.
175. Halling, P. J., Solvent selection for biocatalysis in mainly organic systems: predictions of effects on equilibrium position, *Biotechnol. Bioeng.*, 35, 691–701, 1990.
176. Janssen, A. E. M., van der Padt, A., van Sonsbeek, H. M., and van't Riet, K., The effect of organic solvents on the equilibrium position of enzymatic acylglycerol synthesis, *Biotechnol. Bioeng.*, 41, 95–103, 1993.
177. Gupta, M. N., Batra. R., Tyagi, R., and Sharma, A., Polarity index: the guiding solvent parameter for enzyme stability in aqueous-organic cosolvent mixtures, *Biotechnol. Prog.*, 13, 284–288, 1997.
178. Wang, Z. L., Hiltunen, K., Orava, P., Seppala, J., and Linko, Y. Y., Lipase-catalyzed polyester synthesis, *Pure Appl. Chem.*, A33, 599–612, 1996.
179a. Lide, D. R., *CRC Handbook of Chemistry and Physics*, 73rd ed., CRC Press, Boca Raton, FL, 1992.
179b. Erbeldinger, M., Mesiano, A. J., and Russell, A. J., Enzymatic catalysis of formation of Z-aspartame in ionic liquid — an alternative to enzymeatic catalysis in organic solvents. *Biotechnol. Prog.*, 16, 1129–1131, 2000.
179c. Madeira Lau, R., van Rantwijk, F., and Sheldon, R. A. Lipase-catalyzed reactions in ionic liquids. *Org. Lett.*, 2, 4189–4191, 2000.
180. Griebenow, K. and Klibanov, A. M., On protein denaturation in aqueous-organic mixtures but not in pure organic solvents, *J. Am. Chem. Soc.*, 118, 11695–11700, 1996.
181. Sarney, D. B., Barnard, M. J., Virto, M., and Vulfson, E. N., Enzymatic synthesis of sorbitan esters using a low-boiling-point azeotrope as a reaction solvent, *Biotechnol. Bioeng.*, 54, 351–356, 1997.
182. Almarsson, O. and Klibanov, A. M., Remarkable activation of enzymes in nonaqueous media by denaturing organic cosolvents, *Biotechnol. Bioeng.*, 49, 87–92, 1996.
183. Colombie, S., Tweedell, R. J., Condoret, J. S., and Marty, A., Water activity control: a way to improve the efficiency of continuous lipase esterification, *Biotechnol. Bioeng.*, 60, 362–368, 1998.
184. Janssen, A. E. M., van der Padt, A., and van't Riet, K., Solvent effects on lipase-catalyzed esterification of glycerol and fatty acids, *Biotechnol. Bioeng.*, 42, 953–962, 1993.
185. Wescott, C. R. and Klibanov, A. M., Thermodynamic analysis of solvent effect on substrate specificity of lyophilized enzymes suspended in organic media, *Biotechnol. Bioeng.*, 56, 340–344, 1997.

186. Rubio, E., Fernandez-Mayorales, A., and Klibanov, A. M., Effect of the solvent on enzyme regioselectivity, *J. Am. Chem. Soc.*, 113, 695–696, 1991.
187. Fitzpatrick, P. A. and Klibanov, A. M., How can the solvent affect enzyme enantioselectivity? *J. Am. Chem. Soc.*, 113, 3166–3171, 1991.
188. Tawaki, S. and Klibanov, A. M., Inversion of enzyme enantioselectivity mediated by the solvent, *J. Am. Chem. Soc.*, 114, 1882–1884, 1992.
189. Chaudhary, A. K., Kamat, S. V., Beckman, E. J., Nurok, D., Kleyle, R. M., Hajdu, P., and Russell, A. J., Control of subtilisin substrate specificity by solvent engineering in organic solvents and supercritical fluoroform, *J. Am. Chem. Soc.*, 118, 12891–12901, 1996.
190. Cao, L., Fischer, A., Bornscheuer, U. T., and Schmid, R. D., Lipase-catalyzed solid phase synthesis of sugar fatty acid esters, *Biocat. Biotrans.*, 14 269–283, 1997.
191. Gorman, L. A. S. and Dordick, J. S., Organic solvents strip water off enzymes, *Biotechnol. Bioeng.*, 39, 392–397, 1992.
192. Valivety, R. H., Halling, P. J., and Macrae, A. R., Reaction rate with suspended lipase catalyst shows similar dependence on water activity in different organic solvents, *Biochim. Biophys. Acta*, 1118, 218–222, 1992.
193. Wehtje, E. and Adlercreutz, P., Water activity and substrate concentration effects on lipase activity, *Biotechnol. Bioeng.*, 55, 798–806, 1997.
194. Bell, G., Janssen, A. E. M., and Halling, P. J., Water activity fails to predict critical hydration level for enzyme activity in polar organic solvents: interconversion of water concentrations and activities, *Enzyme Microb. Technol.*, 20, 471–477, 1997.
195. de la Casa, R. M., Sanchez-Montero, R. M., and Sinisterra, J. V., Water adsorption isotherm as a tool to predict the preequilibrium water amount in preparative esterification,. *Biotechnol. Lett.*, 18, 13–18, 1996.
196. Parker, M. C., Moore, B. D., and Blacker, A. J., Measuring enzyme hydration in nonpolar organic solvents using NMR, *Biotechnol. Bioeng.*, 46, 452–458, 1995.
197. Condoret, J. S., Vankan, S., Joulia, X., and Marty, A., Prediction of water adsorption curves for heterogeneous biocatalysis in organic and supercritical solvents, *Chem. Engr. Sci.*, 52, 213–220, 1996.
198. Ghatorae, A. S., Guerra, M. J., Bell, G., and Halling, P. J., Immiscible organic solvent inactivation of urease, chymotrypsin, lipase, and ribonuclease: separation of dissolved solvent and interfacial effects, *Biotechnol. Bioeng.*, 44, 1355–1361, 1994.
199. Mensah, P., Gainer, J. L., and Carta, G., Adsorptive control of water in esterification with immobilized enzymes. II. Fixed-bed reactor behavior, *Biotechnol. Bioeng.*, 60, 445–453, 1998.
200. Rockland, L. B., Saturated salt solutions for static control of relative humidity between 5° and 40°C, *Anal. Chem.*, 32, 1375–1376, 1960.
201. Greenspan, L., Humidity fixed points of binary saturated aqueous solutions, *J. Res. Natl. Bureau Stand. A. Phys. Chem.*, 81A, 89–96, 1997.
202. Ferreira-Dias, S. and da Fonseca, M. M. R., Production of monoglycerides by glycerolysis of olive oil with imobilized lipases: effect of the water activity, *Bioproc. Eng.*, 12, 327–337, 1995.
203. Zhang, X. and Hayes, D. G., Improvement of the rate and extent of lipase-catalyzed saccharide-fatty acid esterification by control of reaction medium, *J. Am. Oil Chem. Soc.*, 76, 1495–1500, 1999.
204. Jonsson, A. A., Wehtje, E., and Adlercreutz, P., Low reaction temperature increases the selectivity in an enzymatic reaction owing to substrate solvation effects, *Biotechnol. Lett.*, 19, 85–88 1997.

205. Kim, J. J., Han, J. J., Yoon, J. H., and Rhee, J. S., Effect of salt hydrate pair on lipase-catalyzed regioselective monoacylation of sucrose, *Biotechnol. Bioeng.*, 57, 121–125, 1998.
206. Zacharis, E., Omar, I. C., Partridge, J., Robb, D. A., and Halling, P. J., Selection of salt hydrate pairs for use in water control in enzyme catalysis in organic solvents, *Biotechnol. Bioeng.*, 55, 367–374, 1997.
207. Luisi, P. L., Giomini, M., Pileni, M. P., and Robinson, B. H., Reverse micelles as hosts for proteins and small molecules, *Biochim. Biophys. Acta*, 947, 209–246, 1988.
208. Sjoblom, J., Lindberg, R., and Friberg, S. E., Microemulsions—phase equilibria characterization, structures, applications and chemical reactions, *Adv. Coll. Interf. Sci.*, 95, 125–287, 1996.
209. Oldfield, C., Enzymes in water-in-oil microemulsions ("reversed micelles"): principles and applications, in *Biotechnology and Genetic Engineering Reviews*, Tombs, M.P., Ed., Vol. 12, Intercept Ltd., Andover, U.K., 255–327, 1994.
210. Walde, P., Giuliani, A. M., Boicelli, C. A., and Luisi, P. L., Phospholipid-based reverse micelles, *Chem. Phys. Lipids*, 53, 265–288, 1990.
211. Waks, M., Proteins and peptides in water-restricted environments, *Prot. Struct. Func. Genet.*, 1, 4–15, 1986.
212. Hayes, D. G. and Gulari, E., 1-Monoglyceride production from lipase-catalyzed esterification of glycerol and fatty acid in reverse micelles, *Biotechnol. Bioeng.*, 38, 507–517, 1991.
213. Morgado, M. A. P., Cabral, J. M. S., and Prazeres, D. M. F., Hydrolysis of lecithin by phospholipase A_2 in mixed reversed micelles of lecithin and sodium dioctyl sulphosuccinate, *J. Chem. Tech. Biotechnol.*, 63, 181–189, 1995.
214. Subramani, S., Dittrich, N., Hirche, F., and Ulbrich-Hofmann, R., Characteristics of phospholipase D in reverse micelles of Triton X-100 and phosphatidylcholine in diethyl ether, *Biotechnol. Lett.*, 18, 815–820, 1996.
215. Backlund, S., Rantala, M., and Molander, O., Characterization of lecithin-based microemulsions used as a media for a cholesterol oxidase-catalyzed reaction, *Coll. Polym. Sci.*, 272, 1098–1103, 1994.
216. Rees, G. D., Robinson, B. H., and Stephenson, G. R., Macrocyclic lactone synthesis by lipases in water-in-oil microemulsions, *Biochim. Biophys. Acta*, 1257, 239–248, 1995.
217. Rao, A. M., John, V. T., Gonzalez, R. D., Akkara, J. A., and Kaplan, D. L., Catalytic and interfacial aspects of enzymatic polymer synthesis in reversed micellar systems, *Biotechnol. Bioeng.*, 41, 531–540, 1993.
218. Vos, K., Laane, C., and Visser, A. J. W. G., Spectroscopy of reversed micelles, *Photochem. Photobiol.*, 45, 863–878, 1987.
219. Escamilla, E., Contreras, M., Escobar, L., and Ayala, G., Biological electron transfer in reverse micellar systems: from enzymes to cells, in *Biomolecules in Organic Solvents*, CRC Press, Boca Raton, FL, 219–239, 1992.
220. Yang, F. and Russell, A. J., Two-step biocatalytic conversion of an ester to an aldehyde in reverse micelles, *Biotechnol. Bioeng.*, 43, 232–241, 1994.
221. Hagen, A. J., Hatton, T. A., and Wang, D. I. C., Protein refolding in reversed micelles, *Biotechnol. Bioeng.*, 35, 955–965, 1990.
222. Hagen, A. J., Hatton, T. A., and Wang, D. I. C., Protein refolding in reversed micelles: interactions of the protein with micelle components, *Biotechnol. Bioeng.*, 35, 966–975, 1990.

223. Hashimoto, Y., Ono, T., Goto, M., and Hatton, T. A., Protein refolding by reversed micelles utilizing solid-liquid extraction technique, *Biotechnol. Bioeng.*, 57, 620–623, 1998.
224. Towey, T. F., Rees, G. D., Steytler, D. C., Price, A. L., and Robinson, B. H., Winsor-II microemulsion systems in bioseparations—a model reaction for extractive bioconversions, *Bioseparations*, 4, 139–147, 1994.
225. Russell, A. J. and Komives, C., May the centrifugal force be with you! A novel centrifugal reactor system is used with reversed micelle-enzyme catalysts to biodegrade organophosphates, *CHEMTECH*, 26–32, 1994.
226. Rees, G. D., Robinson, B. H., and Stephenson, G. R., Preparative-scale kinetic resolutions catalyxed by microbial lipases immobilised in AOT-stabilised microemulsion-based organogels: cryoenzymology as a tool for improving enantioselectivity, *Biochim. Biophys. Acta*, 1259, 73–81, 1995.
227. Tata, M., John, V.T., Waguespack, Y. Y., and McPherson, G. L., Microstructural characterization of novel phenolic organogels through high-resolution NMR spectroscopy, *J. Phys. Chem.*, 98, 3809–3817, 1994.
228. Jonsson, A. A., Adlercreutz, P., and Mattiasson, B., Effects of subzero temperatures on the kinetics of protease catalyzed dipeptide synthesis in organic media, *Biotechnol. Bioeng.*, 46, 429–436, 1995.
229. Van Sonsbeek, H. M., Beeftink, H. H., and Tramper, J., Two-phase bioreactors, *Enzyme Microbiol. Technol.*, 15, 722–729, 1993.
230. Marty, A., Dossat, V., and Condoret, J. S., Continuous operation of lipase-catalyzed reactions in nonaqueous solvents: influence of the production of hydrophilic compounds, *Biotechnol. Bioeng.*, 56, 232–237, 1997.
231. Edmundo, C., Valerie, D., Didier, C., and Alain, M., Efficient lipase-catalyzed production of tailor-made emulsifiers using solvent engineering coupled to extractive processing, *J. Am. Oil Chem. Soc.*, 75, 309–313, 1998.
232. Jeong, J. C. and Lee, S. B., Enzymatic esterification reaction in organic media with continuous water stripping: effect of water content on reactor performance and enzyme agglomeration, *Biotechnol. Tech.*, 11, 853–858, 1997.
233. Ergan, F. and Trani, M., Effect of lipase specificity on triglyceride synthesis, *Biotechnol. Lett.*, 13, 19–24, 1991.
234. Boyer, J. L., Gilot, B., and Guiraud, R., Heterogeneous enzymatic esterification: analysis of the effect of water, *Appl. Microbiol. Biotechnol.*, 33, 372–376, 1990.
235. Napier, P. E., Lacerda, H. M., Rosell, C. M., Valivety, R. H., Vaidya, A. M., and Halling, P. J., Enhanced organic phase enzymatic esterification with continuous water removal in a controlled air-bleed evacuated-headspace, *Biotechnol. Prog.*, 12, 47–50, 1996.
236. Kwoon, S. J., Song, K. M., Hong, W. H., and Rhee, J. S., Removal of water produced from lipase-catalyzed esterification in organic solvent by pervaporation, *Biotechnol. Bioeng.*, 46, 393–395, 1995.
237. Mensah, P., Gainer, J. L., and Carta, G., Adsorptive control of water in esterification with immobilized enzymes. I. Batch reactor behavior, *Biotechnol. Bioeng.*, 60, 434–444, 1998.
238. Monot, F., Borzeix, F., Bardin, M., and Vandecasteele, J. P., Enzymatic esterification in organic media: role of water and organic solvent in kinetics and yield of butyl butyrate synthesis, *Appl. Microbiol. Biotechnol.*, 35, 759–765, 1991.
239. Bloomer, S., Adlercreutz, P., and Mattiasson, B., Facile synthesis of fatty acid esters in high yields, *Enzyme Microbiol. Technol.*, 14, 546–552, 1992.

240. Ducret, A., Giroux, A., Trani, M., and Lortie, R., Enzymatic preparation of biosurfactants from sugars or sugar alcohols and fatty acids in organic media under reduced pressure, *Biotechnol. Bioeng.*, 48, 214–221, 1995.
241. Wehtje, E., Svensson, I., Adlercreutz, P., and Mattiasson, B., Continuous control of water activity during biocatalysis in organic media, *Biotechnol. Techn.*, 7, 873–878, 1993.
242. Rossell, C. M., Vaidya, A. M., and Halling, P. J., Continuous *in situ* water activity control for organic phase biocatalysis in a packed bed hollow fiber reactor, *Biotechnol. Bioeng.*, 49, 284–289, 1996.
243. Ujang, Z., Al-Sharbati, A. N., and Vaidya, A. M., Organic-phase enzymatic esterification in a hollow fiber membrane reactor with *in situ* gas-phase water activity control, *Biotechnol. Prog.*, 13, 39–42, 1997.
244. Hayes, D. G., and Kleiman, R., 1,3-specific lipolysis of *Lesquerella fendleri* oil by immobilized and reverse micellar encapsulated lipases, *J. Am. Oil Chem. Soc.*, 70, 1121–1127, 1993.
245. Kuhl, P. and Halling, P. J., Salt hydrates buffer water activity during chymotrypsin-catalysed peptide synthesis, *Biochim. Biophys. Acta*, 1078, 326–328, 1991.
246. Kvittingen, L., Sjursnes, B., Anthonsen, T., and Halling, P. J., Use of salt hydrates to buffer optimal water level during lipase catalysed synthesis in organic media: a practical procedure for organic chemists, *Tetrahedron*, 48, 2793–2802, 1992.
247. Robb, D. A., Yang, Z., and Halling, P. J., The use of salt hydrates as water buffers to control enzyme activity in organic solvents, *Biocatalysis*, 9, 277–283, 1994.
248. Rosell, C. M. and Vaidya, A. M., Twin-core packed-bed reactors for organic-phase enzymatic esterification with water activity control, *App. Microbiol. Biotechnol.*, 44, 283–286, 1995.
249. Henderson, L. A., Svirki, Y. Y., Gross, R. A., Kaplan, D. L., and Swift, G., Enzyme-catalyzed polymerizations of *epsilon*-caprolactone: effects of initiator on product structure, propagation kinetics, and mechanism, *Macromolecules*, 29, 7759–7766, 1996.
250. Isono, Y., Nabetani, H., and Nakajima, M., Interesterification of triglyceride and fatty acid in a microaqueous reaction system using lipase-surfactant complex, *Biosci. Biotechnol. Biochem.*, 59, 1632–1635, 1995.
251. Hosokawa, M., Takahashi, K., Miyazaki, N., Okamura, K., and Hatano, M., Application of water mimics on preparation of eicosapentenoic and docosahexaenoic acids containing glycerolipids, *J. Am. Oil Chem. Soc.*, 72, 421–425, 1995.
252. Blackwood, A. D., Curran, L. J, Moore, B. D., and Halling, P. J., Organic phase buffers' control biocatalyst activity independent of initial aqueous pH, *Biochim. Biophys. Acta*, 1206, 161–165, 1994.
253. Dolman, M., Halling, P. J., and Moore, B. D., Functionalized dendritic polybenzylethers as acid/base buffers for biocatalysis in nonpolar solvents, *Biotechnol. Bioeng.*, 55, 278–282, 1997.
254. McNeill, G. P. and Yamane, T., Further improvements in the yield of monoglycerides during enzymatic glycerolysis of fats and oils, *J. Am. Oil Chem. Soc.*, 68, 6–10, 1991.
255. Bornscheuer, U. T. and Yamane, T., Activity and stability of lipase in the solid-phase glycerolysis of triolein, *Enzyme Microb. Technol.*, 16, 864–869, 1994.
256. Millqvist-Fureby, A., Adlercreutz, P., and Mattiasson, B., Glyceride synthesis in a solvent-free system, *J. Am. Oil Chem. Soc.*, 73, 1489–1495, 1996.
257. Arcos, J. A., Bernabe, M., and Otero, C., Quantitative enzymatic production of 1,6-diacyl sorbitol esters, *Biotechnol. Bioeng.*, 60, 53–60, 1998.

258. van der Padt, A., Keurentjes, J. T. F., Sewalt, J. J. W., van Dam, E. M., van Dorp, L. J., and van't Riet, K., Enzymatic synthesis of monoglycerides in a membrane bioreactor with an in-line adsorption column, *J. Am. Oil Chem. Soc.*, 69, 748–754, 1992.
259. Derksen, J. T. P., Boswinkel, G., van Gelder, W. M. J., van't Riet, K., and Cuperus, F. P., Enzymatic processing of triglycerides and fatty acids. Enhancing reaction kinetics by continuous product removal, in *Developing Agricultural Biotechnology in the Netherlands*, Vuijk, D. H., Dekkers, J. J., and van der Plas, H. C., Ed., Pudoc Scientific Publishers, Wageningen, Netherlands, 220–225, 1993.
260. Chaudhary, A. K., Beckman, E. J., and Russell, A. J., Rational control of polymer molecular weight and dispersity during enzyme-catalyzed polyester synthesis in supercritical fluids, *J. Am. Chem. Soc.*, 117, 3728–3733, 1995.
261. Bjorkling, F., Frykman, H., Godtfredsen, S. E., and Kirk, O., Lipase catalyzed synthesis of peroxycarboxylic acids and lipase mediated oxidations, *Tetrahedron*, 48, 4587–4592, 1992.
262. Warwel, S. and Rusch-Klaa, M., Chemo-enzymatic epoxidation of unsaturated carboxylic acids, *J. Molec. Catal. B*, 1, 29–35, 1995.
263. Selmi, B., Gontier, E., Ergan, F., Barbotin, J. N., and Thomas, D., Lipase-catalyzed synthesis of tricaprylin in a medium solely composed of substrates. Water production and elimination, *Enzyme Microbiol. Technol.*, 20, 322–325, 1997.
264. Khmelnitsky, Y. L., Hilhorst, R., and Veeger, C., Detergentless microemulsions as a media for enzymatic reactions: cholesterol oxidation catalyzed by cholesterol oxidase, *Eur. J. Biochem.*, 176, 265–271, 1988.

9 Proteomics: Difference Gel Electrophoresis

Mustafa Ünlü and Jonathan Minden

CONTENTS

Introduction 228
Of Proteomes and Gels 228
Difference Gel Electrophoresis (DIGE) 229
Of Proteomes and Mass Spectrometry 230
Good Night, Sweet Gel? 231
Usage Guide 231
General Notes 231
Recipes, Apparatus, and Chemicals for Sample Solubilization and Labeling 231
Lysis Buffer 232
Labeling Solution 232
Quenching Solution 232
Isoelectric Focusing 232
Rehydration Buffer 232
Equilibration Buffer I 233
Equilibration Buffer II 233
SDS-Page 233
4× Resolving Gel Buffer 233
30% Monomer Solution 233
4× Stacking Gel Buffer 233
Light Gradient Gel Solution 234
Heavy Gradient Gel Solution 234
Stacking Gel Solution 234
Image Acquisition and Analysis 234
Destain Solution 235
Methods 235
Sample Solubilization 235
Introduction and General Notes 235
Tissue-Cultured 3T3 Mouse Fibroblasts 235
Drosophila Embryos 236
S. cerevisae 236
Sample Labeling 236

0-8493-9453-8/02/$0.00+$1.50

Isoelectric Focusing (IEF) 237
Introduction and General Notes 237
Rehydration 237
Setting up and Running the First Dimension 238
Equilibration of IEF Gels 239
SDS-PAGE 239
Assembling the Gel Cassettes 239
Pouring the 10 to 15% Gradient Gels 239
Setting up and Running the Second Dimension 239
Image Acquisition and Analysis 240
Introduction and General Notes 240
Image Acquisition 241
Image Analysis 241
Notes 241
References 243

INTRODUCTION

Of Proteomes and Gels

Proteomics is a newly coined term that refers to the field of study of the proteome. The term was defined for the first time in 1994 as "the PROTEin products of a genOME."[1] Since its introduction, the term has found widespread acceptance across the breadth of biological research. A simple search of the Medline database *circa* October 2000 indicated an almost exponential increase in the number of published articles about proteomics (see Figure 9.1) since 1995.

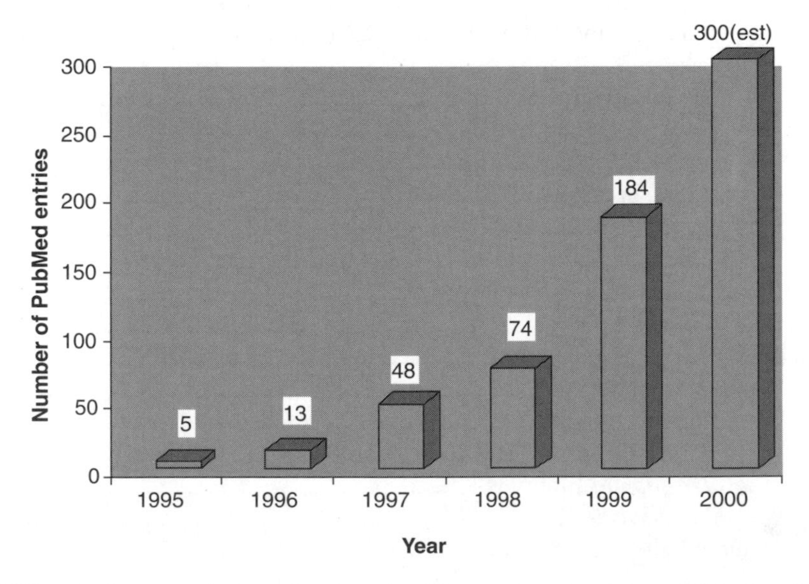

FIGURE 9.1 The meteroic rise of proteomics. The number of articles published each year according to Medline search to find the term "PROTEOM."

The attractiveness of proteomics derives from its promise to uncover changes in global protein expression accompanying many biologically relevant processes, such as development and tumorigenesis. Because proteins are the effector molecules that carry out most cellular functions, studying proteins directly has clear advantages over and (at least in theory) achieves results that go beyond those accessible by genomic analyses. Whereas the information contained in the genome is almost always static, spatio-temporal patterns of protein expression are very complex and dynamic owing to fluctuations in abundance. This complexity is increased several fold by the proteins' ability to be modified functionally after translational.

To detect changes in global protein expression, there is a clear requirement that the methodology employed be able to generate and compare snapshots of the entire protein component of an organism, cell, or tissue type. Two-dimensional gel electrophoresis (2DE) is the key separation technique for complex polypeptide mixtures because it offers the highest practical resolution in protein fractionation. This resolution power derives from orthogonally combining two separations based on two independent parameters: iso-electric focusing (IEF) separates proteins based on charge and SDS-PAGE separates by size. Because separation is achieved under two independent parameters, charge and size, and no dilution effects are incurred, 2DE outperforms all other current separation techniques and yields a two-dimensional gel on which hundreds to a few thousand proteins are fractionated and displayed. Comparison of two or more gels then allows for the detection of the protein species differing between the compared states of the organism, cell, or tissue type under investigation.

2DE was first described simultaneously by several groups in 1975.[2–4] Unfortunately, using 2DE to compare protein extracts to one another has not been very fruitful in terms of significant discoveries. It is instructive to examine the number of published articles on 2DE in the database and then to compare these with the results for proteomics (Figure 9.2). The numbers indicate that, for a technique that has been around for 25 years and promised much, 2D gels have not delivered. This has been due primarily to unavoidable systemic irreproducibilities between different gels. The other salient shortcoming of two-dimensional gels is incomplete proteome display because of the poor performance of the gels with basic and hydrophobic proteins. Some progress has been made with the latter,[5] but most efforts have been concentrated on surmounting the more serious irreproducibility bottleneck. Given that all conventional protein detection methods are nonspecific, running different gels to compare different samples has remained a requirement. Thus, new developments have focused primarily on physically improving the 2DE process[6] and also on computational approaches to better compare several gels.[7] The results of these efforts have been partially successful at best. Side-effects of increased material and time investment complicate a technique that was already considered to be difficult.

Difference Gel Electrophoresis (DIGE)

In order to be able to run two different samples on the same two-dimensional gel,[8] we have developed a detection system that differentiates between two different samples without introducing differences on its own. Difference Gel Electrophoresis (DIGE) uses two fluorescent dyes to label two protein mixtures. The dyes enable samples to be differentiated from each other on the same gel. Dye design was nontrivial: we had to ensure that no electrophoretic mobility differences were imparted

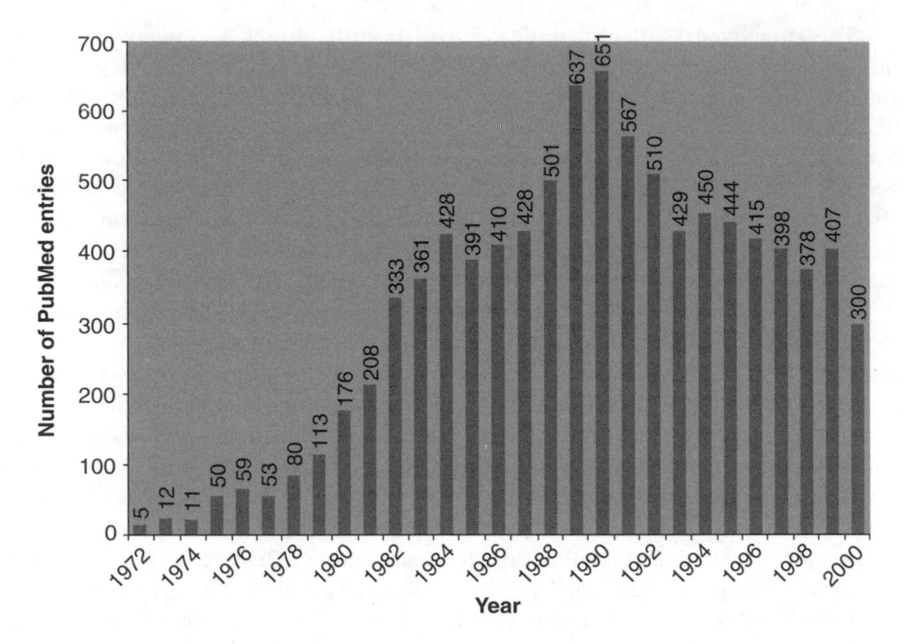

FIGURE 9.2 The rise and fall of two-dimensional gels. The number of articles published each year according to a Medline search to find TWO and DIMENSIONAL and GEL and ELECTROPHORESIS.

to otherwise identical proteins that were differentially tagged and that issues of sensitivity and photostability were properly addressed. We succeeded in this approach and unquestionably demonstrated the reproducibility and feasibility of our modified technique. Thus, in DIGE, every identical protein in one sample superimposes with its differentially labeled counterpart in the second sample, thus eliminating the need to compare different gels and the problems associated with that approach.

Of Proteomes and Mass Spectrometry

Application of mass spectrometry to in-gel enzymic digests of protein spots from two-dimensional gels facilitates the rapid and sensitive identification of proteins of interest.[9] Previously, the only way to identify candidate spots was by slow and less sensitive methods, such as co-expressing or co-electrophoresing known proteins, Western blotting and/or N-terminal sequencing by Edman degradation. An in-depth look at how mass spectrometry has developed and its applications to proteome research is beyond the scope of this chapter; however, suffice it to say that mass spectrometry, in conjunction with genome sequencing projects, has brought high throughput and new levels of sensitivity to the field, literally allowing the word proteomics to be invented.

Widespread use of mass spectrometry to analyze large numbers of spots from two-dimensional gels has also brought out another apparent shortcoming of 2DE. Articles on proteome-wide analyses of *Saccharomyces cerevisiae* point to a growing awareness in the literature and community of another shortcoming of 2DE, namely that no spots with low abundance can be identified from silver stained two-dimensional gels.[10, 11]

Estimates of protein abundance have been linked to codon bias,[12] and the completion of the yeast genome sequence has enabled protein spot abundance to be coupled with quantitative transcript-based analyses.

These recent studies[13] have concluded that there is an inherent limitation of two-dimensional gels, namely that of maximum load compared to the range of protein abundances, which precludes low abundance protein identification. It would appear from these conclusions that, to be useful in detecting any protein with less than moderate abundance, 2DE must be applied to subfractionated or otherwise enriched samples. In fact, the same groups have already started using and advocating mass spectrometry-based techniques for proteome comparison, bypassing completely the 2DE step.[14]

Good Night, Sweet Gel?

Despite the encouraging preliminary results reported from several new promising techniques that are currently in development and that do not require prefractionation of complex protein mixtures, such as ICAT technology,[14] the advantages of separating a mixture of thousands of proteins (such as whole cell lysates) are obvious. 2DE achieves its remarkable resolution by separating proteins via two independent parameters: charge and size. Furthermore, dilution effects during the procedure are minimal compared to column chromotography, thus enabling 2DE to outperform multidimensional chromatography-based approaches.

The question, which we feel is still open to debate, is whether the bells have tolled for 2DE. In other words, is the abundance limitation indeed systemic and well nigh insurmountable, or is it more a problem of the level of detection associated with the staining techniques employed and/or due to losses caused by post-electrophoretic handling before mass spectrometry?

By affording facile and reproducible detection of differences, DIGE has eliminated the requirement for running multiple gels. The fact that all proteins are already fluorescently labeled may take sensitivity of detection to lower levels than is affordable. Also, gels do not need to be stained, reducing the number of post-electrophoretic handling steps. Coupling the fluorescent label to automated spot cutting and digestion will also minimize losses associated with handling further.

USAGE GUIDE

General Notes

All references to H_2O should be read as double distilled H_2O unless stated otherwise. In recipes, the information given in each line corresponds to the final concentration, the name of the ingredient, and the amount used, in that order.

Recipes, Apparatus, and Chemicals for Sample Solubilization and Labeling

Samples of 40% methylamine in water, urea, thiourea, DTT and CHAPS were obtained from Sigma-Aldrich (Milwaukee, WI). HEPES was obtained from Fisher Scientific (Pittsburgh, PA).

Lysis Buffer

8*M* urea	24.0 g

Alternatively, 6 *M* urea and 2 *M* thiourea may be used. See Note 1.

6 *M* urea	18.0 g
2 *M* thiourea	7.6 g

Make up to 40 ml with HPLC-quality H_2O and dissolve the urea.

2% CHAPS (Sigma-Aldrich)	1.0 g
10 m*M* DTT	500 μl of 1 *M* stock or 0.077 g solid
10 m*M* NaHEPES, pH 8.0	5.0 ml of 100 m*M* stock (add the NaHEPES last; see Notes 2 and 3)

Make up to 50.0 ml with HPLC-quality H_2O and store at –80°C in 1- to 1.5-ml aliquots.

Labeling Solution

We typically label samples in lysis buffer with no further modification. Note that we do not add pharmalytes to this solution (see Note 4).

Quenching Solution

A sample of 5 *M* methyl amine was dissolved in 100 m*M* NaHEPES, at pH 8.0. Dissolve 2.38 g HEPES in 38.8 ml of 40% methyl amine aqueous solution. Slowly add approximately 60 ml concentrated HCl with stirring. Cool the solution on ice and measure the pH as the HCl is added until the pH reaches 8.0.

Isoelectric Focusing

The IEF equipment is a Multiphor II from Amersham-Pharmacia (Peapack, NJ), with the DryStrip kit installed. The teflon membranes are from the YSI (Yellow Springs, OH) model 5793 standard membrane kit for oxygen electrodes. Light paraffin oil was obtained from Amersham-Pharmacia.

Rehydration Buffer

2% CHAPS	0.4 g
8 *M* urea	9.6 g (Same composition as lysis buffer described previously; see Note 1.)
6 *M* urea	7.2 g
2 *M* thiourea	3.0 g
2 m*M* acetic acid	2.7 μl of glacial acetic acid (approximately 17 *M*)
10 m*M* DTT	200 μl of 1 *M* stock or 0.031 g solid
1% pharmalyte	500 μl of 40% pharmalyte stock solution*

* Use the appropriate IPG solution that corresponds to the pH range of the IEF gels. See the section on equilibration of IEF gels.

Make up to 20 ml with H_2O; and store at 4°C (see Notes 5 and 6).

Equilibration Buffer I

1× stacking gel buffer	25 ml of 4× solution
1% SDS	20 ml of 10% SDS solution
8.7% glycerol	10 ml of 87% solution
5 m*M* DTT	500 μl of 1 *M* solution or 0.076 g

Bring up to 100 ml with H_2O and store at 4°C (see Note 5).

Equilibration Buffer II

Same as in the section on equilibration buffer I except 2% iodoacetamide (add 2.0 g) replaces DTT, and a trace amount of bromophenol blue is added.

SDS-PAGE

SDS was obtained from Fisher Scientific. The gradient maker, acrylamide and bis-acrylamide of the highest purity were obtained from Bio-Rad (Hercules, CA). The gel equipment was a Hoefer SE-660 18 × 24 cm apparatus from Amersham-Pharmacia. The 0.2 μm filters were from Nalgene (Rochester, NY)

4× Resolving Gel Buffer

36.3 g Tris
Add 150 ml H_2O.
Adjust to pH 8.6 with 6 *N* HCl.
Make up to 200 ml with H_2O.

Filter sterilize through a 0.2-μm filter and store at 4°C.

30% Monomer Solution

60.0 g acrylamide
1.6 g bis-acrylamide
Make up to 200 ml with H_2O

Filter sterilize and store as discussed in the previous section.

4× Stacking Gel Buffer

3.0 g Tris
Add 40 ml H_2O.
Adjust pH to 6.8 with 6 *N* HCl.
Make up to 50 ml with H_2O.

Filter sterilize as above and store at –20°C.

Light Gradient Gel Solution

10% acrylamide	8.25 ml of 30% monomer solution
0.375 *M* Tris	6.25 ml of 4× resolving gel buffer
H_2O	10 ml
0.1% SDS	250 μl of 10% SDS solution

Add these immediately before pouring the gel:

APS	82.5 μl of 10% stock solution
TEMED	8.25 μl

Heavy Gradient Gel Solution

15% acrylamide	12.25 ml of 30% monomer solution
0.375 *M* Tris	6.25 ml of 4× resolving gel buffer
H_2O	3.8 ml
0.1% SDS	250 μl of 10% SDS solution
Sucrose	3.75 g

Add these immediately before pouring the gel:

APS	82.5 μl of 10% stock solution
TEMED	8.25 μl

Stacking Gel Solution

3.5% acrylamide	400 μl of 30% monomer solution
0.175 *M* Tris	800 μl of 4× stacking gel solution
H_2O	3.8 ml
0.1% SDS	33 μl of 10% SDS solution

Add these two immediately before pouring the gel:

APS	16.7 μl of 10% stock solution
TEMED	1.7 μl

Image Acquisition and Analysis

The cooled CCD camera is a 16-bit, series 300 model purchased from Photometrics/ Roper Scientific (Tucson, AZ). It is fitted with a standard 105-mm macro lens from Nikon, available from most photographical suppliers. Used for illumination are two 250-W quartz-tungsten-halogen lamps from Oriel (Stratford, CT). Single bandpass excitation filters (2.5 cm diameter) from Chroma Technology (Brattleboro, VT), are used to excite 545 ± 10 nm and 635 ± 15 nm for Cy3 and Cy5, respectively. A multi-wavelength bandpass emission filter from Chroma Technology is used to image the gels at 587.5 ± 17.5 nm and 695 ± 30 nm for Cy3 and Cy5, respectively. The imager housing was constructed in-house from black plexiglass. Image acquisition is semi-automated and is controlled by a Silicon Graphics (Mountain View, CA) O2 computer workstation.

Destain Solution

1% acetic acid	10 ml glacial acetic acid
40% methanol	400 ml
H_2O	590 ml

METHODS

Sample Solubilization

Introduction and General Notes

Because the number of different cells or tissue types which will be typically analyzed by 2DE is quite large, it is hard to describe a single protocol that will be applicable for all cases. Most samples have merely needed lysis buffer to be added to extract protein. Some examples of this kind are almost all prefractionated or prepurified proteins, human brain slices, *Drosophila* testes, mouse embryonic genital ridges, and pellets of *E. coli*. Some preparations have required only slightly more vigorous disruption (i.e., with a ground glass homogenizer). Other preparations have required even more severe extractions (i.e., using sonication or glass beads for efficient disruption). Yeast and some cultured cells are examples of the latter. Each sample, if it does not simply lyse upon buffer addition, will require an empirical approach to determine the most efficient preparation method. In general, all samples should be kept as cold as possible. All steps leading to and including lysis should be performed on ice. As soon as lysis is complete, samples may be stored indefinitely at –80°C. Repeated thawing and refreezing does not seem to have a deleterious effect. Because the lysis buffer contains urea, provisions of Note 2 apply. (See also Note 7.) The following are three sample extraction protocols that may be used as a basis for developing further protocols.

Tissue-Cultured 3T3 Mouse Fibroblasts

- Grow one 150-mm dish of cells per condition.
- Wash three times with 5 s ml of DMEM culture medium without serum.
- Add 2 ml medium to the plate.
- Remove cells by scraping and transfer to a 15-ml conical tube.
- Repeat the preceding with an additional 2 ml, and transfer to the same conical tube.
- Centrifuge for 5 min at 5000 *g*.
- Discard supernatant and resuspend cells in 1 ml medium.
- Count cells.
- Transfer cells to a 1.5-ml Eppendorf tube on ice.
- Centrifuge for 5 s at maximum speed in a tabletop nanofuge.
- Vortex lightly to loosen pellet.
- Add 50 µl of lysis buffer for every 10^6 cells.

- Stir with pipette tip to remove clumps, if any.
- Centrifuge in the cold room for 5 to 10 min at 13,000 *g*.
- Store at –80°C.

This extract is expected to have between 4 to 8 mg/ml protein.

Drosophila Embryos

Embryos are dechorionated and observed in eggwash solution under a dissection microscope in order to determine their age. Embryos that are at the right stage are removed and rinsed once with ethanol to remove the eggwash solution. They are immediately transferred into ethanol over dry ice. Several days of collections may be accumulated, with the embryos being kept in ethanol at –80°C. After accumulating enough embryos (assume a yield of 1- to 1.5-µg protein yield per embryo), transfer the embryos into a cold ground glass homogenizer on ice, add 1 µl of lysis buffer per embryo and briefly homogenize. Spin in a tabletop microfuge for 5 to 10 s, take the supernatant and discard any solid material. Store at –80°C. This extract should have between 1 to 2 mg/ml protein.

S. cerevisiae

- Obtain a 250-ml yeast culture of 0.5 OD^{600}.
- Centrifuge the culture at 5000 *g* for 15 min at 4°C.
- Decant the supernatant.
- Wash the cells as follows:
  - Suspend the cells in 5 ml of 100 m*M* NaHEPES, pH 8.0.
  - Centrifuge the cells for 10 min at 5000 *g* at 4°C (see Note 15).
  - Remove the supernatant using an aspirator.

Resuspend in 5 ml of 100 m*M* NaHEPES, pH 8.0, and repeat the preceding steps.

- Suspend the pellet of washed cells in 750 µl of lysis buffer.
- Transfer to a 15-ml conical tube containing 1 g acid-washed glass beads.
- Vortex for 3 min in the cold room.
- Spin down and transfer the supernatant to fresh tubes.

Store the extract at –80°C. This extract has an expected protein concentration between 1.5 to 4 mg/ml.

Sample Labeling

Measure the protein concentration in the extracts with the method of choice (e.g., Bradford). To obtain matched fluorescence images, use equal amounts of protein for each sample. Anywhere between 50 to 250 µg of protein per sample may be loaded on a single IEF strip, making the total maximum load 0.5 mg when two samples are being compared. There is very little to no loss of resolution even at the highest level of loading; and there is no reason not to try to achieve it. In fact, the greater load will allow for the detection of lower abundance proteins and permit a better chance of success in the eventual identification of the spots of interest. It is best to run two gels for each comparison where the order of labeling is reversed. This ensures that the observed differences are sample dependent and not dye dependent. Dye-dependent changes are rarely seen when the labeling reaction is done correctly. However, errors in labeling or quenching are often indicated by many dye-dependent changes.

- Bring the desired amount of each sample to up to 48 μl with lysis buffer (see Note 8).
- Add 1 μl of the appropriate dye to each sample (see Note 9).
- Incubate on ice for 15 min.
- Add 1 μl of quenching solution, followed by 0.5 μl of the appropriate pharmalyte solution.
- Incubate on ice for 30 min.
- Immediately load and run on the first dimension.

Isoelectric Focusing (IEF)

Introduction and General Notes

This is arguably the most complex and problematic step in the whole procedure. Notable difficulties include, but are not limited to, sample leakage, gel sparking, and insufficient or incomplete focusing. The first dimension gels are purchased as dry strips and come in a variety of sizes and pH ranges. In the case of Pharmacia Dry Strips, the instructions that come with the gels should be followed in general; however, do note that some of the solutions and procedures recommended by Pharmacia have been modified here. When there is a conflict, use this version for the best results.

Rehydration

The rehydration cassette is made up of two glass plates: one with a 0.5-mm rubber U-gasket, the other plain. Wash both plates with H_2O, then 95% EtOH, and air dry. Hold gel strip at either end and peel off protecting thin layer of plastic. The thicker piece has the gel cast on it. Discard the protecting plastic. The gel is bonded to special plastic called PAG bond plastic. The back side of the strip opposite the gel is hydrophobic. Drop just a few drops of water on this hydrophobic side (see Note 10). Lay the plain glass plate face down and place the gel on it, wet (hydrophobic) side down. Make sure the acidic (pointed) end is near the bottom of plate.

After all gels are in place, roll them flat with a Teflon roller. There is no need to press hard; all that is needed is that the strips do not fall down when the glass plate is inverted. Remove any excess water that might have seeped out.

Make sure the gasket around the edge of top glass plate is clean and dry, then place over bottom plate and clamp two plates together (two clips on each long side and one at the bottom).

Using the special 25-ml syringe included with the kit, fill the cassette with rehydration buffer by injecting from the bottom until the top of the gels are just covered. Try to avoid air bubbles. Lay the cassette assembly flat and rehydrate overnight (see Note 11).

Setting Up and Running the First Dimension

Remove the rehydration buffer from the IEF strips and save at 4°C for future re-use (see Note 6).

To make the contact wicks that will go between the gel and the electrodes, cut the filter paper supplied with the Immobiline apparatus to approximately 0.5 × 0.3 mm rectangles (see Note 12). The wicks are used for ensuring good contact between the electrodes and the gel. They also act as a buffer space for salts that migrate to the electrodes. Cut enough wicks so that there are a few more than twice the number of strips. Immerse all wicks in HPLC quality H_2O for about a minute. Then place all wicks on a paper towel so that they cluster around each other to create a damp patch on the paper. They should be placed close to, but not overlapping, each other. Then start placing the IEF strips on the gel apparatus. By the time the gels are in place and ready to receive the wicks, they are usually at the right dampness (see Note 13).

Lay down a thin line of light paraffin oil into each groove on the strip aligner that will receive an IEF strip. Make sure to use a minimal amount of oil. It is not desirable to have any excess oil on top of the gel, especially over the location of the sample cup. Place one IEF strip over each oil streak with the acidic end pointing toward the positive, red electrode, and making sure that it is parallel to, and aligned with, the other gels using the lines on the flatbed.

Apply one wick at each (basic and acidic) end of each gel. Make sure all the wicks are lined up with each other if there is more than one IEF strip on the plate. Place both electrodes over their respective wicks. Take care to orient the electrodes correctly. They are color coded and the instructions are included in the kit. Incorrect orientation of the electrodes will result in no electrophoresis. Samples will be applied close to the anode, so the sample cup holder should next be placed as close to the positive electrode as possible.

Cover the gel strips with Teflon membranes, except the areas where the sample cups are going to be applied (see Note 14). Dab the gel area around the prospective sample cup location with a paper towel if there is liquid over it; otherwise, leave it alone. After putting the cups on, lightly press on the cup to make sure it sits flatly on the gel. This is also a quite tricky step. Any misalignment of the cups will cause the sample to leak out. Observe the cups from several angles to make sure that they are seated correctly (see Note 16). Apply the sample, taking care not to disturb the sample cup.

Start the run at between 300 to 1000 V for 1 h or as long as the current stays above 0.5 mA. It is important not to allow the current to rise too high, especially during the initial stages of the electrophoresis. As the current drops through the run, slowly raise the voltage up to the maximum allowed by your power supply. At least 2000 V should be achieved: the higher the voltage, the better the focusing. The total amount of volt hours delivered over the run time should be 30 kVh to 50 kVh (see Note 17).

Equilibration of IEF Gels

At the end of the IEF run, equilibrate the gels in Equilibration Buffer I for 15 min, rinse briefly with H_2O to remove DTT, and equilibrate in Buffer II for another 15 min. After equilibration, the gels either should be run immediately on the second dimension or stored at –80°C. They will last almost indefinitely at –80°C (see Note 18).

SDS-Page

Assembling the Gel Cassettes

Rinse the glass plates with H_2O, then with 95% ethanol, and air dry. Rinse clamps and spacers in H_2O and dry. Make sure the insides of clamps are dry and the spacers are free of particulate debris.

Align glass plates and spacers so that all edges are flush, especially at the bottom. Place a large thumb-screw clamp toward the top and a small one at the bottom on each of the long sides of the glass plates. Make sure the bottom of the cassette is sticking out slightly (1 to 2 mm) from the clamps. This drives the gel cassette into the stand and seals the bottom when the cams are inserted and twisted in opposite directions. If desired, check seal by squirting water on the outside of sandwich assembly.

Assemble the equipment for pouring, with the gradient maker on a stir plate, over the cassette(s) (see Note 19).

Pouring the 10 to 15% Gradient Gels

- Pour the heavy solution in front chamber of gradient maker; add stirbar.
- Pour light solution in back chamber of gradient maker, add stirbar balancer.
- Open the mixing channel and the front stopcock at the same time.
- Pour gel and overlay with *n*-butanol.
- Allow gel to polymerize (about 30 min).
- Pour off *n*-butanol, rinse top of gel with H_2O twice and pour about 0.5 cm of the stacking gel.
- Overlay with *n*-butanol again.
- Allow to polymerize for at least 8 h (see Note 20).

Setting Up and Running the Second Dimension

Microwave 10 ml of autoclaved 1% low melting agarose in a 1× stacking gel buffer until it starts to boil. (We also add a trace amount of bromophenol blue). Place an IEF strip on top of the second-dimension gel plastic side down, acidic side facing left. Then rotate the IEF strip so that the plastic backing is parallel to the face of the back gel plate and slide it down vertically while making sure the plastic side, not the gel side, is contacting to the glass plate (see Note 21).

Push the strip down until it is in firm contact with the top of the SDS gel. Add the agarose until it barely covers the top of the strip. Make sure to add evenly from both sides and to avoid air bubbles. Burst bubbles with gel loading pipettor tips after the agarose hardens. Make sure to mark the second dimension gel with sample ID.

Electrophorese in the cold room. For a run of about 8 to 10 h, use 20 mA per gel at constant voltage. For a run of about 16 h, use 8 to 10 mA per gel.

Image Acquisition and Analysis

Introduction and General Notes

The layout of the imaging system is vital for the success of the DIGE. The technique has unique requirements, so that all the hardware and most of the software for image acquisition must be built from scratch. Our system is experimental and transitory in nature. At the time of this writing, the apparatus is under constant revision and development and is not commercially available in its current form. Thus, rather than give the exact minutiae of the image acquisition and analysis process, we list the nature and aim of the operations that are performed to arrive at the final result. Hopefully, this will assist those who are interested in either building or acquiring their own imaging system.

At the writing of this manuscript, commercially available DIGE imagers are just beginning to be developed. For those who might consider either building or acquiring an imager, we list the absolute minimum requirements that the hardware must meet:

- In addition to the obvious requirement that the gel physically not move during imaging, no changes in protein spot position owing to optically induced deformations should occur while switching between channels. Thus, a multiple band-pass emission filter must be employed.
- The imaging hardware needs to be sensitive enough to detect minimally labeled proteins. Because a cooled, coupled charge device (CCD) had been used previously to image fluorescently labeled proteins in the gel,[15] we decided to use a scientific grade CCD camera. CCD cameras without cooling do not have the requisite sensitivity of low noise capabilities.
- The imaging cabinet must be light-tight and illumination must be filtered through the appropriate filter sets.

In the current incarnation of the DIGE imager, the gels are placed flat on a black plexiglass surface at the bottom of the cabinet. The camera is mounted vertically over the gel at about 30 cm away. Illumination is provided by two halogen lamps with fiberoptic leads mounted on top of the cabinet at ±60° incident angles to the bottom to provide an even field of illumination. The standard image acquired by the imager is a 4 × 4 cm square made up of 256 × 256 pixels, each storing 65,000 gray levels as unsigned short integers.

Image Acquisition

After the second dimension run, the gels are removed from their cassettes and incubated in destain solution for a minimum of 1 h. Gels are placed under either destain solution or in 1% acetic acid in H_2O during imaging. Two images are acquired from the central region of the gel, one with each excitation filter. A few spot intensities are compared and used to normalize the acquisition times for the two channels in tile mode. The tile mode is where the entire gel is imaged into one file by stitching together a total of 30 4 × 4 cm squares to generate a single 20 × 24 cm image. Then, two such images are generated, one for each channel. Each image corresponds to one of the dyes and, thus, represents one of the samples that were run on the gel. Acquisition times typically vary between 10 to 180 s per square, which translates to 10 to 180 min total acquisition time per gel.

Image Analysis

The two images from a single gel are inverted and then normalized to each other so that the most abundant spots appear at the same level of intensity. The images are then converted to byte format and placed into a two-frame quicktime movie. Playing this movie in a continuous loop allows for the visual detection of differences. The images are normalized at several different gray scale levels. Normalizing for high values allows for the detection of the abundant protein changes and normalizing at lower values covers the lower abundant protein changes.

NOTES

1. A mixture of urea and thiourea has been reported to aid in the solubilization of membrane proteins.[5]
2. Never warm a protein lysate in urea; always endeavor to keep at least on ice, if not frozen. Minimize the time samples spend away from –80°C. This is because at high temperature and pH urea spontaneously breaks down to yield cyanate, which modifies lysine residues and leads to carbamylated charge trains in the IEF dimension. Low temperature slows down but does not stop this process.[16]
3. Usually the pH of HEPES is adjusted with KOH; however, if SDS is used in any subsequent step (such as running lysate directly on SDS-PAGE), KDS will form, which is insoluble in water. Use NaOH. When making up this buffer, make sure to add the HEPES last (see Note 2).
4. The presence of pharmalytes in the lysis buffer interferes with labeling because some contain amines that react with and inactivate the dye. Similarly, the presence of any other primary amine-containing compound (such as Tris) should be avoided.
5. Similar to Note 2, there is a danger of cyanate formation in this buffer. It is thus preferable to add the acetic acid at the same time as the urea. The lower pH will slow down the breakdown process.

6. This buffer may be reused up to three times or until the DTT is no longer detectable by its smell, whichever comes first.
7. We typically do not use protease inhibitors. The combination of the lysis buffer with its reducing ability, the chaotropic effects of the urea and the surfactant, and the cold temperature seems to inactivate proteolytic activity. We also do not perform any steps requiring room temperature or protein activity (such as the DNAse–RNAse treatment found in some protocols). Furthermore, the presence of the inhibitors may sometimes interfere with the fluorescent labeling.
8. The sample cups on the IEF gel have about 100-μl maximum capacity. However, if necessary, more volume can be handled by ordering more sample cup holder bars separately from the dry strip kit and used to spread one sample between several cups. Because IEF is a focusing technique, the sample does not necessarily have to be applied in exactly the same spot.
9. The dye synthesis is detailed elsewhere.[8] The dyes are not commercially available as of the time of this writing.
10. Try not to get any water on the gel side because this makes the gel very sticky, which causes problems when the strip is being rolled onto the glass plate. If the gel side does get wet, sometimes the situation can be saved by dabbing the water gently with a paper towel. Take care not to touch the paper to the gel. If you know you have sticky gels, then hold down the strips at the basic end while rolling.
11. Rehydration should continue for a minimum of 8 h. The gels should not be used after 24 h in the rehydration solution.
12. Getting the wicks to the exact size is not that important—eyeball accuracy is good enough. Also, although Pharmacia supplies and sells special wick paper, we have found that 3-mm Whatman chromatography paper will work just as well.
13. This is probably one of the trickier aspects of the IEF procedure. It is important not to get the wicks too dry or to leave them too wet. Sparking is usually caused by the wicks being at the wrong dampness. Following these steps usually works, but work pace can play a role as can the ambient humidity. One way of checking to see if the wicks are correct is to remove one from the damp patch and touch it lightly to a dry area on the paper. If it leaves behind a slight imprint, then it is most likely at the right dampness. If too wet, place firmly and dry a little more before placing on the gel. If too dry, rehydrate the wicks and repeat, waiting for one half the time as the first try.
14. The Teflon membranes replace the paraffin oil that is normally used to cover the gels to isolate and keep the gels dry. We have found that using the oil produces a host of problems of its own. The membranes work just as well and are easy to use and remove without the messiness of the oil.
15. It is preferable to use a swing-out rotor to get a more compact pellet and minimize loss of yeast cells. We use a benchtop centrifuge with a swing-out rotor from Sorvall Instruments.
16. Do not press on the sample cups too hard. The gel is quite delicate and making a dent or hole in it is as deleterious as not having the sample cup

seated properly. Even if the sample does leak, this is not as great a tragedy as it might seem at first. As long as the sample stays in the same groove as the gel, and the current is started promptly, most of the sample will still go in.

17. For 13-cm-long dry strips, 20 kVh total also produces acceptable results in most cases.
18. A small disposable petri dish makes a great receptacle for equilibration. Wrap the gel around the inside of the dish, with the gel side pointing in. The dishes may also be marked to keep track of the identities of multiple gels easily.
19. Before starting to pour the gradient gels, make sure the gradient maker chamber is free of polyacrylamide pieces. The tube and mixing channel can easily be blocked by small pieces of gel. Working fast, four gels can be poured one after the other. No washing of the chamber in between gels is necessary.
20. The 8 h is to allow the polymerisation reaction to go to completion. In a crunch, this time may be shortened to as little as 1 h, but be aware that side-chain and N-terminal modification by acrylamide then becomes more of a problem. Gels may be allowed to polymerize as long as overnight provided that the butanol layer is increased so that it does not dry out. We also have stored gels at 4°C for up to 24 h.
21. If the IEF gel has dried, this may cause a problem because the gel has a higher tendency to stick to the glass and to get torn. Should this start to happen, try wetting the IEF strip a little with equilibration buffer II.

REFERENCES

1. Wasinger, V. C., Cordwell, S. J., Cerpa-Poljak, A., Yan, J. X., Gooley, A. A., Wilkins, M. R., Duncan, M. W., Harris, R., Williams, K. L., and Humphery-Smith, I. (1995) Progress with gene-product mapping of the Mollicutes: *Mycoplasma genitalium*, *Electrophoresis*, 16, 1090–1094.
2. Klose, J. (1975) Protein mapping by combined isoelectric focusing and electrophoresis in mouse tissues. A novel approach to testing for induced point mutations in mammals, *Humangenetik*, 26, 231–243.
3. O'Farrell, P. H. (1975) High resolution two-dimensional electrophoresis of proteins, *J. Biol. Chem.*, 250, 4007–4021.
4. Scheele, G. A. (1975) Two-dimensional gel analysis of soluble proteins. Characterization of guinea pig exocrine pancreatic proteins, *J. Biol. Chem.*, 250, 5375–5385.
5. Rabilloud, T., Adessi, C., Giraudel, A., and Lunardi, J. (1997) Improvement of the solubilization of proteins in two-dimensional electrophoresis with immobilized pH gradients, *Electrophoresis*, 18, 307–316.
6. Görg, A., Postel, W., and Günther, S. (1988) The current state of two-dimensional electrophoresis with immobilized pH gradients, *Electrophoresis*, 9, 531–546.
7. Wilkins, M. R., Hochstrasser, D. F., Sanchez, J. C., Bairoch, A., and Appel, R. D. (1996) Integrating two-dimensional gel databases using the Melanie II software, *Trends Biochem. Sci.*, 12, 496–497.
8. Ünlü, M., Morgan, M. E., and Minden, J. S. (1997) Difference gel electrophoresis: a single gel method for detecting changes in protein extracts, *Electrophoresis*, 18, 2071–2077.

9. Shevchenko, A., Jensen, O. N., Podtelejnikov, A. V., Sagliocco, F., Wilm, M., Vorm, O., Mortensen, P., Shevchenko, A., Boucherie, H., and Mann, M. (1996) Linking genome and proteome by mass spectrometry: large-scale identification of yeast proteins from two dimensional gels, *PNAS,* 25, 14440–14445.
10. Futcher, B., Latter, G. I., Monardo, P., McLaughlin, C. S., and Garrels, J. I. (1999) A sampling of the yeast proteome, *Mol. Cell. Biol.*, 11, 7357–7368.
11. Gygi, S. P., Rochon, Y., Franza, B. R., and Aebersold, R. (1990) Correlation between protein and mRNA abundance in yeast, *Mol. Cell. Biol.*, 3, 1720–1730.
12. Coghlan, A. and Wolfe, K. H. (2000) Relationship of codon bias to mRNA concentration and protein length in *Saccharomyces cerevisiae*, *Yeast*, 12, 1131–1145.
13. Gygi, S. P., Corthals, G. L., Zhang, Y., Rochon, Y., and Aebersold, R. (2000) Evaluation of two-dimensional gel electrophoresis based proteome analysis technology, *PNAS*, 97, 9390–9395.
14. Gygi, S. P., Rist, B., Gerber, S. A., Turecek, F., Gelb, M. H., and Aebersold, R. (1999) Quantitative analysis of complex protein mixtures using isotope-coded affinity tags, *Nat. Biotechnol.*, 10, 994–999.
15. Urwin, V. E. and Jackson, P. (1993) Two-dimensional polyacrylamide gel electrophoresis of proteins labeled with the fluorophore monobromobimane prior to the first-dimensional isoelectric focusing: imaging of the fluorescent protein spot patterns using a cooled charge-coupled device, *Anal. Biochem.*, 209, 57–62.
16. Hagel, P., Gerding, J. J. T., Fieggen, W., and Bloemendal, H. (1971) Cyanate formation in solutions of urea, *Biochim. Biophys. Acta*, 243, 366–379.

Index

A

Abbreviations, 119
Ab initio folding, 135
Acetylation, 112-113
Acid liability, 114
Acylation, 98
Adducts, 88
Affinity tags, 11
α-helix, 129, 130
Alexanderov studies, 73
Algorithms, 85, 128, 135
Alignment method web sites, 129
Alkaline and lysis method, modified, 168-169
Allen studies, 107
Alpha-cyano-4-dihydroxycinnamic acid, 75
Amide hydrogen exchange rates, 65
Amido Black, 107
Amino acids, 88, 104-105, 110-111
α-amino alkylation, 116-117
Analyte concentrations, 81-82
Anisotropy, 42
Anomalous dispersion, 2, 24-28
Anomalous scattering
    dispersion methods, 26-28
    isomorphous replacement, 23
    phase problem, 20
    x-ray sources and detectors, 14
Apolipoprotein B (apoB), 160-163
Apomyoglobin (equine), 88
Argand diagram, 19, 22
Arginine, 110-111
Asn-Gly, 114-115
Asp-Pro bonds, 114
Assignments
    large proteins, 57-59
    moderate sized labeled proteins, 56-57
    resonance, 52-53
    side-chain, 59
    small unlabeled proteins, 53-56
    very large proteins, 59-60
Asymmetric units, 9, 30
Atherogenicity, 162
Atherosclerosis, 161-162
Atomic positions, 31
Atomic resolution, 8
Atoms
    anomalous dispersion effects, 24
    fast bombardment (FAB), 3, 72
    heavy, 21-24, 26
    scattering power, 24
Attomole protein identification, 3
Autolysis, 95
Automated sequencers, 104
Average vs. exact monoisotopic molecular weights, 86-87

B

BAC, *see* Bacterial artificial chromosomes (BAC)
Bacterial artificial chromosomes (BAC)
    designing transgenes, 155
    gene regulation, 162
    miniprep sample protocol, 168-169
    P1 sample protocol, 170-172
Ban studies, 7
Base peak, 89
Batch method, 13
Bax, Tjandra and, studies, 63-64
Bax and Davis studies, 54
Bax and Grzesiek studies, 58
Bax and Lerner studies, 46
Bax studies, 46
van Beek, Rossmann and, studies, 18
Berendzen, Terwilliger and, studies, 21-22
Berman studies, 32
Bhoun studies, 107
Biocatalysts, 181-193
Bio-imprinting, 186
B-ions, 91
Birdsall, Feeney and, studies, 50
BLAST database searches, 108, 128, 129
Blocked N-termini, 112-117
BLOCKS database, 126
Bloomer studies, 205
Bolotovsky studies, 18
Boltzmann equations, 40
Book of Fourier, 8
Bovine serum albumin (BSA), 88, 94, 97
Bradford method, 236
Bragg studies, 7
Bravais lattices, 9

Brunger studies, 28, 31
β -strands, 130
β-turns, 130
Buffers, 82, 232-233
Bulk dialysis, 13

C

Caenorhabditis elegans, 146
Calibration, mass accuracy, 87
Capillaries, 16, 84, 94
Capillary electrophoresis (CE), 84
Capillary-LC-MS, 83-84, 88, 94
Carboxypetidase-Y, 92
Carter and Carter studies, 13
CASP, *see* Critical Assessment of Techniques for Protein Structure Prediction (CASP)
CATH database, 134
CCD systems, *see* Charge-coupled device (CCD) systems
CE, *see* Capillary electrophoresis (CE)
Cell dimensions, 21
Cell proteins analysis, 166
Centroid, 88
Chain tracing, 29-30
CHAPS, 232
Charge-coupled device (CCD) systems, 15
Charged molecular ions, 84-86
Charge state, 84-86
Chemical change, 49-51
Chemical exchange, NMR, 49-51
Chemical kinetics, 60-61
Chemical modifications, Edman chemistry, 111-112
Chemical shift, 40-41, 42, 61
Chen studies, 20, 27, 28
Chou-Fasman method, 130
CID, *see* Collision-induced dissociation (CID)
Clarke and Itzhaki studies, 65
Cleavages, CNBr, 113-114, 115
Clore and Gronenborn studies, 46, 62
CNBr cleavages, 113-114
CNS program, 31
Coiled coils study, 127
COILS program, 127
Collision-activate dissociation, *see* Collision-induced dissociation (CID)
Collision-induced dissociation (CID), 79, 89, 94
Comparative modeling, 132, 133-134, 136
COMPOSER software, 134
Computer modeling, *see also* Model building
  comparative modeling, 133-134, 137
  de novo folding, 136, 137-
  fold recognition, 134-136, 137
  fundamentals, 123-124, 138
  motifs, 125-127
  multiple sequence alignment, 129
  pairwise alignment, 127-129
  result comparisons, 136-138
  secondary structure prediction, 129-132
  sequence alignment, 127-129
  sequence analysis, 125-129
  substitution matrices, 127
  tertiary structure prediction, 132-136
  transmembrane proteins, 131-132
Concentration dialysis, 13
Continuous recovery of product, 209
Coomassie R-250, 107
Covalent modification of enzymes, 193
Critical Assessment of Techniques for Protein Structure Prediction (CASP), 136
Crown ethers, 186
Cryoelectron microscopy, 2
Cryogenic temperatures, 16
Cryo-loops supplies, 16
Cryo-protectants, 16
Crystallization, 11-13
Crystallography
  anomalous dispersion methods, 24-28
  crystal selection and mounting, 15-17
  data collection and processing, 17-19
  detectors, 13-15
  diffraction, 9-11
  isomorphous replacement methods, 21-24
  model building, 29-30
  molecular replacement, 28-29
  phase problem, 19-29
  refinement, 30-32
  sample preparation, 11-13
  validation, 32
  x-ray sources, 13-15
Crystals
  growth, 12
  lattices, 9, 18
  mounting, 15-16
  nucleation, 12
  orientation, 18
  seeding, 13
  selection and mounting, 15-17
  systems, 9
  unit cells, 9
Crystal-to-detector distance, 17
Current, 238
Cysteine, 110
Cytochrome-C (equine), 88

D

DABITC, *see* Dimethylaminoazobenzene isothiocyanate (DABITC)
Dahlquist, McIntosh and, studies, 57
Dansyl-Edman technique, 104
Databases, protein sequence, 125-127
Data collection and processing, 15-19
Data-dependent scanning, 98
Data interpretation, 108-109
Data reduction, 18
Dauter studies, 27
Davis, Bax and, studies, 54
Dayhoff matrix, 127, 128
Dayie studies, 65
DDBJ, *see* DNA Database of Japan (DDBJ)
Dead-end elimination, 133
Dean-Stark trap, 205
Dehydrogenases, 186
Delayed extraction, 79
De novo folding prediction, 135, 137
De novo sequencing, 90
Detection limits, 84, 92
Detectors
  data collection, 17-18
  image plate, 17
  saturation, 17
  sensitivity, 82
  swing, 17
  x-ray sources, 13-15
Detergents, sample preparation, 82
Deuteration, 60, 63
Dialysis, 13
Difference Fourier electron density maps, 31
Difference gel electrophoresis (DIGE), 4, 229-230
Diffraction pattern
  Bragg's law, 10
  crystals, 9-11, 16
  data collection, 18
  protein crystallography, 8
Diffraction tutorials, 8
Diffusion cell, *see* Vapor diffusion
DIGE, *see* Difference gel electrophoresis (DIGE)
Digitization rate, 87
Dihydroxy-benzoic acid, 75
Dimers, 89
Dimethylaminoazobenzene isothiocyanate (DABITC), 105
Dipolar coupling, 2, 42, 46
Direct infusion, 92, 94
Disrupting genes, 147
Distant DNA elements, 162
DNA, *see also* Sample protocols for yeast
  fragments for pronuclear injection, 169-170
  microarrays, 154
  micro-injection, 158
  PCR analysis, 172-173
  rapid preparation, 165-166
  southern analysis, 172-173
DNA Database of Japan (DDBJ), 125
Dohmeier and Jorgenson studies, 73
Dole studies, 73
Doolittle, Kyte and, studies, 132
Dordick studies, 182
Dotsch and Wagner studies, 62
Double labeled proteins, 58
Double shutoff technique, 150-151
Douglas studies, 99
Dynamic programming, 127-128, 135

E

EBI, *see* European Bioinformatics Institute (EBI)
Edman chemistry
  acetylation, 112-113
  acid liability, 114
  amino-acids, 104-105, 110-111
  -amino alkylation, 116-117
  Asn-Gly, 114-115
  Asp-Pro bonds, 114
  blockages, 113-116
  chemical modifications, 111-112
  cleavage at Trp, 115
  CNBr cleavages, 113-114
  coupling, 3
  data interpretation, 108-109
  degradation, 3-4, 114, 230
  fundamentals, 104, 106-117, 118
  glycosylation, 118
  history, 104-106
  in-gel digestion, 115-116
  internal residues, 117-118
  methionine, 113-114
  N-terminus, 112-117
  peptide sequencing, 92
  phosphorylation, 117-118
  post-translational modifications, 117-118
  pyroglutamate, 113
  ragged ends, 116
  reactions, 112
  salt, 109-110
  samples, 105-108
  sequencing, 112
  site-directed labeling, 111-112
Edmonds and Smith studies, 11
Educational Internet software, 93
Electromagnetic spectrum, 8
Electron density, 20

Electron density maps
anomalous dispersion methods, 26
model building, 28-29
phase problem, 20
refinement of models, 31
Electrophoresis
capillary electrophoresis (CE), 84
difference gel electrophoresis (DIGE), 4
gel electrophoresis, 85, 171-172
SDS-PAGE 2-D-Gel electrophoresis, 92, 229-230
two-dimensional electrophoresis (2DE), 229
Electrospray ionization (ESI), 3, 73-75, 81-82
EMBL, *see* European Molecular Biology Laboratory (EMBL)
Embryos, 236
Emission filter, 240
Eng and Yates studies, 98
Engh and Huber studies, 31, 32
Enzymes
covalent modification, 193
effect of solvents, 194-198
immobilization, 187-190
nonaqueous modification, 198-201, 204-208
organic solvents, 180
peptide sequencing, 88-92
reactions, 202-211
salt hydrate pairs, 206
screening experiments, 202-203
water mimics, 208
Enzymes-lipid aggregates, 190
Enzymology, 180
Epitope tags, 147, 151
Equilibrium constants, 60-61
Equine apomyoglobin, 88
Equine cytochrome-C, 88
ESI, *see* Electrospray ionization (ESI)
European Bioinformatics Institute (EBI), 128
European Molecular Biology Laboratory (EMBL), 125
Exact vs. average monoisotopic molecular weights, 86-87
Excitation filter, 241
Exposure time, data collection, 17
Expression plasmid, 155

F

FAB, *see* Fast atom bombardment (FAB)
Fast atom bombardment (FAB), 3, 72
Fast exchange, nuclear magnetic resonance, 49-51
FD, *see* Field desorption (FD)
Feeney and Birdsall studies, 50
Femtomole level, 81
Feng and Konish studies, 99
Fenn studies, 73
Ferentz and Wagner studies, 62
Fibroblasts, mouse, 235-236
FID, *see* Free induction decay (FID)
Field desorption (FD), 72
First dimension, proteomics, 238-239
FLAG tags, 151
Flash freezing techniques, 16
Fluorescence spectroscopy, 38
Fluorescent dyes, 229
Fluorescent images, 236-237
Fluorodinitrobenzene, 104
Fohlman, Roepstorff and, studies, 91
Folding, 98-99, 134, 137
Founder mice, 159
Fourier transform ion cyclotron (FT) mass spectrometer, 80-81
Fourier transform methods, 46
Free induction decay (FID), 47
Free interface diffusion, 13
Free *R* factor, 31
Freeze-drying, 182-185
Friedel pairs, 25
Frozen approximation, 135
FSSP, 134
Full width at half maximum, 87
Function homology, 27-28
Furey and Swaminathan studies, 23-24
Fusion proteins, 151

G

Gap penalty, 128
Gardner and Kay studies, 60
Garman studies, 16
Gel cassettes, 239
Gels
advances, 3, 4
digest, 93
electrophoresis, 86, 171-172
proteomics, 228-229, 231, 239
SDS-PAGE 2-D-Gel electrophoresis, 92, 229-230
two-dimensional limitations, 231
GenBank, 125
Gene disruption, 149-150, 153
Gene expression measurements, 152
Genetic analysis
fundamentals, 146
gene disruption, 149-150
global transcription measurement, 152-153
Med2 subunit studies, 153-154
multiprotein complexes, 146-149

protocols, 163-174
Saccharomyces cerevisiae, 146-155
tagging yeast proteins, 150-151
transgenic mice, 155-163
Genome-wide expression analysis, 152
Georgalis and Saenger studies, 11
Giessmann, Hillenkamp, Karas and, studies, 75
Glass loop, freezing, 16
Glide planes, 9
Global transcription measurement, 152-153
Glossaries, 68-70
Glutamine residues, 115
Glutathione-S transferase (GST) domain, 151
Glycosylation, 98, 118
Goniometer head, 16
GOR method, 130
Goto and Kay studies, 60
Goto studies, 192
Green fluorescent protein, 151
Green studies, 21
Gronenborn, Clore and, studies, 46, 62
Grzesiek, Bax and, studies, 58
Guntert studies, 62

H

Hajduk studies, 62
Halling studies, 199, 207
Hansen studies, 64
Hassel studies, 9
H/D, *see* Hydrogen/deuterium (H/D)
Heavy atoms, 21-24, 26
Heavy metals, amino acid residue affinity, 24
Hendrickson and Teeter studies, 27
Hendrickson studies, 20, 27
Heteronuclear NOE, 65
Heteronuclear single quantum correlation (HSQC), 47-49, 57, 62
Higher magnetic field strengths, 2
High resolution triple quadrupole mass spectrometers, 79, 80
High salt concentrations, 82, 85
Hillenkamp, Karas, and Giessmann studies, 75
His tag, 151
Histidine, 110-111
Hitscherich studies, 11
HNCA assignment experiments, 56, 59
Holo-TFIID, 150-151
Homologous recombination, 149
Homology modeling, 134
Homomyoglobin, 99
Hooft studies, 32
Hope studies, 16
HPLC, 105, 115
HSQC, *see* Heteronuclear single quantum correlation (HSQC)
Huber, Engh and, studies, 31
Human genome sequence, 98
Hydrogen/deuterium (H/D), 99
Hydrophobicity, scale of, 132
Hydrophobic proteins, 93

I

IEF, *see* Isoelectric focusing (IEF)
Ikura studies, 58
Illumination, 234
Image acquisition and analysis, 234-235, 240-241
Image plate detectors, 17
Imaging, protein analysis, 99
Immobilization, 187-190
Information theory, 130
In-gel digests, 94, 115-116, 230
Instrumentation, mass spectrometry, 73-81
Interleukin-8, 87
Internal motions, 65
Internal residues, 117-118
Intrinsic resolution, 87
Inverse protein folding, 134
Ions
charged molecular, 84-86
electrospray ionization, 73-75
ionization advances, 3
matrix-assisted laser desorption ionization, 75-77
sources, 73-77
Ion trap, 79
ISAS, *see* Iterative single-wavelength anomalous scattering (ISAS) process
Isoelectric focusing (IEF), 237-239
Isomorphous replacement methods, 21-24
Isotope distribution, 87
Isotopic labeling
deuteration, 60
double labeled proteins, 58
nuclear spins, 38
resonance assignment, 53
very large protein assignments, 60
Iterative single-wavelength anomalous scattering (ISAS) process, 20, 26-27
Itzhaki, Clarke and, studies, 65

J

Jancarik and Kim studies, 13
Jardine studies, 11
Jeener studies, 46

Jenkins studies, 50
Jergenson, Dohmeier and, studies, 73
Jones studies, 30

K

Karas, and Giessmann, Hillenkamp, studies, 75
Karl-Fischer titration, 211
Kay, Gardner and, studies, 60
Kay, Goto and, studies, 60
Kay studies, 65
Keratin contamination, 95
Kim, Jancarik and, studies, 13
Kinetic rates, 60-61
Klibanov studies, 208
Konermann studies, 99
Konish, Feng and, studies, 99
Kyte and Doolittle studies, 132

L

Labeling schemes, 2, 232
Ladder sequencing, peptide, 92
Lane and Lefevre studies, 65
Large-scale protein preparation, 163-164
Lasers, 75, *see also* Matrix-assisted laser desorption ionization (MALDI)
Laskowski studies, 32
Lefevre, Lane and, studies, 65
Lerner, Bax and, studies, 46
Leslie studies, 18
Lian and Roberts studies, 50
Lian studies, 50
Ligands, 61-62, 186
Light sources for researchers, 14
Lim method, 130
Linbro plates, 13
Linear discriminant analysis, 131
Linear mode (TOF mass spectrometer), 79
Liquid bridge, 13
Liquid chromatography, capillary, 83-84
Liquid volume requirements, 81
Lithium acetate, 164-165
Liu studies, 26
Loop freezing, 16
Loop structures, 133
Low-resolution reflections, 17
Lusatti plot, 32
Lyophilization, 182-187
Lyoprotectants, 185
Lysine, 111

M

MAD, *see* Multi-wavelength anomalous dispersion (MAD)
Main chain torsion angles, 31
MALDI, *see* Matrix-assisted laser desorption ionization (MALDI)
Malkin and McPherson studies, 11
Mann, Wilm and, studies, 73
Marqusee, Raschke and, studies, 65
Martin, Scoble and, studies, 11
Martin studies, 91
Mascot software, 93
Mass accuracy, 87-88
Mass analyzers, 77-81
Mass spectroscopy (MS)
  advances, 2-4
  analyte concentrations, 81-82
  average vs. exact monoisotopic molecular weights, 86-87
  capillary electrophoresis ESI-MS, 84
  capillary liquid chromatography MS-MS, 83-84
  charge state, 84-86
  data-dependent scanning, 98
  ESI analysis, 81-82
  femtomole level, 81
  folding, 98-99
  fundamentals, 72-73
  instrumentation, 73-81
  ion sources, 73-77
  liquid volumes, 81
  MALDI-MS analysis, 81-82
  mass accuracy, 87-88
  mass analyzers, 77-81
  modifications, 98-99
  molecular weight determinations, 84-88
  MS-MS sequencing, 88-92
  peptides, 84-88, 88-92, 95
  pipet tip packing bed purification method, 83
  protein identification, 92-98
  protein sequencing, 88-92
  proteomics, 230-231
  quality, 82
  resolution, 87
  sample preparation, 81-84, 93-95, 99
  sequence tag approach, 95
  SEQUEST approach, 98
  technique of choice, 72
  tissue analysis, 98-99
Le Master and Richards studies, 57
Le Master studies, 60
MatchMaker program, 135

Matrix-assisted laser desorption ionization (MALDI)
  advances, 3
  mass spectrometry instrumentation, 72-73
  peptide mapping, 94
  sample preparation, 81-82
Matthews' method, 17
McIntosh and Dahlquist studies, 57
McPherson, Malkin and, studies, 11
McPherson studies, 13, 27
McRee studies, 16, 30
Mediator complex, 154
Med2 subunit studies, 153-154
Merril studies, 11
Metal clusters, 26
Metal ions, 26
Metalloproteins, 26
Methionine, CNBr cleavages, 113-114
Mice, transgenic, *see* Transgenic mice
Micro arrays, 153
Micro-batch method, 13
Micro dialysis, 13
Microemulsions, 201-202
Micro-injection of DNA, 157-159
Micro-pipette tip, 83
Miller indices, 10, 19
Miller studies, 22
Minor, Otwinski and, studies, 18
MIR, *see* Multiple isomorphous replacement (MIR)
MIRAS approach, 23
MMDB, *see* Molecular Modeling Database (MMDB)
Model building, 29-30, 30-31, 32, *see also* Computer modeling
Modeling, computer, *see* Computer modeling
MODELLER software, 134
Moderate sized labeled proteins, 56-57
Modifications
  Edman chemistry, 111
  proteins, 98-99
  steps, 90-91
  transgenes, 155-157
Modified alkaline and lysis method, 168-169
Molecular boundary, 30
Molecular Modeling Database (MMDB), 134
Molecular replacement (MR), 28-29
Molecular weight, 72, 84-88
Monoisotopic molecular weight, 86-87
Monomer buffer, 233
Moore studies, 62
Moqtaderi studies, 149
Morphology, 15
Mosaicity, 16, 17
Motifs, 125-127
Mounting supplies, 17
Mouse fibroblasts, 235-236
MR, *see* Molecular replacement (MR)
MS, *see* Mass spectroscopy (MS)
MS-Digest software, 93
MS-Edman search program, 109
MS-Fit software, 93
MS-MS scan modes, 78-79, 88-92
MS^n, 79
MS-Product software, 93
Multidimensional spectra, 46-49
Multiple isomorphous replacement (MIR), 20-21, 24, 26-27
Multiple sequence alignment, 129
Multiprotein complexes, *S. cerevisiae,* 146-149
Multi-wavelength anomalous dispersion (MAD)
  anomalous dispersion methods, 26-27
  macromolecular structure solutions, 14
  phase problem, 20-21
  sample preparation, 11
Multi-wire area detectors, 15
Multi-wire x-ray area detectors, 15
Multi-wire x-ray area fundamentals, 15
Murshudov studies, 31

N

Nakajima studies, 192
Nanoflow capillary, 94
Navaza studies, 28
NCBI, 125
N-dimensional NMR, 46-49
Nd-YAG laser, 75
Needleman-Wunsch alignment, 128
Neuhaus and Williamson studies, 44
Neural networks, 130, 131
Neutral loss MS-MS scan mode, 78-79
Next generation detectors, 15
Next generation x-ray detectors, 15
Nicotinic acid, 75
Nielsen studies, 162
Nitrogen laser, 75
NMR, *see* Nuclear magnetic resonance (NMR) spectroscopy
NOE, *see* Nuclear overhauser effect (NOE)
NOESY, *see* Nuclear overhauser effect spectroscopy (NOESY)
Nomenclature, peptide fragmentation, 91
Nonaqueous environments
  advances, 4
  biocatalysts, 181-193, 203-204, 210-211
  bio-imprinting, 186
  co-factors, 186
  continuous phase, 194-198

controlling water activity, 198-201, 204-208
covalent modification, 193
crown ethers, 186
enzymatic reactions, 202-211
fundamentals, 180, 211
immobilization, 187-190
ligands, 186
lyophilization, 182-187
lyoprotectants, 185
microemulsions, 201-202
monitoring progress, 210-211
organic phase buffers, 208
pH buffer salts, 185
pH control, 208
polymers, 186-187
product recovery, 209
purification, 181-182
reactions, 209-211
reactor design, 203-204
reversed micelles, 201-202
salts, 185
screening experiments, 202-203
solubilization of enzymes, 190-193
substrate preparation, 193
surfactants, 186
temperature, 203
water mimics, 208
Noncrystallographic symmetry, 30
Nonredundant protein sequence database, 125
Nonvolatile salts, 82
N-terminus, 112-117
Nuclear magnetic resonance (NMR) spectroscopy
advances, 2-4
applications, 51-60
chemical shift, 40-41
degeneracy, 41
equilibrium constants, 60-61
experiments, 60-66
fundamentals, 37-49
higher magnetic field strengths, 2
HSQC, 47-49
kinetic on-rates, 60-61
large proteins, 57-59
ligand binding sites, 61-62
line shape, 49-51
moderate-sized labeled proteins, 56-57
multidimensional, 46-49
nuclear spin exchange, 49-51
relaxation, 41-46, 65-66
resonance assignments, 52-60
sensitivity, 40, 58, 59
small unlabeled proteins, 54-56
solid state, 2
structure determination, 62-65
topological mapping, 61-62
very large proteins, 59-60
Nuclear overhauser effect (NOE), 44, 63, 65
Nuclear overhauser effect spectroscopy (NOESY)
HSQC, 48, 57
multidimensional NMR, 44, 48
small unlabeled protein assignment, 56
structure determination, 63
Nucleation, crystal growth, 12

O

Okahata studies, 192
One-dimensional NMR, 46-49
Open reading frames (ORFs), 146
O program, 30
Order parameter, 65
ORFs, *see* Open reading frames (ORFs)
Organic phase buffer agents, 208
Organic phase buffers, 208
Oscillation step, 17
Otwinski and Minor studies, 18
OWL database, 125

P

Packing material, 83
PAIRCOIL analysis, 127
Pairwise alignment, 127-129
Palmer studies, 65
PAM matrices, 127
Paraffin oil, 232
Partial orientation, structure determination, 63
Patterson function
isomorphous replacement methods, 21, 24
model building, 30
molecular replacement, 28
Paul, Wolfgang, 79
Paul trap, 79
PCR analysis, 172-174
PCR-mediated gene disruption, 164
PD, *see* Plasma desorption (PD)
PDB, *see* Protein Data Bank (PDB)
PepFrag software, 93
Peptide fragmentation nomenclature, 91
Peptide ladder sequencing, 88-92
Peptide mapping, 95
Peptide Search software, 93
Petsko studies, 24
Pflugrath studies, 18
Phase ambiguity problem, 23-24, 26, 27
Phase angle, isomorphous replacement methods, 22
Phase buffers, 208

Phase problem
anomalous dispersion methods, 24-28
crystallography overview, 19-21
isomorphous replacement methods, 21-24
molecular replacement, 28-29
Phase triangle relationship, 25
Phasing, anomalous dispersion, 2
PH buffer salts, 185
PH control, nonaqueous environments, 208
PHD method, 130-131
Phenylthiohydantoin (PTH), 104
Phosphate buffer, 82
Phosphorylated residue, 117-118
Phosphorylation, 98, 117-118
Pipet tip packing bed purification method, 83
PIR, *see* Protein identification resource database (PIR)
PITC, 116
Plasma desorption (PD), 72
Point groups, 9
Polarities of solvents, 196
Polymers, 186-187
Ponceau S, 107
P1 and BAC miniprep protocol, 168-169
P1 bacteriophage sample protocol, 170-172
Post source decay (PSD), 80
Post-translational modifications, 98, 117-118
Potassium, 85
Powers studies, 65
Precursor mass, 89
Precursor MS-MS scan mode, 78-79
Predictions of transmembrane helices, 132
Prestegard studies, 63
PRINTS database, 126
PROCHECK program, 32
Product MS-MS scan mode, 78-79
ProFound software, 93
Programs, *see* Software; Web sites
Pronuclear injection, sample protocols, 169-170, 170-172
PROSITE database, 125-127
Protein abundance, 231
Protein crystallography, *see* Crystallography
Protein Data Bank (PDB), 32, 129
Protein database search engine, 92 93, 96, 97
Protein identification resource database (PIR), 125
Proteins
adsorption, 81
disulfide bonds, 93
dynamics, 38, 65
identification, 81, 92 98, 95
ligand interactions, 38
protein interactions, 149
regions, model building, 29
structure, nuclear magnetic resonance, 38
Protein sequence
databases, 125 127
mass spectrometry, 88 92
model building, 30
Proteomics
advances, 4
buffers, 232-233
DIGE, 229-230
first dimension, 238-239
fundamentals, 228-231
gel cassettes, 239
gels, 228-229, 231, 239
image acquisition and analysis, 234-235, 240-241
isoelectric focusing, 232-234, 237-239
mass spectrometry, 230-231
monomer buffer, 233
mouse fibroblasts, 235-236
notes and information, 231, 235, 237, 240, 241-243
rehydration, 237-238
sample solubilization and labeling, 231-232, 235-237
SDS-Page, 233, 239
second dimension, 239-240
solutions, 232, 234, 325
usage guide, 231-235
Proton bound dimers and trimers, 89
PROWL software, 93
PSD, *see* Post source decay (PSD)
PTH, *see* Phenylthiohydantoin (PTH)
PTH derivative, 118
Pulsed-field gel electrophoresis, 171-172
Purification of biocatalysts, 181-182
Purity, sample preparation, 82
PVDF, 107
Pyroglutamate, 113

Q

QDFM, *see* Quantitative DNA fiber mapping (QDFM)
Quadrupole ion trap (QIT) mass spectrometer, 79
Quadrupole time-of-flight (Qq-TOF) mass spectrometer, 80
Quantitative DNA fiber mapping (QDFM), 157
Quenching solution, 232

R

RAD, *see* Resolved anomalous phasing (RAD)
Ragged ends, 116

Ramachandran plot, 31, 32
Ramakrishnan and Ramachandran studies, 31
Rapid preparation of DNA, 165-166
Rapid screening systems, 13
RARE, *see* RecA-assisted restriction endonuclease (RARE) cleavage technique
Raschke and Marqusee studies, 65
Reactions, enzymatic, 202-211
Reactor design, nonaqueous environments, 203-204
Reagent kits, 13
RecA-assisted restriction endonuclease (RARE) cleavage technique, 156, 168
Refinement of models, 30-32
Reflection mode (TOF mass spectrometer), 79
REFMAC program, 31
Rehydration, 237-238
Relaxation measurements, 65-66
Resolution, 18, 63-64, 87
Resolution range, 29-30
Resolved anomalous phasing (RAD), 27
Resolving power, 87
Resonance assignments
  fundamentals, 52-53
  general approach, 53-54
  large proteins, 57-59
  moderate-sized labeled proteins, 56-57
  small unlabeled proteins, 54-56
  very large proteins, 59-60
Resonance lines, 49-51
Reversed micelles, 201-202
R_{free} calculation, 31
Richards, Le Master and, studies, 57
Richardson and Richardson studies, 115
RNA, undegraded isolation, 166-167
RNA polymerase II, 147, 153
RNase contamination, 167
Roberts, Lian and, studies, 50
Roberts studies, 62
Robotics, 2
Roepstorff and Fohlman studies, 91
Rossmann and van Beek studies, 18
Rotating anode generators, 13-14
Rotational correlation time, 45
Rotation function, 28, 30
Rubin, Young and, studies, 160
Russell studies, 209

S

Saccharomyces cerevisiae
  gene disruption, 149-150
  global transcription measurement, 152-153
  mediator complex, 153-154
  Med2 subunit studies, 153-154
  multiprotein complexes, 146-149
  proteomics, 230
  sample solubilization, 236
  tagging proteins, 150-151
SAD, *see* Single-wavelength anomalous dispersion (SAD)
Saenger, Georgalis and, studies, 11
Salt hydrate pairs, 206, 207
Salts, 109-110, 185
Sambrook studies, 11
Sample preparations
  crystallization, 11-13
  Edman chemistry, 107-108
  improvements, 105-106
  MALDI and ESI, 81-84
  multi-wavelength anomalous dispersion (MAD), 11
  peptide ladder sequencing, 92
  protein identification, 93-95
  quality and purity, 82, 99, 108
Sample protocols for yeast, *see also* Transgenic mice
  DNA fragments, 169-170
  large-scale protein preparation, 163-164
  lithium acetate, 164-165
  modified alkaline and lysis method, 168-169
  P1 and BAC miniprep, 168-169
  P1 bacteriophage, 170-172
  PCR analysis, 172-174
  PCR-mediated gene disruption, 164
  pronuclear injection, 169-170, 170-172
  rapid preparation of DNA, 165-166
  RARE cleavage of large DNA, 168
  total cell proteins analysis, 166
  undegraded RNA isolation, 166-167
Sample solubilization and labeling, 231-232, 235-237
Sanger's reagent, 104
SA refinement, *see* Simulated annealing (SA) refinement
Saturation of detectors, 17
Scalar coupling, 46, 64
Scale of hydrophobicity, 132
Scoble and Martin studies, 11
SCOP database, 134
Screw axes, 9
SDS, *see* Sodium-dodecyl-sulfate (SDS)
SDS-Page, 233, 239
SDS-PAGE 2-D-Gel electrophoresis, 92, 229-230
Secondary ionization mass spectrometry (SIMS), 72
Secondary ionization microscopy, 3
Secondary structure prediction method, 129-132

Second dimension, proteomics, 239-240
Seeding, crystal, 13
Self-rotation function, 30
Sensitivity, NMR spectroscopy, 40, 58, 59-60
Sequencing
alignment, 127 129, 133
analysis, 125 129
data, 108
databases, 125
homology, 28
limitations, 112
tags, 95, 147
SEQUEST software, 93, 98
Serum albumin (bovine) (BSA), 88, 94, 96
Shake and Bake program, 22
Sheldrick studies, 31
SHELX program, 31
Shieh studies, 13
Shizuya studies, 168
Side-chain assignments, 59
Signal to noise ratio
anomalous dispersion methods, 26
data collection, 17, 18
detection limits, 92
very large protein assignments, 59
SIMS, *see* Secondary ionization mass spectrometry (SIMS)
Simulated annealing (SA) refinement, 31
Sinapinic acid, 75
Single isomorphous replacement (SIR), 20-21, 24, 26
Single-wavelength anomalous dispersion (SAD), 14, 20-21, 26-28
SIR, *see* Single isomorphous replacement (SIR)
SIRAS approach, 23
Site-directed labeling, 111-112
Slow exchange, NMR, 49-51
Small fiber loops, 16
Small unlabeled protein assignment, 53-56
Smith, Edmonds and, studies, 11
Smith-Waterman algorithm, 128
Sodium, 85
Sodium-dodecyl-sulfate (SDS), 82
Software, *see also* Web sites
CATH database, 134
CNS, 31
COMPOSER, 134
educational on the Internet, 93
MatchMaker, 135
MODELLER, 134
O, 30
PROCHECK, 32
REFMAC, 31
SCOP database, 134
SHELX, 31
THREADER, 135
TOPITS, 135
wARP, 30
WHAT_CHECK, 32
X-PLOR, 31
XtalView, 30
Solid phase extraction (SPE), 84
Solid state NMR, 2
Solubilization, 235-236
Solubilization of enzymes, 190-193
Solutions, proteomics, 232, 234, 325
Solvent flattening, 23-24, 27
Solvent molecules, 31
Solvent regions, 29
Solvents, 194-198
SOLVE program, 21-22
Space groups
crystals and diffraction, 9
data collection, 18
isomorphous replacement methods, 21
model building, 29-30
SPE, *see* Solid phase extraction (SPE)
Spectral density function, 45, 65
Spectroscopy, *see* Mass spectroscopy (MS); Nuclear magnetic resonance (NMR) spectroscopy
Spin-lattice relaxation, 42-43, 65
Spinning-cup instruments, 105
Spin-spin relaxation, 45-46, 65
S-SAD, *see* Sulfur SAD
Steller studies, 18
Stereochemical properties, 133
Stereochemistry, 31, 32
Stimulated emission, 42
Stone studies, 65
Storage-phorphor-based imaging plate, 15
Structural genomics, 1
Structure determination, 2, 62-65
Structure factor, 19, 22
Structure-function studies, 160-161
Substitution matrices, 127
Sulfation, 98
Sulfur SAD, 27
Surfactants, 186
Sutra studies, 13
Swaminathan, Furey and, studies, 23-24
SWISS-MODEL web site, 134
SWISS-PROT database, 125
Symmetry (space group), 21
Synchrotron data collection, 15
Synchrotron sources, 2, 14, 20
Systematic absences, 185

T

TAFs, *see* TBP-associated factors (TAFs)
Tagging yeast proteins, 150-151
TBP-associated factors (TAFs), 150-151
Teeter, Hendrickson and, studies, 27
Teflon membranes, 232
Temperatures, 16, 203
Teng studies, 16
Tertiary structures, 99, 132-136, 137
Terwilliger and Berendzen studies, 21-22
Thiourea, 232
THREADER program, 135, 137
Threading, 134
Time-of-flight (TOF) mass spectrometer, 79
Tissue analysis, 98-99
Tjandra and Bax studies, 63-64
Tjandra studies, 63
TOCSY, *see* Total correlation spectroscopy (TOCSY)
Tolman studies, 63
TOPITS program, 135
Topological mapping, NMR spectroscopy, 61-62
Topology of proteins, 134
Torsion angles, 31, 64
Total correlation spectroscopy (TOCSY)
  large proteins, 59
  ligand binding sites, 62
  moderate sized labeled protein assignment, 57
  small unlabeled protein assignment, 54-55
Transformation of yeast, 164-165
Transgenes, 155-57, 160
Transgenic mice, *see also* Sample protocols for yeast
  analysis, 159-160
  atherogenicity of ApoB, 161-162
  designing transgenes, 155
  distant DNA elements, 162
  DNA preparation, 157
  fundamentals, 155
  gene regulation, 162
  lipoprotein heart secretions, 163
  micro-injections, 158-159
  modification of trangenes, 155-157
  structure function studies, 160-161
Transmembrane proteins, 131-132
Trimers, 89
Triple resonance experiments, 58
Triple-stage quadrupole (QqQ) mass spectrometer, 78-79
Tryptic peptides, 95, 97
Two-dimensional NMR, 46-49
Two-hybrid method, 149, 153

U

Undegraded DNA, isolation, 166-167
Unit cells, 9-11, 15, 18
Urea, 108, 232
Usage guide, proteomics, 231-235

V

Validation of models, 32
Vapor diffusion, 12-13
Virtual Mass Spectrometry Laboratory, 93
Viscosity, 51

W

Wagner, Dotsch and, studies, 62
Wagner, Ferentz and, studies, 62
Wang studies, 20, 23-24
WARP software, 30
Water activity, controlling, 198-201, 204-208
Water mimics, 208
Web sites, *see also* Software
  alignment methods, 129
  AMAS, 129
  anomalous dispersion, 25
  BLAST database, 108, 128, 129
  BLOCKS, 126
  Book of Fourier, 8
  CCD systems, 15
  Clustal W, 129
  DDBJ, 125
  detectors, 18
  diffraction tutorials, 8
  DNA preparation, mini, 172-173
  DNA quick preparation, 173-174
  EMBL, 125
  enzyme suppliers and technology, 182
  FASTA, 129
  GenBank, 125
  isolation of DNA fragments, 169-170
  light sources for researchers, 14
  mounting supplies, 17
  MS-Edman search program, 109
  OWL, 125
  PIR, 125
  PRINTS, 126
  PROSITE, 126
  Protein Data Bank, 32
  protein identification, 93
  rapid screening systems, 13
  reagent kits, 13
  rotating anode generators, 14

secondary structure prediction methods, 131
sequence databases, 125
sequence motif databases, 127
Shake and Bake program, 22
SOLVE program, 21-22
storage-phorphor-based imaging plate, 15
SWISS-MODEL, 134
SWISS-PROT, 125
XtalView, 30
WHAT_CHECK program, 32
Wider studies, 62
Williamson, Neuhaus and, studies, 44
Wilm and Mann studies, 73
Ω-loops, 129
Woodward studies, 65
Wu studies, 27

X

X-PLOR program, 31
X-ray crystallography, 38
X-rays, 2, 13-15, 18
XtalView program, 30

Y

YAC gene-targeting approach, 157
Yates, Eng and, studies, 98
Yeast, *see Saccharomyces cerevisiae*
Y-ions, 91
Young and Rubin studies, 160